Frontiers in Clinical Drug Research-Anti Infectives

(Volume 5)

Edited by

Atta-ur-Rahman, *FRS*
Honorary Life Fellow,
Kings College, University of Cambridge, Cambridge, UK

Frontiers in Clinical Drug Research-Anti-Infectives

Volume # 5

Editor: Atta-ur-Rahman

ISSN (Online): 2352-3212

ISSN (Print): 2452-3208

ISBN (Online): 978-1-68108-637-8

ISBN (Print): 978-1-68108-638-5

BENTHAM SCIENCE PUBLISHERS LTD.

End User License Agreement (for non-institutional, personal use)

This is an agreement between you and Bentham Science Publishers Ltd. Please read this License Agreement carefully before using the ebook/echapter/ejournal (**"Work"**). Your use of the Work constitutes your agreement to the terms and conditions set forth in this License Agreement. If you do not agree to these terms and conditions then you should not use the Work.

Bentham Science Publishers agrees to grant you a non-exclusive, non-transferable limited license to use the Work subject to and in accordance with the following terms and conditions. This License Agreement is for non-library, personal use only. For a library / institutional / multi user license in respect of the Work, please contact: permission@benthamscience.net.

Usage Rules:

1. All rights reserved: The Work is the subject of copyright and Bentham Science Publishers either owns the Work (and the copyright in it) or is licensed to distribute the Work. You shall not copy, reproduce, modify, remove, delete, augment, add to, publish, transmit, sell, resell, create derivative works from, or in any way exploit the Work or make the Work available for others to do any of the same, in any form or by any means, in whole or in part, in each case without the prior written permission of Bentham Science Publishers, unless stated otherwise in this License Agreement.
2. You may download a copy of the Work on one occasion to one personal computer (including tablet, laptop, desktop, or other such devices). You may make one back-up copy of the Work to avoid losing it.
3. The unauthorised use or distribution of copyrighted or other proprietary content is illegal and could subject you to liability for substantial money damages. You will be liable for any damage resulting from your misuse of the Work or any violation of this License Agreement, including any infringement by you of copyrights or proprietary rights.

Disclaimer:

Bentham Science Publishers does not guarantee that the information in the Work is error-free, or warrant that it will meet your requirements or that access to the Work will be uninterrupted or error-free. The Work is provided "as is" without warranty of any kind, either express or implied or statutory, including, without limitation, implied warranties of merchantability and fitness for a particular purpose. The entire risk as to the results and performance of the Work is assumed by you. No responsibility is assumed by Bentham Science Publishers, its staff, editors and/or authors for any injury and/or damage to persons or property as a matter of products liability, negligence or otherwise, or from any use or operation of any methods, products instruction, advertisements or ideas contained in the Work.

Limitation of Liability:

In no event will Bentham Science Publishers, its staff, editors and/or authors, be liable for any damages, including, without limitation, special, incidental and/or consequential damages and/or damages for lost data and/or profits arising out of (whether directly or indirectly) the use or inability to use the Work. The entire liability of Bentham Science Publishers shall be limited to the amount actually paid by you for the Work.

General:

1. Any dispute or claim arising out of or in connection with this License Agreement or the Work (including non-contractual disputes or claims) will be governed by and construed in accordance with the laws of the U.A.E. as applied in the Emirate of Dubai. Each party agrees that the courts of the Emirate of Dubai shall have exclusive jurisdiction to settle any dispute or claim arising out of or in connection with this License Agreement or the Work (including non-contractual disputes or claims).
2. Your rights under this License Agreement will automatically terminate without notice and without the

Bentham Science Publishers Ltd.
Executive Suite Y - 2
PO Box 7917, Saif Zone
Sharjah, U.A.E.
Email: subscriptions@benthamscience.net

CONTENTS

PREFACE

The 5th volume of **Frontiers in Clinical Drug Research – Anti Infectives** comprises six chapters that cover a variety of topics including prolonging antibiotic life, biofilms in medical devices and various antiviral drugs.

In chapter 1, Kurtböke *et al.* present an overview of new bioactive compound and marine actinomycetes that might be further exploited for their potential as novel and potent drug candidates. In chapter 2, Zárate-Bladés *et al.* have reviewed the developments in the field of pathogenesis of *Clostridium difficile* infections and clinical and experimental studies on the therapeutic effects of gut microbial transportation (GMT) against recurrent *C. difficile* infections (CDI). The methodology of gut microbial transportation (GMT) and applications in the context of the mechanisms of microbiota effects on the immune system are also discussed.

Chapter 3 by Filazi & Yurdakok-Dikmen presents an insight into antimicrobials, particularly against direct-food microbes, as well as other alternative products such as plant-derived compounds, bacteriophage and phage lysins, and antimicrobial peptides. They also discuss novel approaches applicable in the field. In chapter 4, Ahmad *et al.* explain the current status of different nanoantibiotic (nAbts) loaded systems, their release mechanisms, key targets, formulations and modes of action. Important features of nanoantibiotics (nAbts) such as size, surface charge, hydrophobicity/philicity, biofilm formation, stimuli-receptive and functionalization against multidrug-resistant (MDR) pathogens are also described.

In chapter 5, Monroy-Torres & Medina-Jiménez present primary objectives of cranberry and other functional foods in urinary tract infections in women. Chapter 6 by Hameed & Fatima discusses insights into the association of Magnesium (Mg2+) in survival of *Myobacterium Tuberculosis* (Mtb) and how it can be exploited as an anti-mycobacterial drug target.

I would like to thanks all the authors for their excellent contributions that will be of great interest. I would also like to thank the editorial staff of Bentham Science Publishers, particularly Mr. Zain Rehman, Mr. Shehzad Naqvi and Mr. Mahmood Alam for their support.

Prof. Atta-ur-Rahman, *FRS*
Honorary Life Fellow,
Kings College,
University of Cambridge,
Cambridge
UK

List of Contributors

Ayhan Filazi	Department of Pharmacology and Toxicology, Faculty of Veterinary Medicine, Ankara University, Ankara, Turkey
Amjad Khalil	Department of Life Sciences, King Fahd University of Petroleum and Minerals (KFUPM), Dhahran, Saudi Arabia
Aline Machiavelli	Laboratory of Immunoregulation, iREG, Department of Microbiology, Immunology and Parasitology, Federal University of Santa Catarina, UFSC, Florianopolis, 88040-900, SC, Brazil Laboratory of Applied Immunology, LIA, Department of Microbiology, Immunology and Parasitology, Federal University of Santa Catarina, UFSC, Florianopolis, 88040-900, SC, Brazil
Anthony Richard Carroll	Environmental Futures Research Institute, Griffith University, Gold Coast Campus, QLD, Australia
Aguinaldo R. Pinto	Laboratory of Applied Immunology, LIA, Department of Microbiology, Immunology and Parasitology, Federal University of Santa Catarina, UFSC, Florianopolis, 88040-900, SC, Brazil
Ali Talha Khalil	Department of Eastern Medicine and Surgery, Qarshi University, Lahore, Pakistan
Ana Karen Medina-Jiménez	University Observatory of Food and Nutritional Security of the State of Guanajuato/ Observatorio Universitario de Seguridad Alimentaria y Nutricional del Estado de Guanajuato (OUSANEG), Mexico
Begum Yurdakok-Dikmen	Department of Pharmacology and Toxicology, Faculty of Veterinary Medicine, Ankara University, Ankara, Turkey
Catherine W. Wippel	Hospital Infection Control Service, SCIH, University Hospital "Polydoro Ernani de Sao Thiago", Federal University of Santa Catarina, UFSC, Florianopolis, 88040-900, SC, Brazil
Caetana P. Zamparette	Laboratory of Applied Molecular Microbiology, MIMA, Department of Clinical Analysis, Federal University of Santa Catarina, UFSC, Florianopolis, 88040-900, SC, Brazil
Carlos R. Zárate-Bladés	Laboratory of Immunoregulation, iREG, Department of Microbiology, Immunology and Parasitology, Federal University of Santa Catarina, UFSC, Florianopolis, 88040-900, SC, Brazil
Daniela C. Tartari	Laboratory of Applied Molecular Microbiology, MIMA, Department of Clinical Analysis, Federal University of Santa Catarina, UFSC, Florianopolis, 88040-900, SC, Brazil
D. İpek Kurtböke	GeneCology Research Centre and the Faculty of Science, Health, Education and Engineering, University of the Sunshine Coast, Maroochydore DC, Australia
Irshad Ahmad	Department of Life Sciences, King Fahd University of Petroleum and Minerals (KFUPM), Dhahran, Saudi Arabia
Jéssica de Andrade	Hospital Infection Control Service, SCIH, University Hospital "Polydoro Ernani de Sao Thiago", Federal University of Santa Catarina, UFSC, Florianopolis, 88040-900, SC, Brazil

Larissa Buedenbender	Environmental Futures Research Institute, Griffith University, Gold Coast Campus, QLD, Australia
Muhammad Ayaz	Department of Pharmacy, University of Malakand, Khyber Pakhtunkhwa, Pakistan
Muhammad Ovais	CAS Center for Excellence in Nanoscience, CAS Key Laboratory for Biomedical Effects of Nanomaterials and Nanosafety, National Center for Nanoscience and Technology (NCNST), Beijing, P.R. China University of Chinese Academy of Sciences, Beijing, P.R. China
Nashmia Zia	Department of Pharmacy, University of Peshawar, Peshawar, Pakistan
Patricia A. Vanny	Hospital Infection Control Service, SCIH, University Hospital "Polydoro Ernani de Sao Thiago", Federal University of Santa Catarina, UFSC, Florianopolis, 88040-900, SC, Brazil Laboratory of Immunoregulation, iREG, Department of Microbiology, Immunology and Parasitology, Federal University of Santa Catarina, UFSC, Florianopolis, 88040-900, SC, Brazil
Rebeca Monroy-Torres	Environmental Nutrition and Food Security Laboratory, Medicine and Nutrition Department, Health and Science Division, University of Guanajuato, Guanajuato, Mexico
Saif Hameed	Amity Institute of Biotechnology, Amity University Haryana, Gurugram (Manesar), India
Thaís C.M. Sincero	Laboratory of Applied Molecular Microbiology, MIMA, Department of Clinical Analysis, Federal University of Santa Catarina, UFSC, Florianopolis, 88040-900, SC, Brazil
Zeeshan Fatima	Amity Institute of Biotechnology, Amity University Haryana, Gurugram (Manesar), India

CHAPTER 1

Integrated Approaches for Marine Actinomycete Biodiscovery

Larissa Buedenbender[1], **Anthony Richard Carroll**[1] and **D. İpek Kurtböke**[2,*]

[1] *Environmental Futures Research Institute, Griffith University, Gold Coast Campus, QLD, 4222, Australia*

[2] *GeneCology Research Centre and the Faculty of Science, Health, Education and Engineering, University of the Sunshine Coast, Maroochydore DC, QLD4558, Australia*

Abstract: Since the discovery of penicillin in 1928, microbial natural products have been exploited as an unexhausted resource for biodiscovery by the pharmaceutical industry. Unlike primary metabolites such as amino acids, carbohydrates and fatty acids that maintain function and utilized for the growth of an organism; secondary metabolites are specific to its producer but not essential for survival. However, the structural complexity of these natural products is closely linked to the ecological role of the producing organism that supports their survival in their niche. Sessile or slow-moving organisms thus rely more heavily on bioactive secondary metabolites, which act as defences, antimicrobials, allelochemicals, signalling molecules, UV protectants or feeding deterrents, thus often form symbiotic associations with microorganisms that produce such metabolites. Technological advancements and the advent of new tools such as nuclear magnetic resonance (NMR) and mass spectrometry (MS) have enhanced our understanding of the bioactivity of these natural products and aided the discovery of numerous biologically active lead structures, drug candidates, and drugs to treat various diseases. This chapter will thus overview the microbiological and chemical techniques currently used to maximize the discovery of new bioactive compounds in particular, the ones from marine actinomycetes that might be further exploited for their potential as novel and potent drug candidates.

Keywords: Actinomycetes, Actinobacteria, Actinomycetales, Anti-Plasmodial Activity, Ascidians, Chemical Diversity, Drug Discovery, Integrated Approaches to Biodiscovery, Marine Natural Products, Microbial Metabolome, Natural Products.

* **Corresponding author D. İpek Kurtböke:** GeneCology Research Centre and the Faculty of Science, Health, Education and Engineering, University of the Sunshine Coast, Maroochydore DC, QLD 4558, Australia; Tel: +61754302819; Fax: +61(07)54302881; E-mail: IKurtbok@usc.edu.au

Atta-ur-Rahman (Ed.)

INTRODUCTION

Alexander Fleming's discovery of penicillin (**1**) from *Penicillium notatum* in 1928 and its subsequent development into an antibiotic substance in the 1940s was probably the most significant breakthrough in modern medicine, providing the foundation of drug discovery from microorganisms [1]. The commercial production of penicillin allowed for the treatment of bacterial infections, which until then were often fatal. Between 1940 and 2013, 156 natural products were sourced from microorganisms, mainly from the actinobacterial order *Actinomycetales* (commonly termed actinomycetes), have been approved by the U. S. Food and Drug Administration (FDA) as antimicrobials (*i.e.* rifamycin, **2**), anticancer drugs (*i.e.* actinomycin, **3**) and immunosuppressive agents (*i.e.* cyclosporin, **4**) Fig. (**1**) [2].

Of all drugs newly approved between 1981 and 2014, 26% were natural products or natural product derived and 25% were natural product inspired synthetic drugs [3]. Of the natural product derived drugs, 30% were isolated from microorganisms [2] and these findings highlight the importance of natural products in drug discovery, particularly those from microbial sources.

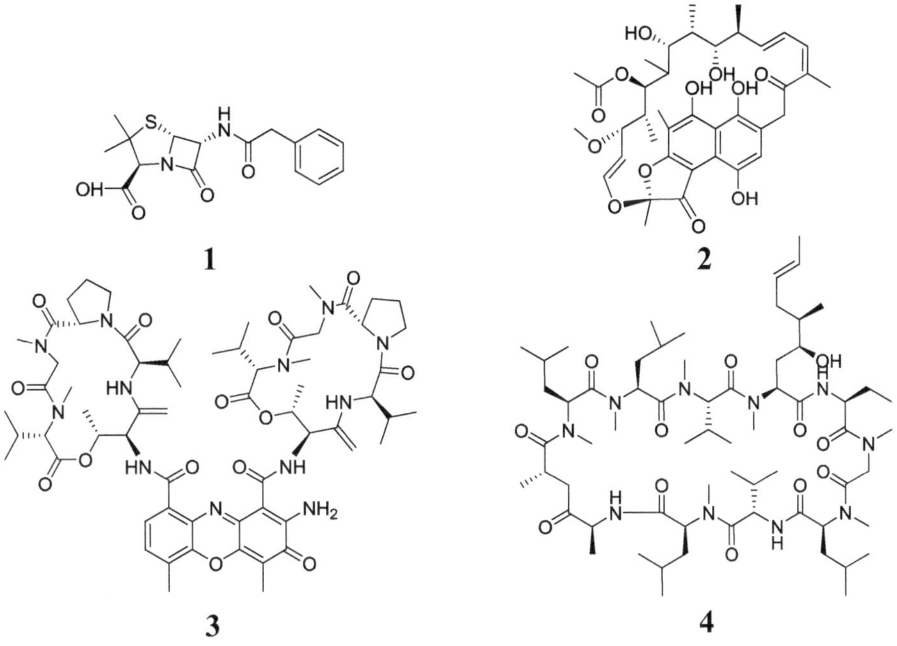

Fig. (1). Examples of potent microbial drugs: penicillin G (**1**), rifamycin SV (**2**), actinomycin (**3**) and cyclosporin (**4**).

However, the time it takes from the discovery of a natural product until it becomes a marketed drug is lengthy and the rediscovery of already known natural products has become increasingly recurrent [4]. In the 1990's, high throughput screening (HTS) of combinatorial compound libraries became favoured by pharmaceutical companies to overcome the rediscovery problem as they are generally easier and cheaper to develop [5]. Yet, the surge did not last long and the number of Food and Drug Administration (FDA) approved drugs decreased once pharmaceutical companies began to focus on purely synthetic drugs [6]. Combinatorial compounds only occupy a very small area of chemical space, while natural products cover a much larger area of chemical space, which aligns well with that of the marketed drugs [7]. Consequently, natural products and their derivatives remain of immense importance for new drug discovery and development. Therefore, innovative approaches increasing chemical diversity are needed to continue with the successful exploitation of natural products for drug development.

One successful approach to enhance chemical diversity is to target organisms inhabiting extreme environments that have not previously been investigated. Examples include target-directed search for rarely isolated actinomycetes such as acidophilic *Catenulispora* and *Actinospica* [8]; or the use of highly selective isolation techniques such as the use of phage battery to remove the common bacterial taxa on isolation plates to recover slow growing rare members of the actinomycetes [9 - 11]. Moreover, the high rediscovery rate of already known natural products from terrestrial bacteria again has directed a focus to microorganisms from different ecological niches, particularly the marine environment. Marine environments cover 70% of the earth's surface and thus provide a vast and highly biodiverse resource for biodiscovery [12]. The underlying hypothesis in marine biodiscovery has been that biological diversity correlates with chemical diversity [13, 14]. Accordingly, in recent years increased research has targeted bacterial species from marine environments, that are taxonomically distinct to terrestrial organisms [12]. The physiochemical properties of the marine environment, including pH, temperature, osmolarity, and pressure, as well as the presence of uncommon halogenated functional groups were suggested to result in the biosynthesis of natural products with enhanced bioactivities and quite different properties compared to the terrestrial environments [12, 15, 16]. Several cases have been reported where natural products isolated from marine invertebrates, particularly from sponges and ascidians, closely resembled analogues from terrestrial microorganisms, suggesting that many marine invertebrate metabolites are synthesised by microbial symbionts. Consequently, a new focus in biodiscovery is the targeting of marine invertebrate symbionts. However, culturing of obligate symbionts in the laboratory still poses a major challenge.

Another way to minimise rediscovery has been to implement sophisticated dereplication approaches early in the drug discovery efforts. An example of this has been pre-screening of microbial strains and extracts with either molecular or spectrometric techniques to avoid repeatedly isolating the same microbial natural products [17]. A different approach to maximise the immense potential of microorganisms for natural product drug discovery has also been to diversify the laboratory conditions. For the target-directed isolation of new microbial species with different properties, media based on marine organic matter that more closely mimic the natural environment in the oceans were used. Examples include the use of sponge extract agar by Webster and co-workers [18] to culture previously uncultured marine organisms. Moreover, since microorganisms produce natural products in response to environmental stresses, together with a range of highly selective culturing media, chemical, physical and biological elicitors that can trigger the biosynthesis of new metabolites were also tested [16, 19]. In addition, bioactive compound secretion was also achieved by Okazaki and co-workers [20] who used a medium containing ‘Kobu Cha’, a Japanese seaweed product, to trigger the production of a benzanthraquinone antibiotic by a marine actinomycete [16, 20].

MARINE MICROBIAL BIODISCOVERY

Culturing Bioactive Marine Bacteria for Biodiscovery

The potential for biodiscovery from marine microorganisms is immense. Extensive microbial 16S rRNA sequence libraries are being established based on metagenomics data that provide location and function of marine microorganisms [21]. Currently, the number of prokaryotic species is estimated to be over a million [22], yet less than 0.1% of those have been cultured in the laboratory [23]. Therefore, it is very likely that the cultured microorganisms do not reflect the actual environmental diversity [22]. The establishment of pure microbial cultures poses the major challenge for microbiologists, but to date, it is still the ultimate goal for biodiscovery in order to examine the physiology, ecology, and metabolism of bioactive compound producing microorganisms [22].

The physiochemical properties of the marine environment are highly unstable, which makes it difficult to recreate appropriate culturing conditions for marine bacteria in the laboratory [24]. The most obvious marine-specific nutrient for obligate marine microorganisms is sodium. Besides the conventional carbon and nitrogen sources, sediment extracts, sponge extracts, and natural seawater were also used to mimic natural environmental conditions [25]. Cultivation temperature may also play a *key* role, especially during the initial isolation attempts. Not all marine bacteria observed through microscopic counts form cultures on agar.

Typically, the observed bacterial count using selective fluorochrome stains, such as DAPI (4,6-diamidino-2-phenylindole), which interacts with nucleic acids to facilitate bacterial counts that can be used with live cells [26], was at least three orders of magnitude higher than the counts achieved through conventional plating techniques [27]. Thus, enrichment techniques are often used, since artificial nutrient-rich agar plates or liquid cultures generally select for fast-growing species that often outcompete slow growing organisms [28]. Enrichment media can be regarded as selective media because it favours the growth of specific bacteria, but its main purpose is to increase the number of bacteria of interest to a detec*table* level [29]. For instance, recently described novel *Salinispora* species were isolated using successful enrichment methods based on the antibiotic resistances *i.e.* against novobiocin, and the ability of these actinomycetes to degrade recalcitrant chitin [30].

Enrichments mainly yield fast-growing bacteria; however, since rare marine taxa have been reported to produce bioactive compounds, efficient high-throughput culturing (HTC) methods have been established to target novel and slow-growing marine bacteria. HTC approaches use low-nutrient media in microtitre plates and the concept of extinction cultures, which was already adopted by the biotech company Diversa Corporation, USA in the early 1980s [31]. Hereby, the samples are diluted to known minute numbers of bacteria ranging from 1 to 10 cells per well [32]. Bacteria are incubated individually for longer periods without being overgrown by fast-growing bacteria. This approach provides 14- to 1400-fold higher cultivation success than traditional microbiological isolation techniques. Previously uncultured marine Proteobacteria clades could be cultured by HTC, which were previously only known through metagenomic studies [33].

Another promising approach to isolate marine bacteria is the diffusion chamber method, which allows connection between the cultured organisms and their environment [34, 35]. This approach separates the target microorganism through a semi-permeable membrane from their natural environment. Nutrients diffuse through the membrane, and toxic substances can diffuse away [36]. First attempts using this approach were very laborious, but successful high-throughput methods have been developed that increased the microbial recovery by up to 50% [35]. The isolation chip (iChip) is an effective version of diffusion chambers that can also be deployed *in situ* [37]. The iChip consists of hundreds of miniature diffusion chambers of approximately 1 mm diameter in a central plate, which is dipped into a microbial suspension in molten agar [37]. Semi-permeable membranes cover the central plate allowing diffusion of nutrients into the chamber but restrict the movement of the cells to the outside environment. Two supporting plates are screwed to the central plate providing sufficient pressure to seal the isolation [37]. This device allows simultaneous isolation of environmental bacteria. Initially, an environmental sample, *i.e.* sediment, is diluted so that approximately one bacterial cell is delivered to each miniature chamber, the device is then sealed with semi-permeable membranes between central and side plates, and incubated back in the natural environment where the sample was taken from originally. Once a colony of sufficient cells is produced, it is likely that the previously uncultured isolates are able to grow *in vitro*. This isolation approach has recently led to a major breakthrough in microbial research as the novel bacterium *Eleftheria terrae* could be isolated from soil samples [38]. This bacterium is the producer of the antibiotic teixobactin, the first new antibiotic class discovered in the last 30 years without detectable resistance [38]. To date, no marine bacteria have been isolated with the iChip, this was probably due to the so far limited incubation time of only two weeks [39]. However, with longer incubation times it might be probable to isolate marine bacteria with optimised iChip conditions in the future.

INTEGRATED APPROACHES IN MICROBIAL BIODISCOVERY

Unrevealing the Diversity and Genetic Potential of Microorganisms

Recent molecular advances including genomic studies highlighted the differences in the genetic potential of different microorganisms that enable them to produce secondary metabolites [40]. Thus, genes coding for these compounds are not uniformly distributed in nature and in fact, most of the bacterial genomes might be lacking gene clusters specific to code secondary metabolite production [8]. However, naturally gifted actinomycetes have been found to possess more than 20 gene clusters coding the synthesis of secondary metabolites; examples include *Streptomyces coelicolor* [8, 41] and *Streptomyces avermitilis* [8, 42, 43]. Moreover, actinomycetes other than streptomycetes were also found to possess multiple gene clusters for secondary metabolism [44]. Genomic data now align with reoccurring trends of diverse metabolite secretions by rare actinomycetes. Furthermore, phylogenetically distant strains were claimed more likely to possess different genes than the phylogenetically related ones [8], as a result phylogenetically unrelated strains are more likely to be targeted for screening of new antibiotics as possession of different genes would enable them to produce of different metabolites [8]. Additionally, recent advances in DNA sequencing technologies made entire genome sequencing possible in rapid and inexpensive ways [45]. During these investigations, actinomycetes were found to contain genes encoding enzymes that synthesize an immense diversity of potential secondary metabolites. Investigations into the homologous and heterologous expression of these often "silent" cryptic secondary metabolite-biosynthetic genes under ordinary laboratory fermentation conditions, led to the discovery of novel secondary metabolites [46].

Metagenomics and Genome Mining in Natural Product Discovery

Diverse marine environments ranging from tropical to polar waters with their adapted marine microflora offer untapped sources for marine biodiscovery. Metagenomics allows for high-throughput analysis of the microbial diversity and distributions in the environment without the need of culturing [47]. Novel molecular advances in the field of metagenomics have been revealing an unprecedented microbial diversity established through sequence-based approaches and function-based approaches that enabled the detection of novel gene clusters from these ocean metagenomes [48, 49]. In the sequence-based approach, the DNA is extracted from environmental samples, such as marine sediment, seawater, or from marine macroorganisms such as sponges and other marine invertebrates to explore their symbionts. Then, generally a short region of the 16S gene is amplified to generate sequences that can be searched in databases and new bioinformatics tools such as the Quantitative Insights Into Microbial Ecology (QIIME) toolbox facilitate assessment of the microbial communities [50]. Metagenomic analyses have so far displayed the sheer number of 'unculturable' microorganisms in the environment including uncovering of a new group of low GC and ultra-small marine Actinobacteria [51].

Functional metagenomics approaches aim to identify gene clusters that encode for bioactive metabolites have also facilitated biodiscoveries [49]. The increased knowledge of biosynthetic pathways, specifically of polyketide synthesis and non-ribosomal peptide synthesis, has opened the door for new development of molecular approaches to natural product drug discovery. Examples of these approaches have been recently been covered by Lane and Moore [40], Pimentel-Elardo *et al.* . [52], Sun *et al.* [53], Trindade-Silva *et al.* [54] and Zotchev *et al.* [49].

One Strain Many Compounds (OSMAC) Approach

Whole genome sequencing has revealed the true genetic potential of microorganisms; however, only a fraction of the biosynthetic genes are transcribed under laboratory conditions and many biosynthetic genes remain silent [55, 56]. Considering the functions of natural products in microorganisms, it is assumed that every natural product is a result of interactions of the organism with its environment [57, 58]. Manipulation of culture conditions with chemical or physical elicitors can exert stresses on the microbial culture and as a result, lead to enhanced production of secondary metabolites. In 2002, Bode and co-workers [59] reported that through slight changes such as media composition, aeration, culture vessel or the addition of enzyme inhibitors, large effects on the secondary metabolite production could be observed. Twenty different metabolites could be isolated from just one strain and the Zeek group gave this approach the term "one stain – many compounds (OSMAC)" [59, 60]. Although this approach was first adapted by Hans Zähner in 1977 and had been in use by the antibiotic industry from the mid-1960s [61]. Nowadays, this approach is a widespread practice and has resulted in the isolation of natural products with enhanced chemical diversity [62 - 65].

Elicitation of the Microbial Metabolome: Community Cultures

In natural environments, microorganisms usually exist in diverse microbial communities. Inter- and intra-specific interactions of microorganisms may stimulate and enhance natural product synthesis [57]. The specific mechanisms of these interactions have not been fully understood yet, although four different mechanisms have been proposed by Abdelmohsen and co-workers [66]. Metabolite synthesis may be triggered through (a) physical cell-to-cell interactions, (b) small molecule mediated interactions, (c) enzymes produced by one species that activate the metabolite precursor of another species. Alternatively, metabolite production could be made possible through gene transfer between two different species [66]. Community cultures (co-cultures) of two or more different microorganisms intend to mimic such interactions in the laboratory [56]. Even though, this approach is perceived as a recent concept; Martin and co-workers [67] already reported on the application of a 5-chambered diffusion apparatus 'EcoLogen' in 1974 and the device has been used by the group for co-cultivation of two organisms that were separated through a diffusive membrane [68]. Several studies have since exploited co-cultivations on solid agar or mixed liquid fermentations with or without diffusion cells to induce biosynthesis of bioactive natural products that are not expressed in standard pure cultures [69 - 73]. Cueto and co-workers [72] demonstrated that mixed fermentations stimulated the production of a new compound, pestalone (**5**, Fig. (**2**)), in a marine *Pestalotia* species when co-cultured with an unidentified antibiotic-resistant marine bacterium. Pestalone, which is active against *Staphylococcus aureus* and *Enterococcus faecium*, was not detected in monocultures of the *Pestalotia* strain [72].

While these techniques allow interactions between different members of the cultured communities that could trigger the production of promising lead structures for biomedical research, co-cultures are highly dynamic and therefore difficult to reproduce. Furthermore, this technique still does not provide other variables of the source environment that in nature stimulate the production of secondary metabolites [56].

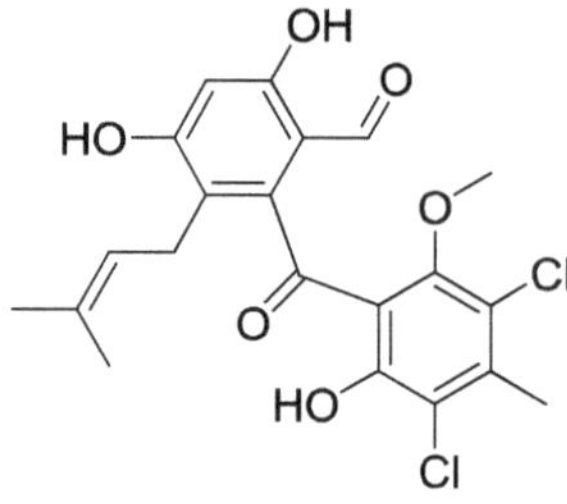

Fig. (2). Pestalone (**5**), a new natural product triggered through co-cultivation.

OVERCOMING THE SUPPLY ISSUES ENCOUNTERED IN MARINE BIODISCOVERY

The re-supply of bioactive compounds derived from natural sources poses a major challenge, as natural populations of marine invertebrates are too small and often only minute amounts of the natural products are produced by these organisms [74]. Nonetheless, marine natural product drug discovery is now an interdisciplinary field, which combines traditional natural products chemistry, synthetic chemistry, microbial- and molecular biology, metabolomics and toxicology to maximize the rates of biodiscovery. The union of these disciplines has resulted in several success stories and delivered new drugs and examples are provided below.

One of the most significant true marine drugs is the ascidian-derived anti-tumour agent trabectedin (**6**), marketed as Yondelis®. This compound was the first drug to be directly sourced from the marine environment. However, the producer *Ecteinascidia turbinata* only yielded 0.0001% of this compound [75]. Five grams of the natural product were needed for clinical trials, and to produce this amount of compound, an unsustainably large quantity of tunicate biomass (5 tonnes) was required [76]. Such extensive harvesting of the marine invertebrates can be restricted due to the general shortage of the marine organisms and can have adverse effects on the environment [74]. In the most extreme case, extinction of the target species could result [24]. Therefore, aquaculture and mariculture were implemented to retrieve more biomass of that ascidian *Ecteinascidia turbinata* [75]. However, yields of in-sea culturing are affected by environmental factors and often low; furthermore, diseases can spread easily in the farm environment; consequently, the pharmaceutical company PharmaMar needed to develop a new approach for full-scale production of the compound [77]. The structure of trabectedin was inherently similar to the base structures of safracins (**7**) and saframycines (**8**) both derived from terrestrial bacteria, indicating that trabectedin was of microbial origin Fig. (**3**) [78, 79]. PharmaMar established a semisynthetic process starting with the bacterial metabolite safracin B [75]. Total synthesis was also described, but the semisynthetic approach provided higher yields as well as additional related compounds [75, 80]. In 2007, Yondelis® was approved as a treatment for soft tissue sarcoma [81] - 38 years after the anti-cancer activity was first detected in the *Ecteinascidia turbinata* extract [82] and 21 years after the structure of the active compound was formally characterised [83,84]. Using metagenomics sequencing of the microbial DNA associated with the ascidian host later identified the biosynthetic gene cluster encoding for trabectedin, which could be linked to the yet uncultured gamma-proteobacterium termed '*Candidatus* Endoecteinascidia frumentensis' based on complete genome sequencing [85].

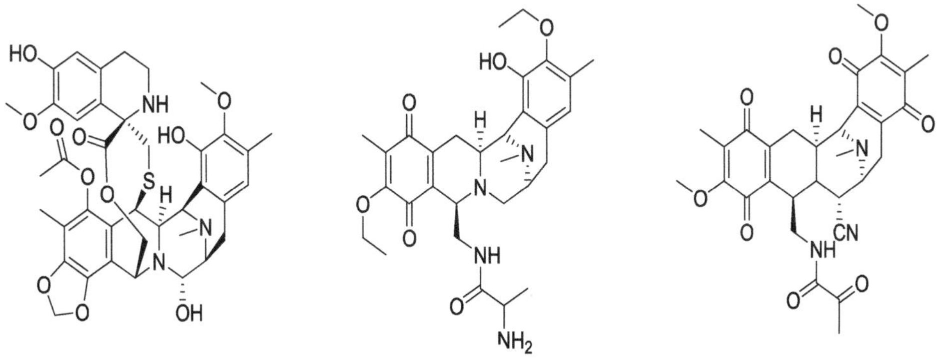

6 **7** **8**

Fig. (3). Trabectedin (**6**) and related safracin A (**7**) and saframycine A (**8**).

As it becomes more evident that symbiotic microorganisms are synthesising many natural products, cultivation of those microorganisms or heterologous gene expression techniques, where genes are expressed in a host organism, hold valuable alternatives to chemical synthesis or extraction from marine invertebrate sources [24, 55, 86 - 88]. For example, the cytotoxic patellamides A (**9**) and C (**10**) Fig. (**4**), originally produced by the cyanobacterium *Prochloron didemni*, could be produced through heterologous gene expression in *E. coli* [89]. Although this approach seems promising, at present it still has its constraints as not all genes will be expressed in all hosts, and it is likely that the target genes need specific chemical or environmental cues in order to be expressed [55, 90]. The supply problem is thus still the biggest hurdle to drug discovery from marine natural products [74], but examples like that of Yondelis® or the total synthesis of Halaven® bring great hope to the natural product research community [91, 92]. Natural products possess many valuable and unique features and there is a continuous interest in the discovery of new organisms, especially from unusual ecological niches [93]. New analytics and biotechnologies will further impel the field [94].

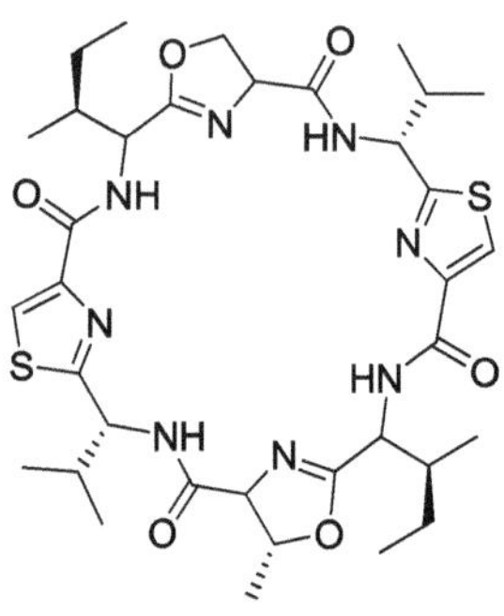

Fig. (4). Patellamides A (**9**) and C (**10**) from ascidian-associated cyanobacteria.

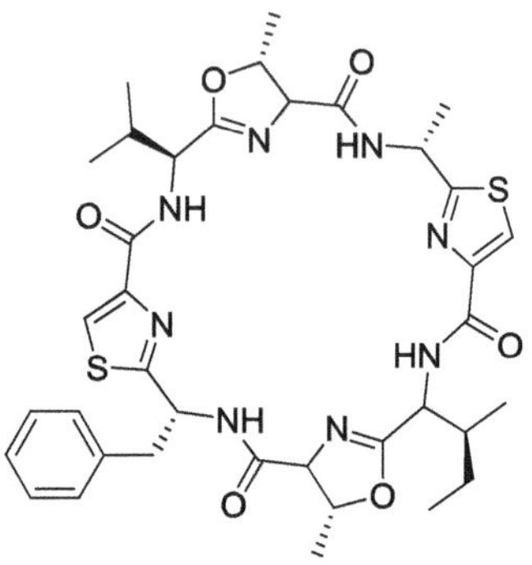

OVERCOMING THE ISSUE OF REDISCOVERY

Dereplication Efforts

Natural products are undoubtedly an important source of new drugs [3]; however, biological screening, large-scale fermentation, isolation and structure elucidation are very resource- and time-intensive stages of the chain of biodiscovery [95]. The frequent rediscovery of already known natural products in the past decades has resulted in a decline in the use of natural products as sources for pharmaceuticals; therefore, new analytical tools were required. To minimise rediscovery, efficient dereplication protocols have to be implemented early in natural product isolation efforts [95, 96]. Dereplication is the identification and elimination of known metabolites within samples that are targeted for new natural product discovery [96]. This is particularly important for microbial natural products, where microbial strains cannot easily be distinguished through morphological features and it is not guaranteed that strains that visually appear different also produce different secondary metabolites [17]. Conversely, strains that display similar morphological features might produce totally distinct metabolomes. Dereplication can either be performed at the microorganismal level or at the chemical extraction stage.

Taxonomic Dereplication

Molecular techniques currently allow dereplication at the genetic level such as the utilization of cluster analysis to identify and dereplicate bacterial species. 16S rRNA sequencing is a widely accepted technique to identify environmental bacteria by using a common gene that possesses the universal primer of all bacteria that is distinct to the small subunit rRNA of eukaryotic organisms [97]. Different microbial species have varying regions within the 16S rRNA sequence that, when matched up against 16S gene databases, can be utilized for identification. Even microorganisms that cannot be identified to the species level will at least be placed in a group of related organisms [97]. The American National Centre for Biotechnology Information (NCBI) offers an extensive open source database of nucleotide sequences and offers a basic local alignment search tool (BLAST) available at https://blast.ncbi.nlm.nih.gov/ that matches input sequences against all deposited sequences [98]. However, the ability to biosynthesise certain metabolites can vary even between different strains of the same taxonomic species [99, 100]. Conversely, facilitated through horizontal gene transfer, taxonomically distinct species may produce the same metabolites [101]. Therefore, culture-dependent methods that pre-screen microbial crude extracts are more commonly utilised by natural product chemists to evaluate the true potential of each bioactive strain.

Chemical Dereplication

Traditionally, sample selection was based on biological pre-screening of the crude extract and subsequent bioassay-guided fractionation was used to identify the bioactive metabolites in the sample [102]. This method is challenged by the repeated isolation of known natural products as it does not provide any chemical information about the active compounds in the extract. Recently, analytical spectroscopic and spectrometric data screening of the entire metabolome of an organism has become more popular to dereplicate samples, and more dereplication databases have emerged [17, 95]. Amongst others, these include the Dictionary of Natural Products, AntiBase, MarinLit, GNPS and DEREP-NP [95, 103 - 106].

The two most common chemical profiling tools are LC-MS/MS and NMR profiling [107]. The advantages of mass spectrometry are the high resolution and sensitivity, as well as the possibility to use MS/MS fragmentation. However, the technique is highly dependent on ionisation of compounds, is difficult to quantify and replicate between different instruments and thus not universally applicable [107]. NMR techniques, on the other hand, have much lower resolution and sensitivity compared to MS, but NMR is non-selective, non-destructive, and reasonably quantitative, it requires minimal sample handling and undeniably its greatest advantage is that it provides structural information about the constituents of the extracts [107].

Metabolomics was originally aimed at studying the total metabolomic processes within organisms and mostly applied in the human health sector, but also plant research has advanced the field of metabolomics [108]. More recently its application has become appreciated in natural products drug discovery, particularly for dereplication and mode of action studies [108]. Chemometric-metabolomics profiling approaches provide an overview of the expressed metabolites in extracts, reduce redundancy and can be used for large datasets without comprehensive fractionation. In chemometric-metabolomics approaches, the chemical profiles are subsequently analysed using multivariate statistics, such as Principal Component Analysis (PCA) or Partial Least Squares Discriminant Analysis (PLS-DA) [109, 110]. PCA is an unsupervised method that reduces multivariate data to a few principal components, while PLS-DA is a supervised technique that determines the variation within a dataset based on classification labels [111]. These techniques produce a scores plot that allows visualisation of the spectral variations of the samples [111]. Samples containing the same or similar chemical profiles will cluster together while differing profiles will cluster away from each other [109 - 111].

Hou and co-workers [112] analysed a microbial collection of 47 samples using LC-MS and utilised PCA to prioritise chemically unique microbial strains for natural product discovery. They noted that one of the limiting factor to this analysis is that it is not readily applicable for large datasets and rather proposed a successive analysis of a smaller number of microbial isolates (20– 50 at a time) [112]. Even in their analysis of 47 isolate samples, over 25000 bucket variables derived from these samples and 74 principal components were needed to explain 98% of the variation in the dataset, indicating that the first two principal components most likely only accounted for a very small percentage of the variation in the data. Nonetheless, the group successfully selected a microbial strain that produced new natural products [113]. MS-based metabolomics approaches are now more commonly used to aid natural product discovery [64, 110, 114 - 117]. LC-MS data is easily matched against databases; however, if a compound is a 'hit' that does not match the database, LC-MS data does not provide any information about the compound except for its molecular mass. More recently, molecular networking based on MS/MS fragmentation patterns that allow assignment of structural classes to unknown compounds has gained a lot of popularity [118, 119].

In recent years, NMR advancement led to faster acquisition through stronger magnets and made this technique more powerful and applicable for metabolomics studies [93, 120]. Yet, only a few natural products studies have used NMR-based metabolomics as a dereplication tool. Chen and co-workers [117] used ^{1}H-NMR as well as LC-MS data for metabolomic profiling of sponge-associated actinomycetes. Thereby, ^{1}H-NMR data of 64 microbial isolates was binned into 0.01 ppm integral regions and the resulting peaks were analysed *via* PCA. Comparison to LC-MS PCA revealed some overlap between the two analytical methods, which both identified a *Streptomyces* sp. to contain unique chemistry [117]. Schroeder and his co-workers [121] used differential analysis of double quantum filtered correlation spectroscopy (DOF-COSY) spectra, which they graphically analysed based on stacking of spectral bitmaps to identify of unique peaks. This was followed by detailed analysis of the unique signals and their corresponding spin-systems, which for a number of compounds was sufficient for their elucidation [121]. For more complex structures, additional HSQC, HMBC, and NOESY spectra were acquired, which identified two new natural products [121].

BIODISCOVERY FROM MARINE ACTINOMYCETES

Actinobacteria are Gram-positive and morphologically diverse bacteria [122]. Currently, intraclass relatedness of the class actinobacteria reveals the presence of nine orders (http://www.bacterio.net/-classifphyla.html, accessed 14-08-2017). Accordingly, in this review “Actinobacteria” will only refer to the “actinomycetes” covering the members of the order *Actinomycetales*. Actinomycetes are often filamentous and present diverse colony morphologies, which according to their function can be divided into areal hyphae, substrate mycelium, and in agar cultures often produce diffusible pigments [123, 124]. They inhabit a wide range of habitats from arid deserts to deep ocean sediments and are responsible for the production of many antibiotics, anticancer agents and immunosuppressive compounds [125, 126]. They are known to have symbiotic associations with plants, insects, and marine invertebrates [12, 127 - 129]. Actinomycetes, among other natural product producing bacteria, are the most prolific suppliers of bioactive compounds to the pharmaceutical industry [130].

Studies of marine actinomycetes date back to the 1940s, when the pioneers of marine microbiology, ZoBell and Rosenfeld, recognised the antimicrobial potential of marine microorganisms [16]. However, only in the last two decades, extensive research in marine microbiology and chemistry has led to the isolation of many marine actinomycete species and the appreciation of their biosynthetic potential for drug discovery. Development of new cultivation techniques allowed isolation of marine actinomycetes, which have produced novel and structurally diverse chemistry [39, 131]. Of the total of over 1000 natural products isolated from marine actinomycetes, 700 were isolated in the last ten years (Fig. **5**).

The marine species described in these studies belong to the genera *Dietzia*, *Streptomyces*, *Salinibacterium*, *Aeromicrobium*, *Williamsia*, *Verrucosispora*

Kocurea, and *Polaribacter*, as well as *Salinispora, Demequina, Salinibacterium, Sciscionella, Serinococcus,* and *Micromonospora*. So far, only a small fraction of actinomycetes can be cultivated from the marine environment. Truly marine-adapted actinomycetes are generally difficult to culture in the laboratory, due to their specific growth requirements [25, 132].

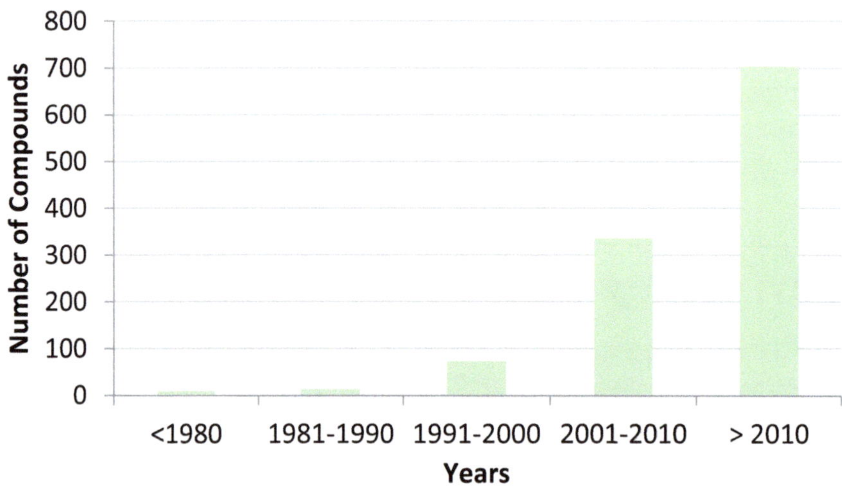

Fig. (5). Number of natural products isolated from marine actinomycetes over the past decades.

However, metagenomic data has revealed 10400 rRNA gene sequences of actinomycetes that are widely distributed throughout the marine environment; 16S rRNA genes have been isolated from the upper water column, different geological sediments, as well as marine sponges, ascidians, and seaweeds [133]. Considering that the marine environment makes up two thirds of the earth's surface, isolations of marine actinomycetes have only been focused on a small range of geographic locations (Fig. **6**). Before the year 2000, natural products had only been reported from twenty different of geographical locations. Once the natural products chemistry community acknowledged the value of marine actinomycetes for biodiscovery, a new focus was placed on new geographical regions, including the Sea of Japan, the China Sea and the Philippines Sea in the 2000s (Fig. **6**). Since then the major focus of marine actinomycete biodiscovery has been the coastlines of the Northern hemisphere. Consulting the mapped distribution of collection locations Fig. (**6**), it becomes evident that there is still immense potential for new biodiscovery from less accessed regions. Particularly, the deep-sea environments remain largely unexplored, most likely due to difficult logistics of accessing these parts of the oceans. However, both cultivation methods and molecular techniques are shedding light into the existence and distribution of marine bacteria providing a broader picture on their biogeography. Such mapping has vital importance for design and development of natural product search and discovery strategies [134].

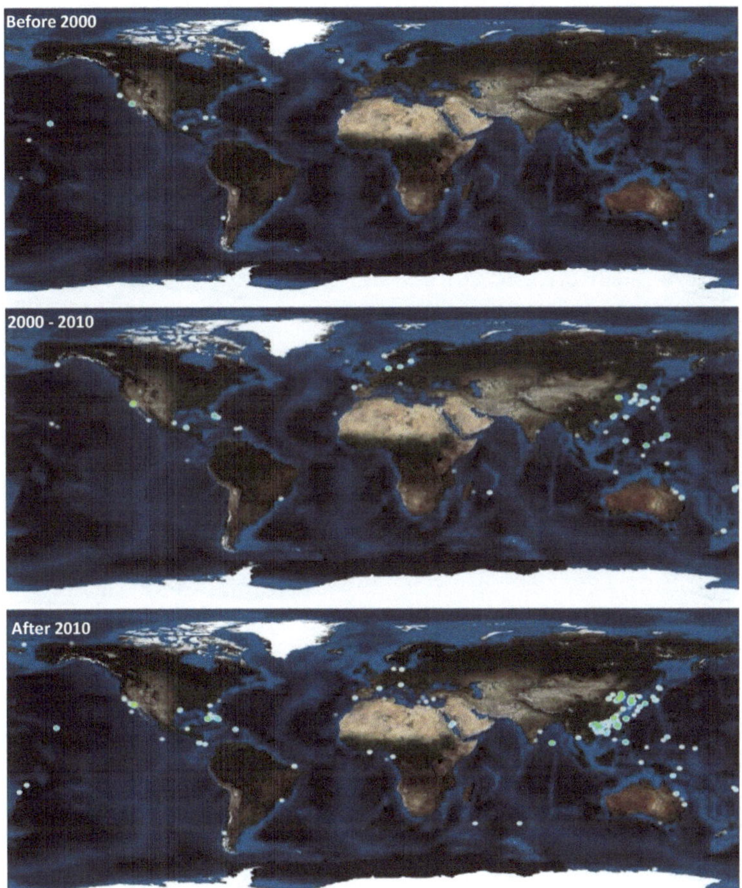

Fig. (6). Collection locations of new compounds isolated from marine actinomycetes per decade (maps generated from data obtained from in MarinLit [103].

Marine Actinomycete Derived Natural Product Diversity

Marine actinomycetes, like their terrestrial counterparts, have an extraordinary large genome, which points towards an exciting potential to biosynthesize secondary metabolites with biological activity [135]. The first full genome sequenced from an obligate marine actinomycete species was that of *Salinispora tropica* strain CNB-440, one of the producers of salinosporamide A [136]. The genome is composed of an average G + C content of 69.5% and 17 secondary metabolic biosynthesis gene clusters, thus devoting approximately 10% of the genome to the production of natural products [136]. Genome sequencing of the marine actinomycete *Salinispora arenicola* revealed 39 biosynthetic loci that produce secondary metabolites [137], including saliniketals (**11**, [138]) and arenamides (**12,** Fig. (**7**)) [139]. However, it has become apparent that most biosynthetic genes are silent under normal laboratory conditions or only low concentrations of natural products are produced [140]. Genome sequences of *Streptomyces coelicolor* and *Streptomyces avermitilis*, two extensively studied species, also revealed interesting biosynthetic gene clusters with the potential to produce new natural products that are not yet expressed under the laboratory conditions [41, 43]. Thus, it seems necessary for marine actinomycete derived drug discovery that new culturing techniques and genome-based natural product mining develop in parallel. The culture-dependent approach is still very important to study the expressed bacterial metabolome, and if successful allows rapid assessment of structure, novelty and bioactivity [141]. In contrast, however, the culture-independent approach will reveal the full biosynthetic potential of marine actinomycetes.

Actinomycetes have yielded several structurally unique and diverse natural products many of which were biologically active. Examples of bioactive marine actinomycete metabolites are: salinipyrones (anti-tumour, **13**), proximicines (anti-tumour, antibiotic, **14**), daryamides (anti-tumour, **15**) or violaceins (anti-protozoal, **16**) Fig. (**7**) [142 - 145]. The most promising discovery was salinosporamide A (**17**) produced by two rare *Salinispora* species. The compound is biosynthesized by an unusual hybrid PKS/NRPS pathway and features a bicyclic ring structure with a chloroethyl group and is functionalised with a cyclohexene [146]. Extracts showed IC_{50} values of less than 2 ng/mL against human colon carcinoma HCT-116 cells and were found to be a highly selective tumour cell growth inhibitor [147, 148]. Salinosporamide A, under the name Marizomib®, is currently sponsored by the biopharmaceutical company Celgene. The U.S. National Institute of Health reports the completion of two Phase I clinical trials for advanced malignancies and one Phase II clinical trial in patients with relapsed/refractory multiple myeloma [149]. Currently, two Phase I clinical trials are recruiting for combination therapy trials for glioblastoma and a Phase III trial of Marizomib® is set to recruit patients with newly diagnosed glioblastoma [149].

We have investigated the chemical diversity of all marine actinomycete natural products that were reported in the literature until the year 2017. Thereof, we generated Self-Organising Maps (SOMs) based on the structural features of the compounds. 'SOMs are a very useful tool to visualise highly multidimensional data in a non-linear fashion against a low-dimensional grid' [150]. Here, we imported the structures of all marine actinomycete compounds into DataWarrior

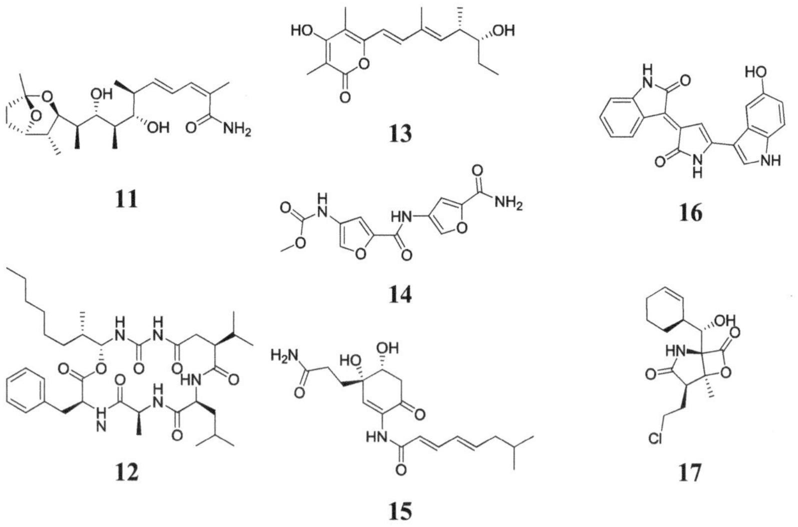

Fig. (7). Examples of bioactive marine actinomycete metabolites: saliniketal (**11**) and arenamide A (**12**) salinipyrone (**13**), proximycin A (**14**), daryamide A (**15**), violacein (**16**), salinosporamide A (**17**).

(http://www.openmolecules.org/datawarrior/), using the 'Create Self-Organising Map' option, SOMs were generated based on certain molecular features such as FragFP and compounds were colour coded according to their genus source (Fig. **8**). From this, it becomes clear that marine actinomycetes produce molecules of great diversity; particularly the genus *Streptomyces* produces the most diverse metabolites (Fig. **8**). This is not surprising, *Streptomyces* from the terrestrial environment are known to be the most prolific natural product producing genus amongst the actinomycetes, and the same appears true for the marine environment where a total of 696 compounds have been isolated from these marine-sourced *Streptomyces*. Interestingly, it appears that all other genera are chemically distinct

a. Diversity of all actinomycete genera
b. Diversity of the genus *Streptomyces*
c. Diversity of actinomycete genera except *Streptomyces*

with little to no overlap in terms of chemical diversity Fig. (**8**).

Between 2015 and 2017, 363 new natural products were derived from marine actinomycetes; this represents about one-third of the isolated compounds (based on data obtained from ESI files associated with the Natural Products Reports reviews on marine natural products [151 - 153]). Of these compounds 211 exhibited activities against a tested biological target, the majority of

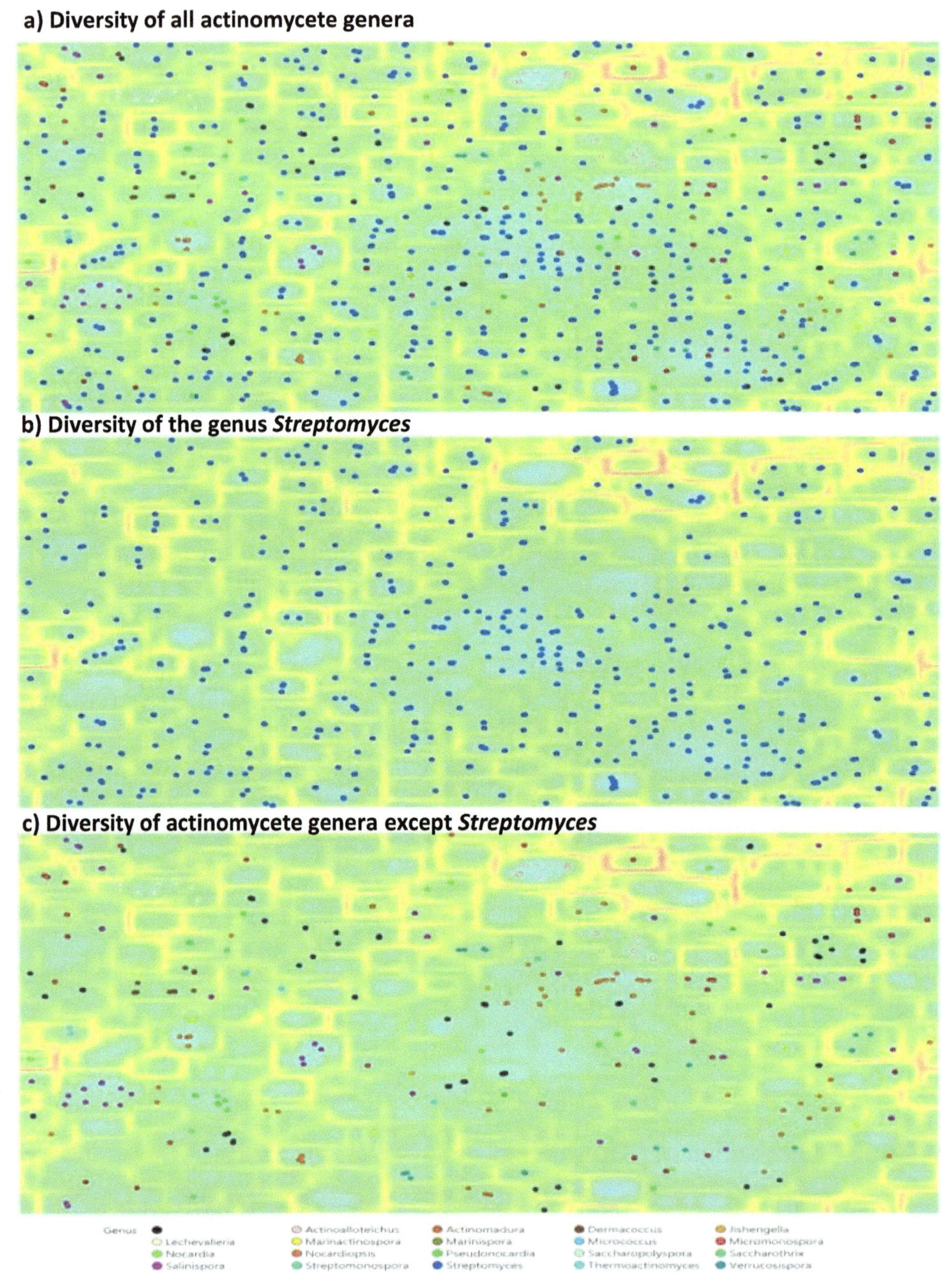

Fig. (8). SOMs illustrating chemical diversity of all isolated marine actinomycete compounds.

actinomycete compounds were cytotoxic and antibacterial (Fig. **9**). Only a few accounts of antimalarial (15), antifungal (10) and antitrypanosomal (5) have been reported to date (Fig. **9**). However, the majority of compounds had not been tested against these targets and it is believed that further screening of these compounds will result in the description of additional bioactivities.

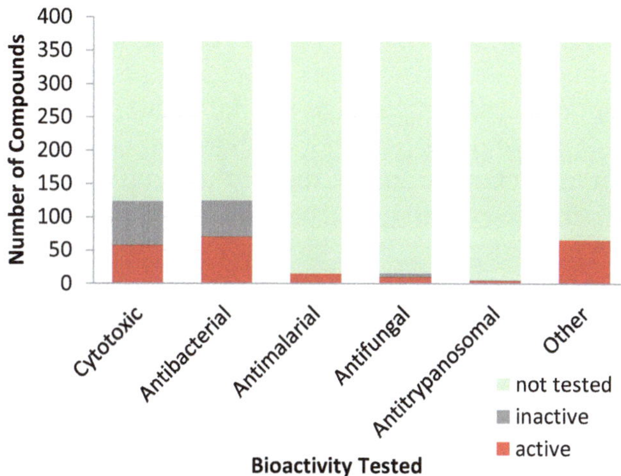

Fig. (9). Tested bioactivities of all of the 363 new marine actinomycete natural products discovered and published in the literature between 2016 and 2018.

Bioactivity Target: The Malaria Parasite

Malaria is a vector-borne infectious disease, caused by a protozoan parasite from the genus *Plasmodium*, with *Plasmodium falciparum* causing the most severe form of human malaria [154]. In 2015, the malaria parasite was responsible for almost half a million human deaths mostly in developing countries of Africa, Latin America and South-East Asia [155]. Humans contract malaria when a female *Anopheles* mosquito feeds on human blood. As the mosquito takes a blood meal, *Plasmodium* sporozoites in its saliva are simultaneously injected into the human body, which then travel to the liver [154]. Here, they bind to negatively charged sugars and asexually reproduce to give rise to thousands of merozoites that invade red blood cells, where the intra-erythrocytic cycle begins [156]. Within the red blood cells, merozoites develop into the early trophozoites, also known as the ring form. The mature trophozoites asexually multiply, ultimately forming 16 - 32 daughter merozoites. The production of daughter merozoites causes the red blood cell to rupture, releasing merozoites into the bloodstream. At this stage, the disease manifests and the human host experiences symptoms of strong fevers and chills [157]. Some merozoites then re-invade new red blood cells, repeating the cycle. Inside the red blood cells, a small portion of merozoites will form schizonts, producing male and female gametocytes. The gametocytes are transmitted back to a mosquito as it takes a blood meal and there they fuse to form oocysts, which divide into sporozoites. These are then transferred to the mosquito's saliva glands and the life cycle of the parasite begins again once the mosquito feeds on a new human host's blood [157].

Most drugs to treat malaria are natural products derived from plants. Quinine (**18**), chloroquine (**19**), artemisinin (**20**) and its derivatives, and antifolates are the most commonly used anti-plasmodial drugs, but resistance is emerging (Fig. **10**). All currently available antimalarial drugs have similar modes of actions: antifolate drugs target the folate biosynthesis pathway, quinoline-type compounds interfere with the enzymatic conversion of toxic ferric haem into to a nontoxic form resulting in cell lysis of the parasite, and artemisinin blocks the production of phosphoinositol [158]. Resistance to chloroquine, the cheapest and safest antimalarial drug, was first reported in Asia in 1957 and by 1970 chloroquine-resistance was spread throughout Asia, Africa, and South America. Therefore, antifolate combination drugs were developed; however, the *Plasmodium* parasite developed resistance soon after their development. Artemisinin brought new hope to treat the disease and artemisinin mono and combination therapy proved effective against the parasite. In 2005, the World Health Organisation reported resistance to artemisinin at the Thailand-Cambodia border and advised withdrawal

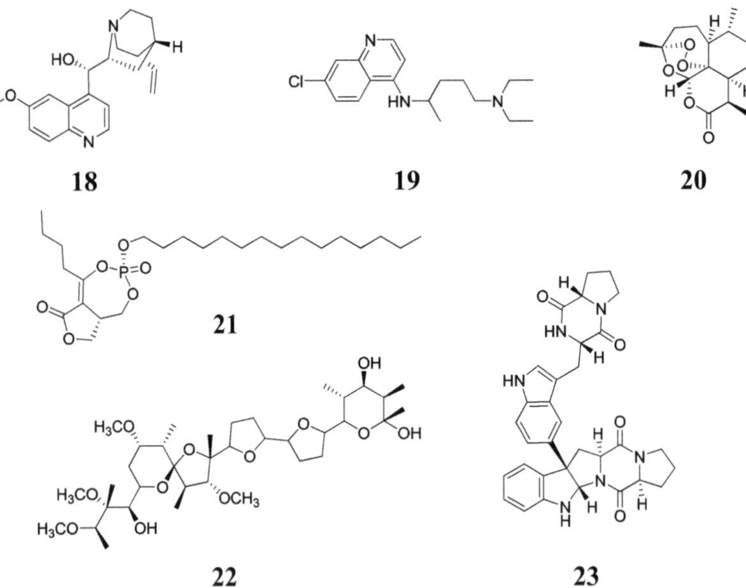

Fig. (10). Anti-plasmodial natural products: quinine (**18**), chloroquine (**19**), artemisinin (**20**), salinipostin A (**21**), a marine *Streptomyces* derived polyether (**22**) and naseseazine C (**23**).

from monotherapy. The spread of artemisinin resistance presents a major threat to millions of people infected with malaria every year. The economic cost of further spread of treatment resistance is estimated at 146 million US$ and 130 million US$ for policy change in the affected areas [159]. A further 385 million US$ associated with productivity loss due to malaria caused morbidity and death have been estimated [159]. Therefore, new drugs with novel targets are urgently needed.

Marine natural products have been again a new area of focus towards finding novel anti-plasmodial agents and marine bacteria have yielded anti-plasmodial natural products [158]. Salinosporamide A (**17**), from the marine actinomycete *Salinispora tropica* demonstrated potent activity against chloroquine-sensitive and resistant *P. faliciparum* clones. In a mouse model, growth inhibition of the parasite was observed at very low doses of 130 µg/kg [160]. The long-chain bicyclic phosphotriester compounds salinipostins A (**21**) - K isolated from the marine-sediment derived *Salinispora* species exhibited anti-plasmodial activity, with salinipostin A having the lowest EC_{50} (50 nM) and high selectivity against mammalian cells [161]. In a study by Na and co-workers [162], a marine *Streptomyces* sp. strain was identified that produced an anti-plasmodial polyether (**22**) (0.145 to 0.29 µM). The diketopiperazine dimer naseseazine C (**23**) was derived from a marine sediment *Streptomyces* species and displayed moderate anti-plasmodial activity of 3.52 µM [163]. Lipinski's 'rule of five', which was based on physiochemical properties of orally available drugs [164], is often the guiding principal for lead-like and drug-like natural product libraries [165,166].Assessment of 126140 unique natural products indicated that 60% had no violation of Lipinski's rule and 80% had less than two violations [166]. Abdelmohsen and co-workers [158] reviewed marine natural products and their drug potential that have biological activity against drug-resistant pathogens. It was found that also most of the marine natural products, including several marine microbial compounds and invertebrate compounds that are likely to be synthesised by a microbial symbiont, exhibit drug-like properties as they adhered to the Lipinski's rule. Fewer anti-plasmodial compounds showed good oral bioavailability; particularly the salinipostins were unfavourable due to their high number of rotatable bonds, thus these characteristics should be assessed thoroughly before pursuing development [158]. However, Lipinski already stated that some orally active natural products do not adhere to the proposed 'rule of five' but can bind to specific transporters [164]. Intravenous administration could also be considered for non-orally active compounds. Thus, new discoveries of natural products from marine environments, specifically from symbiotic actinomycetes of marine macro-organisms might bring new prospects for the treatment of drug-resistant malarial infections.

Untapped Ecological Niche: Ascidian - Actinomycete Associations

Ascidians are sessile, filter-feeding marine invertebrates known to produce highly potent bioactive natural products [167]. These invertebrates are associated with a breadth of associated microorganisms and it has become evident that many of these bioactive metabolites are produced by an associated symbiont [168]. Didemnin B and ecteinascidin 743, two potent anti-tumour agents originally isolated from ascidians, have been proven to be of bacterial origin and similar evidence was found for about another 80 metabolites [169]. The associations between microorganisms and ascidians have been described as species-specific and can be of obligate or facultative nature [89, 170].

Actinomycetes are commonly detected by molecular sequencing techniques in many ascidians [148, 171 - 173]. Most studies have been unable to conclusively confirm any functional role of ascidian-associated actinomycetes. In an ascidian from the *Microcosmus* genus, actinomycetes were associated with the digestive tract most likely through ingestion from the water column through feeding [174]. Actinomycetes in general play a key role in the digestion of recalcitrant organic matter in terrestrial invertebrates [175]. Consequently, through such characteristics, they might have a digestive functional role to play in ascidians that also feed on plankton containing excessive amounts of silica. Furthermore, numerous bioactive compounds isolated from ascidians have remarkable similarities to actinomycete-derived metabolites (Fig. **11**). For instance, the ascidian antitumor compound, namenamicin (**24**), has striking structural similarity to calicheamicins (**25**) and esperamicins (**26**) both synthesised by actinomycetes [176 - 178]. Several staurosporine derivatives (*e.g.* **27**), a structure class known to be synthesised by terrestrial and marine actinomycetes, have repeatedly been isolated from *Eudistoma* ascidians [179], pointing towards a functional role of staurosporine production by an associated actinomycetes strain [180 - 182]. Cultures of ascidian-associated actinomycetes have produced unique and novel chemistry under laboratory conditions; however, most studies were again inconclusive in revealing their functional roles in the ascidian actinomycete association. In the hunt for the true producers of namenamicin, He and co-workers [181] isolated a halophilic *Micromonspora* species, which produced lomaiviticins A (**28**) and B, two novel dimeric diazobenzofluorene glycosides that exhibit potent antitumor properties. A *Streptomyces* sp. strain isolated from *Aplidium lenticulum* at Heron Island, Australia, resulted in the discovery of the aromatic spiroketal polyketide griseorhodin A (**29**) [183] and *Streptomyces* sp. YM14-060 from an unidentified ascidian from Palau produced the new cytotoxic antibiotics, piericidins C7 and C8 (**30**) [184]. Another ascidian derived *Micromonospora* sp. yielded a new antimicrobial alkaloid, diazepinomicin (**31**, [185]); and peptidolipins B – F (**32**) with antibacterial properties were isolated from an ascidian-derived *Nocardia* species [186]. Recently, Zhang and co-workers [113] isolated the two new compounds micromonohalimanes A and B (**33**) from a *Micromonospora* sp. associated with the ascidian, *Symplegma brakenhielmi*; this is the first account of such halimane-type diterpenoids from *Micromonospora*.

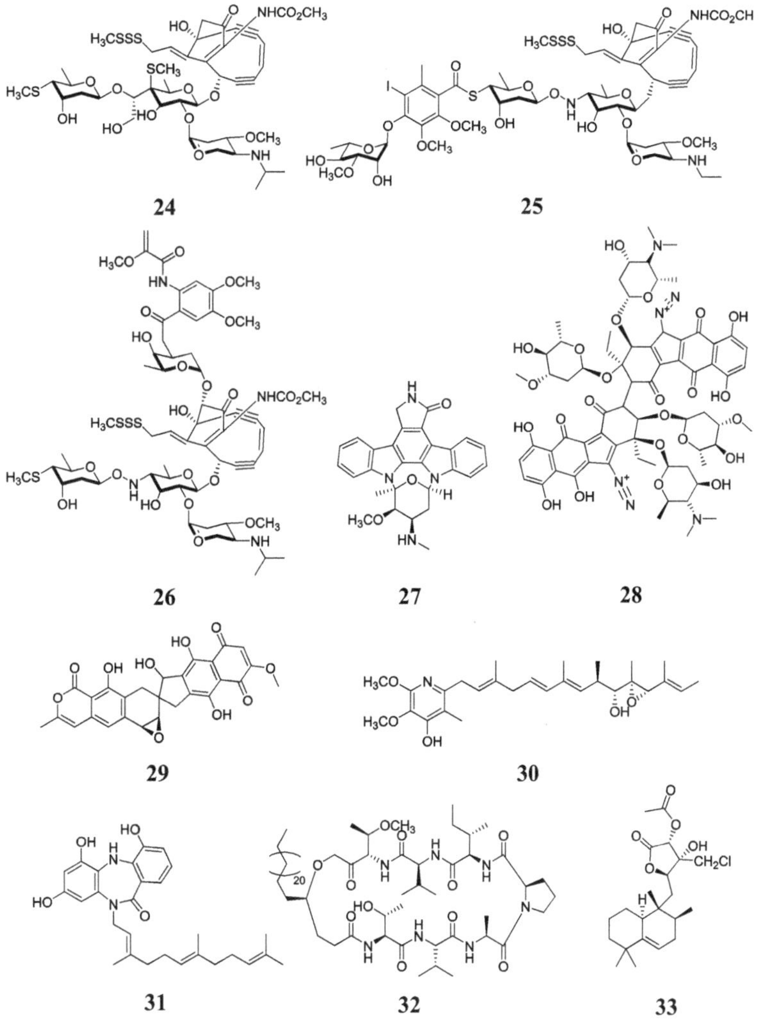

Fig. (11). Diverse secondary metabolites derived or potentially derived from ascidian-associated actinomycetes.

Integrated Approaches to Maximize the Discovery Rate of Marine Actinomycetes as a Source of New Anti-Plasmodial Compounds: *An Australian Example*

In the search for new chemical diversity and anti-plasmodial compounds, actinomycetes were isolated from three different marine ascidians from Hastings Point in Northern New South Wales, Australia. This marine environment is still largely unexplored and the ascidians belonged to the species *Aplidium solidum*, *Polyclinum vasculosum* and *Symplegma rubra*, which had not previously been studied for marine actinomycete biodiscovery [187] (Fig. **12**). Metagenomic 16S rRNA profiling was initially utilised to identify the presence of actinomycete DNA in these ascidian samples. Subsequently, conventional techniques were employed to isolate a total of 120 diverse actinomycetes, belonging to 78 distinct OTUs [187]. These techniques were specifically efficient for the isolation of *Streptomyces* and *Micromonospora* species. After culturing the actinomycete collection in liquid ASW-A media for 2 weeks at 28°C, the cultures were separated from the media through centrifugation and extracted in methanol. The extracts were screened for anti-plasmodial activity and chemical diversity using LC-MS/MS-based molecular networking.

Out of the 120 tested isolates, 63 exhibited inhibition against the *Plasmodium falciparum* (3D7) when tested at a concentration of 0.2 mg/mL Fig. (**13a**). At a lower dose (0.2 µg/mL) 18 isolates still showed inhibitory activity against the malaria parasite. The MS-based chemometric screening approach revealed the immense chemical diversity of the ascidian-associated isolates, and many genus-specific clusters were observed in the GNPS network [187].

Fig. (12). Images of the three ascidians collected from Hastings Point, NSW, Australia (a) *Symplegma rubra* (b) *Aplidium solidum* (c) *Polyclinum vasculosum* (adapted from [187]).

This further confirmed the notion that the chemical diversity is genus-specific as observed in the SOMs (Fig. **8**). Five known natural product MS ions could be identified against the GNPS database [106]. However, most detected ions remained undescribed, indicating that these are potentially new metabolites. Because mass spectrometry does not give any indication about the concentrations nor about the types of molecules present in the extract, we also employed HSQC-TOCSY NMR fingerprinting of the extracts that demonstrated anti-plasmodial activity [188].

The 2D NMR spectrum of the ascidian-associated *Streptomyces* species (USC-16018, (Fig. **13b**)) indicated the presence of a broad diversity of polyketide-associated cross-peaks. Subsequent large-scale fermentation and preparative HPLC resulted in the purification of a new ansamycin polyketide, as well as the known natural products, elaiophylin, cyclo-L-Pro-L-Leu, cyclo-L-Pro-L-Phe, cyclo-L-Pro-L-Val and cyclo-L-Pro-L-Tyr [189]. It was previously reported that elaiophylin exhibited anti-plasmodial activity (IC_{50} 0.22 μg/mL) [190]. Ansamycin polyketides have been identified from terrestrial and marine actinomycetes, with the 17-allylamino-geldanamycin derivative (17AAG) proceeding to advanced clinical trials as an anti-tumour drug [191]. Many ansamycin compounds act as Hsp90 inhibitors. When the *Plasmodium* parasite changes host, heat shock proteins including Hsp90 are highly expressed in the parasite and Hsp90 inhibitors have thus been proposed as a good drug target for anti-malarial drugs [156]. Therefore, the new compound was tested against a chloroquine-sensitive and chloroquine-resistant *P. falciparum* strain and caused inhibition (> 75%) of both parasite strains at a tested concentration of 40 μM [189].

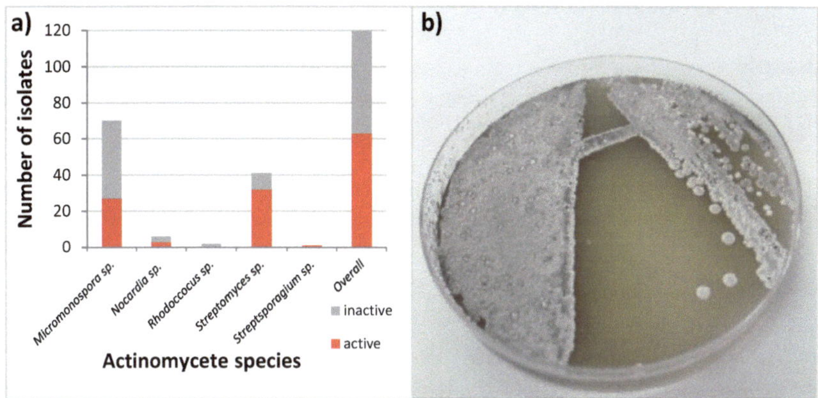

Fig. (13). a) Anti-plasmodial activity tested for 120 ascidian-associated actinomycetes at 0.2 mg/mL crude concentration; b) selected *Streptomyces* sp. USC-16018 (on oatmeal agar) isolated from *S. rubra.*

This study exemplified how integrated approaches can be employed to tackle the issue of rediscovery [189]. Implementation of molecular, microbiological, mass spectrometry and NMR techniques highlighted the abundance and diversity of bioactive compound producing actinomycetes associated with marine ascidians and their immense chemical potential as scaffolds for new drug structures.

Furthermore, chemometric profiling allowed informed selection of the most bioactive actinomycete isolate and resulted in the discovery of a new natural product.

CONCLUSIONS AND PROSPECTS

Natural products present the most successful ecologically adapted metabolites that have evolved to interact with specific biological targets and therefore make effective leads for drug discovery. Microbial natural products, specifically from marine actinomycetes, offer an unmatched structural diversity and can provide a sustainable source of supply. In the past numerous actinomycete metabolites have made their way as drugs to the pharmaceutical market. However, the isolation and elucidation of natural products is time consuming and rediscovery of already known compounds is a major challenge. Consequently, interdisciplinary approaches that connect molecular and microbiology, bioinformatics, metabolomics and analytical chemistry are needed for successful natural product drug discovery as well as in-depth understanding in marine ecology and functional metagenomics of marine bacteria [192]. As stated by Baltz [193] the achievements stemming from genomic sciences, coupled with advancements in genetic engineering of actinomycete secondary metabolite synthesis will result in the second Renaissance of new and potent antibiotic discovery.

CONSENT FOR PUBLICATION

Not applicable.

CONFLICT OF INTEREST

The authors confirm that they have no conflict of interest to declare for this publication.

ACKNOWLEDGEMENTS

The authors acknowledge Professor John Blunt from the University of Canterbury, NZ for provision of all compound structures of marine-derived actinomycetes and their collection locations prior to 2014 and MarinLit for all data on compounds after 2014. We further acknowledge Sandra Duffy and Vicky Avery for biological activity testing of actinomycete crude extracts. L. Buedenbender acknowledges the provision of an Australian Postgraduate Award and Griffith University International Postgraduate Research Scholarship.

REFERENCES

[1] Cragg GM, Grothaus PG, Newman DJ. New horizons for old drugs and drug leads. J Nat Prod 2014; 77(3): 703-23.

[http://dx.doi.org/10.1021/np5000796] [PMID: 24499205]

[2] Patridge E, Gareiss P, Kinch MS, Hoyer D. An analysis of FDA-approved drugs: natural products and their derivatives. Drug Discov Today 2016; 21(2): 204-7. [http://dx.doi.org/10.1016/j.drudis.2015.01.009] [PMID: 25617672]

[3] Newman DJ, Cragg GM. Natural products as sources of new drugs from 1981 to 2014. J Nat Prod 2016; 79(3): 629-61. [http://dx.doi.org/10.1021/acs.jnatprod.5b01055] [PMID: 26852623]

[4] Baltz RH. Marcel Faber Roundtable: is our antibiotic pipeline unproductive because of starvation, constipation or lack of inspiration? J Ind Microbiol Biotechnol 2006; 33(7): 507-13. [http://dx.doi.org/10.1007/s10295-005-0077-9] [PMID: 16418869]

[5] Rouhi AM. Rediscovering natural products. Chem Eng News 2003; 81(41): 77. [http://dx.doi.org/10.1021/cen-v081n041.p077]

[6] Li JW-H, Vederas JC. Drug discovery and natural products: end of an era or an endless frontier? Science 2009; 325(5937): 161-5. [http://dx.doi.org/10.1126/science.1168243] [PMID: 19589993]

[7] Feher M, Schmidt JM. Property distributions: differences between drugs, natural products, and molecules from combinatorial chemistry. J Chem Inf Comput Sci 2003; 43(1): 218-27. [http://dx.doi.org/10.1021/ci0200467] [PMID: 12546556]

[8] Busti E, Monciardini P, Cavaletti L, *et al.* Antibiotic-producing ability by representatives of a newly discovered lineage of actinomycetes. Microbiology 2006; 152(Pt 3): 675-83. [http://dx.doi.org/10.1099/mic.0.28335-0] [PMID: 16514148]

[9] Kurtböke DI. Actinophages as indicators of actinomycete taxa in marine environments. Antonie van Leeuwenhoek 2005; 87(1): 19-28. [http://dx.doi.org/10.1007/s10482-004-6535-y] [PMID: 15726287]

[10] Kurtböke Dİ. Exploitation of phage battery in the search for bioactive actinomycetes. Appl Microbiol Biotechnol 2011; 89(4): 931-7. [http://dx.doi.org/10.1007/s00253-010-3021-5] [PMID: 21120467]

[11] Kurtböke DI. Biodiscovery from rare actinomycetes: an eco-taxonomical perspective. Appl Microbiol Biotechnol 2012; 93(5): 1843-52. [http://dx.doi.org/10.1007/s00253-012-3898-2] [PMID: 22297430]

[12] Kurtböke Dİ, Grkovic T, Quinn RJ. Marine Actinomycetes in biodiscovery Springer Handbook of Marine Biotechnology. Springer 2015; pp. 663-76.

[13] Salomon CE, Magarvey NA, Sherman DH. Merging the potential of microbial genetics with biological and chemical diversity: an even brighter future for marine natural product drug discovery. Nat Prod Rep 2004; 21(1): 105-21. [http://dx.doi.org/10.1039/b301384g] [PMID: 15039838]

[14] Williams PG. Panning for chemical gold: marine bacteria as a source of new therapeutics. Trends Biotechnol 2009; 27(1): 45-52. [http://dx.doi.org/10.1016/j.tibtech.2008.10.005] [PMID: 19022511]

[15] Bhatnagar I, Kim S-K. Immense essence of excellence: marine microbial bioactive compounds. Mar Drugs 2010; 8(10): 2673-701. [http://dx.doi.org/10.3390/md8102673] [PMID: 21116414]

[16] Fenical W. Chemical studies of marine bacteria: developing a new resource. Chem Rev 1993; 93(5): 1673-83. [http://dx.doi.org/10.1021/cr00021a001]

[17] Ito T, Masubuchi M. Dereplication of microbial extracts and related analytical technologies. J Antibiot (Tokyo) 2014; 67(5): 353-60.

[http://dx.doi.org/10.1038/ja.2014.12] [PMID: 24569671]

[18] Webster NS, Wilson KJ, Blackall LL, Hill RT. Phylogenetic diversity of bacteria associated with the marine sponge Rhopaloeides odorabile. Appl Environ Microbiol 2001; 67(1): 434-44. [http://dx.doi.org/10.1128/AEM.67.1.434-444.2001] [PMID: 11133476]

[19] Knight V, Sanglier J-J, DiTullio D, *et al.* Diversifying microbial natural products for drug discovery. Appl Microbiol Biotechnol 2003; 62(5-6): 446-58. [http://dx.doi.org/10.1007/s00253-003-1381-9] [PMID: 12838377]

[20] Okazaki T, Kitahara T, Okami Y. Studies on marine microorganisms. IV. A new antibiotic SS-228 Y produced by Chainia isolated from shallow sea mud. J Antibiot (Tokyo) 1975; 28(3): 176-84. [http://dx.doi.org/10.7164/antibiotics.28.176] [PMID: 1126873]

[21] Simon C, Daniel R. Metagenomic analyses: past and future trends. Appl Environ Microbiol 2011; 77(4): 1153-61. [http://dx.doi.org/10.1128/AEM.02345-10] [PMID: 21169428]

[22] Joint I, Mühling M, Querellou J. Culturing marine bacteria - an essential prerequisite for biodiscovery. Microb Biotechnol 2010; 3(5): 564-75. [http://dx.doi.org/10.1111/j.1751-7915.2010.00188.x] [PMID: 21255353]

[23] Cragg GM, Newman DJ. Natural products: a continuing source of novel drug leads. Biochim Biophys Acta 2013; 1830(6): 3670-95. [http://dx.doi.org/10.1016/j.bbagen.2013.02.008] [PMID: 23428572]

[24] Leal MC, Sheridan C, Osinga R, *et al.* Marine microorganism-invertebrate assemblages: perspectives to solve the “supply problem” in the initial steps of drug discovery. Mar Drugs 2014; 12(7): 3929-52. [http://dx.doi.org/10.3390/md12073929] [PMID: 24983638]

[25] Jensen PR, Gontang E, Mafnas C, Mincer TJ, Fenical W. Culturable marine actinomycete diversity from tropical Pacific Ocean sediments. Environ Microbiol 2005; 7(7): 1039-48. [http://dx.doi.org/10.1111/j.1462-2920.2005.00785.x] [PMID: 15946301]

[26] Roane TM, Pepper IL, Maier RM. Microscopic techniques.Environmental microbiology Chapter 9. 2nd ed. San Diego: Academic Press 2009; pp. 157-72. [http://dx.doi.org/10.1016/B978-0-12-370519-8.00009-2]

[27] Jensen PR, Fenical W. Strategies for the discovery of secondary metabolites from marine bacteria: ecological perspectives. Annu Rev Microbiol 1994; 48(1): 559-84. [http://dx.doi.org/10.1146/annurev.mi.48.100194.003015] [PMID: 7826019]

[28] Pepper IL, Gerba CP. Cultural methods.Environmental microbiology Chapter 10. 2nd ed. San Diego: Academic Press 2009; pp. 173-89. [http://dx.doi.org/10.1016/B978-0-12-370519-8.00010-9]

[29] Tortora GJ, Funke BR, Case CL. Gut microbiota composition correlates with diet and health in the elderly. Nature 2012; 488: pp. ((7410))178-84.

[30] Mincer TJ, Fenical W, Jensen PR. Culture-dependent and culture-independent diversity within the obligate marine actinomycete genus Salinispora. Appl Environ Microbiol 2005; 71(11): 7019-28. [http://dx.doi.org/10.1128/AEM.71.11.7019-7028.2005] [PMID: 16269737]

[31] Newman D. Screening and identification of novel biologically active natural compounds. F1000 Res 2017; 6: 783. [http://dx.doi.org/10.12688/f1000research.11221.1] [PMID: 28649374]

[32] Connon SA, Giovannoni SJ. High-throughput methods for culturing microorganisms in very-lo--nutrient media yield diverse new marine isolates. Appl Environ Microbiol 2002; 68(8): 3878-85. [http://dx.doi.org/10.1128/AEM.68.8.3878-3885.2002] [PMID: 12147485]

[33] Cho J-C, Giovannoni SJ. Cultivation and growth characteristics of a diverse group of oligotrophic marine Gammaproteobacteria. Appl Environ Microbiol 2004; 70(1): 432-40.

[http://dx.doi.org/10.1128/AEM.70.1.432-440.2004] [PMID: 14711672]

[34] Steinert G, Whitfield S, Taylor MW, Thoms C, Schupp PJ. Application of diffusion growth chambers for the cultivation of marine sponge-associated bacteria. Mar Biotechnol (NY) 2014; 16(5): 594-603. [http://dx.doi.org/10.1007/s10126-014-9575-y] [PMID: 24838766]

[35] Nichols D, Cahoon N, Trakhtenberg EM, *et al.* Use of iChip for high-throughput *in situ* cultivation of "uncultivable" microbial species. Appl Environ Microbiol 2010; 76(8): 2445-50. [http://dx.doi.org/10.1128/AEM.01754-09] [PMID: 20173072]

[36] Bollmann A, Lewis K, Epstein SS. Incubation of environmental samples in a diffusion chamber increases the diversity of recovered isolates. Appl Environ Microbiol 2007; 73(20): 6386-90. [http://dx.doi.org/10.1128/AEM.01309-07] [PMID: 17720826]

[37] Lewis K, Epstein S, D'Onofrio A, Ling LL. Uncultured microorganisms as a source of secondary metabolites. J Antibiot (Tokyo) 2010; 63(8): 468-76. [http://dx.doi.org/10.1038/ja.2010.87] [PMID: 20648021]

[38] Ling LL, Schneider T, Peoples AJ, *et al.* A new antibiotic kills pathogens without detectable resistance. Nature 2015.

[39] Hameş-Kocabaş EE, Uzel A. Isolation strategies of marine-derived actinomycetes from sponge and sediment samples. J Microbiol Methods 2012; 88(3): 342-7. [http://dx.doi.org/10.1016/j.mimet.2012.01.010] [PMID: 22285852]

[40] Lane AL, Moore BS. A sea of biosynthesis: marine natural products meet the molecular age. Nat Prod Rep 2011; 28(2): 411-28. [http://dx.doi.org/10.1039/C0NP90032J] [PMID: 21170424]

[41] Bentley SD, Chater KF, Cerdeño-Tárraga A-M, *et al.* Complete genome sequence of the model actinomycete Streptomyces coelicolor A3(2). Nature 2002; 417(6885): 141-7. [http://dx.doi.org/10.1038/417141a] [PMID: 12000953]

[42] Ikeda H, Nonomiya T, Ōmura S. Organization of biosynthetic gene cluster for avermectin in Streptomyces avermitilis: analysis of enzymatic domains in four polyketide synthases. J Ind Microbiol Biotechnol 2001; 27(3): 170-6. [http://dx.doi.org/10.1038/sj.jim.7000092] [PMID: 11780788]

[43] Ōmura S, Ikeda H, Ishikawa J, *et al.* Genome sequence of an industrial microorganism Streptomyces avermitilis: deducing the ability of producing secondary metabolites. Proc Natl Acad Sci USA 2001; 98(21): 12215-20. [http://dx.doi.org/10.1073/pnas.211433198] [PMID: 11572948]

[44] Sosio M, Bossi E, Bianchi A, Donadio S. Multiple peptide synthetase gene clusters in Actinomycetes. Mol Gen Genet 2000; 264(3): 213-21. [http://dx.doi.org/10.1007/s004380000336] [PMID: 11085259]

[45] Ochi K, Tanaka Y, Tojo S. Activating the expression of bacterial cryptic genes by rpoB mutations in RNA polymerase or by rare earth elements. J Ind Microbiol Biotechnol 2014; 41(2): 403-14. [http://dx.doi.org/10.1007/s10295-013-1349-4] [PMID: 24127067]

[46] Corre C, Challis GL. New natural product biosynthetic chemistry discovered by genome mining. Nat Prod Rep 2009; 26(8): 977-86. [http://dx.doi.org/10.1039/b713024b] [PMID: 19636446]

[47] Brady SF, Simmons L, Kim JH, Schmidt EW. Metagenomic approaches to natural products from free-living and symbiotic organisms. Nat Prod Rep 2009; 26(11): 1488-503. [http://dx.doi.org/10.1039/b817078a] [PMID: 19844642]

[48] Kennedy J, Flemer B, Jackson SA, *et al.* Marine metagenomics: new tools for the study and exploitation of marine microbial metabolism. Mar Drugs 2010; 8(3): 608-28. [http://dx.doi.org/10.3390/md8030608] [PMID: 20411118]

[49] Zotchev S, Sekurova O, Kurtböke DI. Metagenomics of marine actinomycetes: From functional gene diversity to biodiscovery. CRC Press 2017; pp. 165-86.

[50] Caporaso JG, Kuczynski J, Stombaugh J, *et al.* QIIME allows analysis of high-throughput community sequencing data. Nat Methods 2010; 7(5): 335-6. [http://dx.doi.org/10.1038/nmeth.f.303] [PMID: 20383131]

[51] Ghai R, Mizuno CM, Picazo A, Camacho A, Rodriguez-Valera F. Metagenomics uncovers a new group of low GC and ultra-small marine Actinobacteria. Sci Rep 2013; 3(1): 2471. [http://dx.doi.org/10.1038/srep02471] [PMID: 23959135]

[52] Pimentel-Elardo SM, Buback V, Gulder TA, *et al.* New tetromycin derivatives with anti-trypanosomal and protease inhibitory activities. Mar Drugs 2011; 9(10): 1682-97. [http://dx.doi.org/10.3390/md9101682] [PMID: 22072992]

[53] Sun W, Peng C, Zhao Y, Li Z. Functional gene-guided discovery of type II polyketides from culturable actinomycetes associated with soft coral Scleronephthya sp. PLoS One 2012; 7(8): e42847. [http://dx.doi.org/10.1371/journal.pone.0042847] [PMID: 22880121]

[54] Trindade-Silva AE, Rua CP, Andrade BG, *et al.* Polyketide synthase gene diversity within the microbiome of the sponge Arenosclera brasiliensis, endemic to the Southern Atlantic Ocean. Appl Environ Microbiol 2013; 79(5): 1598-605. [http://dx.doi.org/10.1128/AEM.03354-12] [PMID: 23275501]

[55] Wilkinson B, Micklefield J. Mining and engineering natural-product biosynthetic pathways. Nat Chem Biol 2007; 3(7): 379-86. [http://dx.doi.org/10.1038/nchembio.2007.7] [PMID: 17576425]

[56] Marmann A, Aly AH, Lin W, Wang B, Proksch P. Co-cultivation--a powerful emerging tool for enhancing the chemical diversity of microorganisms. Mar Drugs 2014; 12(2): 1043-65. [http://dx.doi.org/10.3390/md12021043] [PMID: 24549204]

[57] Scherlach K, Hertweck C. Triggering cryptic natural product biosynthesis in microorganisms. Org Biomol Chem 2009; 7(9): 1753-60. [http://dx.doi.org/10.1039/b821578b] [PMID: 19590766]

[58] Alain K, Querellou J. Cultivating the uncultured: limits, advances and future challenges. Extremophiles 2009; 13(4): 583-94. [http://dx.doi.org/10.1007/s00792-009-0261-3] [PMID: 19548063]

[59] Bode HB, Bethe B, Höfs R, Zeeck A. Big effects from small changes: possible ways to explore nature's chemical diversity. ChemBioChem 2002; 3(7): 619-27. [http://dx.doi.org/10.1002/1439-7633(20020703)3:7<619::AID-CBIC619>3.0.CO;2-9] [PMID: 12324995]

[60] Schiewe HJ, Zeeck A. Cineromycins, γ-butyrolactones and ansamycins by analysis of the secondary metabolite pattern created by a single strain of streptomyces. J antibioti 1999; 52(7): 635-42.

[61] Zähner H. Some aspects of antibiotics research. Angew Chem Int Ed Engl 1977; 16(10): 687-94. [http://dx.doi.org/10.1002/anie.197706871] [PMID: 412435]

[62] English AL, Boufridi A, Quinn RJ, Kurtböke DI. Evaluation of fermentation conditions triggering increased antibacterial activity from a near-shore marine intertidal environment-associated *Streptomyces* species. Synth Syst Biotechnol 2016; 2(1): 28-38. [http://dx.doi.org/10.1016/j.synbio.2016.09.005] [PMID: 29062959]

[63] Pinedo C, Moraga J, Barua J, *et al.* Chemically induced cryptic sesquiterpenoids and expression of sesquiterpene cyclases in botrytis cinerea revealed new sporogenic (+)-4-epi eremophil-9-en-11-ols. ACS Chem Biol 2016; 11(5): 1391-400. [http://dx.doi.org/10.1021/acschembio.5b00931] [PMID: 26900713]

[64] Abdelmohsen UR, Cheng C, Viegelmann C, *et al.* Dereplication strategies for targeted isolation of

new antitrypanosomal actinosporins A and B from a marine sponge associated-Actinokineospora sp. EG49. Mar Drugs 2014; 12(3): 1220-44.
[http://dx.doi.org/10.3390/md12031220] [PMID: 24663112]

[65] Lin Z, Zhu T, Wei H, Zhang G, Wang H, Gu Q. Spicochalasin a and new aspochalasins from the marine□derived fungus spicaria elegans. Eur J Org Chem 2009; 2009(18): 3045-51.
[http://dx.doi.org/10.1002/ejoc.200801085]

[66] Abdelmohsen UR, Grkovic T, Balasubramanian S, Kamel MS, Quinn RJ, Hentschel U. Elicitation of secondary metabolism in actinomycetes. Biotechnol Adv 2015; 33(6 Pt 1): 798-811.
[http://dx.doi.org/10.1016/j.biotechadv.2015.06.003] [PMID: 26087412]

[67] Martin DF, Kutt EC, Kim YS. Use of a multiple diffusion chamber unit in culture studies. Application to gomphosphaeria aponina. Environ Lett 1974; 7(1): 39-46.
[http://dx.doi.org/10.1080/00139307409437382]

[68] Martin BB, Martin DF. Use of the ecologen to study hydrilla growth inhibitors. J Aquat Plant Manage 1986; 24: 82-4.

[69] Traxler MF, Watrous JD, Alexandrov T, Dorrestein PC, Kolter R. Interspecies interactions stimulate diversification of the Streptomyces coelicolor secreted metabolome. MBio 2013; 4(4): e00459-13.
[http://dx.doi.org/10.1128/mBio.00459-13] [PMID: 23963177]

[70] Slattery M, Rajbhandari I, Wesson K. Competition-mediated antibiotic induction in the marine bacterium Streptomyces tenjimariensis. Microb Ecol 2001; 41(2): 90-6.
[PMID: 12032613]

[71] Zhu F, Chen G, Chen X, Huang M, Wan X. Aspergicin, a new antibacterial alkaloid produced by mixed fermentation of two marine-derived mangrove epiphytic fungi. Chem Nat Compd 2011; 47(5): 767-9.
[http://dx.doi.org/10.1007/s10600-011-0053-8]

[72] Cueto M, Jensen PR, Kauffman C, Fenical W, Lobkovsky E, Clardy J. Pestalone, a new antibiotic produced by a marine fungus in response to bacterial challenge. J Nat Prod 2001; 64(11): 1444-6.
[http://dx.doi.org/10.1021/np0102713] [PMID: 11720529]

[73] Onaka H, Mori Y, Igarashi Y, Furumai T. Mycolic acid-containing bacteria induce natural-product biosynthesis in Streptomyces species. Appl Environ Microbiol 2011; 77(2): 400-6.
[http://dx.doi.org/10.1128/AEM.01337-10] [PMID: 21097597]

[74] Radjasa OK, Vaske YM, Navarro G, *et al.* Highlights of marine invertebrate-derived biosynthetic products: their biomedical potential and possible production by microbial associants. Bioorg Med Chem 2011; 19(22): 6658-74.
[http://dx.doi.org/10.1016/j.bmc.2011.07.017] [PMID: 21835627]

[75] Cuevas C, Francesch A. Development of Yondelis (trabectedin, ET-743). A semisynthetic process solves the supply problem. Nat Prod Rep 2009; 26(3): 322-37.
[http://dx.doi.org/10.1039/b808331m] [PMID: 19240944]

[76] Martins A, Vieira H, Gaspar H, Santos S. Marketed marine natural products in the pharmaceutical and cosmeceutical industries: tips for success. Mar Drugs 2014; 12(2): 1066-101.
[http://dx.doi.org/10.3390/md12021066] [PMID: 24549205]

[77] Dunlap WC, Jaspars M, Hranueli D, *et al.* New methods for medicinal chemistry--universal gene cloning and expression systems for production of marine bioactive metabolites. Curr Med Chem 2006; 13(6): 697-710.
[http://dx.doi.org/10.2174/092986706776055643] [PMID: 16529560]

[78] Ikeda Y, Shimada Y, Honjo K, Okumoto T, Munakata T. Safracins, new antitumor antibiotics. III. Biological activity. J Antibiot (Tokyo) 1983; 36(10): 1290-4.
[http://dx.doi.org/10.7164/antibiotics.36.1290] [PMID: 6358171]

[79] Arai T, Takahashi K, Ishiguro K, Yazawa K. Increased production of saframycin A and isolation of

saframycin S. J Antibiot (Tokyo) 1980; 33(9): 951-60.
[http://dx.doi.org/10.7164/antibiotics.33.951] [PMID: 7440416]

[80] Endo A, Yanagisawa A, Abe M, Tohma S, Kan T, Fukuyama T. Total synthesis of ecteinascidin 743. J Am Chem Soc 2002; 124(23): 6552-4.
[http://dx.doi.org/10.1021/ja026216d] [PMID: 12047173]

[81] Gordon EM, Sankhala KK, Chawla N, Chawla SP. Trabectedin for soft tissue sarcoma: current status and future perspectives. Adv Ther 2016; 33(7): 1055-71.
[http://dx.doi.org/10.1007/s12325-016-0344-3] [PMID: 27234989]

[82] Sigel MM, Wellham LL, Lichter W, *et al.* In Food-Drugs from the sea: Proceedings. Washington, DC: Marine Technology Society 1969; pp. 281-94.

[83] Rinehart KL, Holt TG, Fregeau NL, *et al.* Ecteinascidins 729, 743, 745, 759a, 759b, and 770: Potent antitumor agents from the Caribbean tunicate Ecteinascidia turbinata. J Org Chem 1990; 55(15): 4512-5.
[http://dx.doi.org/10.1021/jo00302a007]

[84] Wright AE, Forleo DA, Gunawardana GP, Gunasekera SP, Koehn FE, McConnell OJ. Antitumor tetrahydroisoquinoline alkaloids from the colonial ascidian ecteinascidia turbinata. J Org Chem 1990; 55(15): 4508-12.
[http://dx.doi.org/10.1021/jo00302a006]

[85] Schofield MM, Jain S, Porat D, Dick GJ, Sherman DH. Identification and analysis of the bacterial endosymbiont specialized for production of the chemotherapeutic natural product ET-743. Environ Microbiol 2015; 17(10): 3964-75.
[http://dx.doi.org/10.1111/1462-2920.12908] [PMID: 26013440]

[86] Schmidt EW. The secret to a successful relationship: lasting chemistry between ascidians and their symbiotic bacteria. Invertebr Biol 2015; 134(1): 88-102.
[http://dx.doi.org/10.1111/ivb.12071] [PMID: 25937788]

[87] Malpartida F, Hopwood DA. Molecular cloning of the whole biosynthetic pathway of a Streptomyces antibiotic and its expression in a heterologous host. Nature 1984; 309(5967): 462-4.
[http://dx.doi.org/10.1038/309462a0] [PMID: 6328317]

[88] Wenzel SC, Müller R. Recent developments towards the heterologous expression of complex bacterial natural product biosynthetic pathways. Curr Opin Biotechnol 2005; 16(6): 594-606.
[http://dx.doi.org/10.1016/j.copbio.2005.10.001] [PMID: 16226455]

[89] Long PF, Dunlap WC, Battershill CN, Jaspars M. Shotgun cloning and heterologous expression of the patellamide gene cluster as a strategy to achieving sustained metabolite production. ChemBioChem 2005; 6(10): 1760-5.
[http://dx.doi.org/10.1002/cbic.200500210] [PMID: 15988766]

[90] Warren RL, Freeman JD, Levesque RC, Smailus DE, Flibotte S, Holt RA. Transcription of foreign DNA in Escherichia coli. Genome Res 2008; 18(11): 1798-805.
[http://dx.doi.org/10.1101/gr.080358.108] [PMID: 18701636]

[91] Newman DJ. Developing natural product drugs: Supply problems and how they have been overcome. Pharmacol Ther 2016; 162: 1-9.
[http://dx.doi.org/10.1016/j.pharmthera.2015.12.002] [PMID: 26706239]

[92] Aicher TD, Buszek KR, Fang FG, *et al.* Total synthesis of halichondrin B and norhalichondrin B. J Am Chem Soc 1992; 114(8): 3162-4.
[http://dx.doi.org/10.1021/ja00034a086]

[93] Molinski TF. All natural: the renaissance of natural products chemistry. Org Lett 2014; 16(15): 3849-55.
[http://dx.doi.org/10.1021/ol501917g] [PMID: 25081565]

[94] Imhoff JF, Labes A, Wiese J. Bio-mining the microbial treasures of the ocean: new natural products.

Biotechnol Adv 2011; 29(5): 468-82.
[http://dx.doi.org/10.1016/j.biotechadv.2011.03.001] [PMID: 21419836]

[95] Zani CL, Carroll AR. Database for rapid dereplication of known natural products using data from MS and fast NMR experiments. J Nat Prod 2017; 80(6): 1758-66.
[http://dx.doi.org/10.1021/acs.jnatprod.6b01093] [PMID: 28616931]

[96] Gaudêncio SP, Pereira F. Dereplication: racing to speed up the natural products discovery process. Nat Prod Rep 2015; 32(6): 779-810.
[http://dx.doi.org/10.1039/C4NP00134F] [PMID: 25850681]

[97] Hildebrand M, Waggoner LE, Lim GE, Sharp KH, Ridley CP, Haygood MG. Approaches to identify, clone, and express symbiont bioactive metabolite genes. Nat Prod Rep 2004; 21(1): 122-42.
[http://dx.doi.org/10.1039/b302336m] [PMID: 15039839]

[98] Altschul SF, Gish W, Miller W, Myers EW, Lipman DJ. Basic local alignment search tool. J Mol Biol 1990; 215(3): 403-10.
[http://dx.doi.org/10.1016/S0022-2836(05)80360-2] [PMID: 2231712]

[99] Ritacco FV, Haltli B, Janso JE, Greenstein M, Bernan VS. Dereplication of Streptomyces soil isolates and detection of specific biosynthetic genes using an automated ribotyping instrument. J Ind Microbiol Biotechnol 2003; 30(8): 472-9.
[http://dx.doi.org/10.1007/s10295-003-0038-0] [PMID: 12687492]

[100] Waksman SA, Schatz A. Strain specificity and production of antibiotic substances. Proc Natl Acad Sci USA 1943; 29(2): 74-9.
[http://dx.doi.org/10.1073/pnas.29.2.74] [PMID: 16588605]

[101] Lawrence JG. Selfish operons and speciation by gene transfer. Trends Microbiol 1997; 5(9): 355-9.
[http://dx.doi.org/10.1016/S0966-842X(97)01110-4] [PMID: 9294891]

[102] Antonio J, Molinski TF. Screening of marine invertebrates for the presence of ergosterol-sensitive antifungal compounds. J Nat Prod 1993; 56(1): 54-61.
[http://dx.doi.org/10.1021/np50091a008] [PMID: 8450321]

[103] MarinLit A database of the marine natural products literature [homepage on the Internet]. Cambridge: Royal Society of Chemistry [updated 20th Apr 2018; cited 20th Apr 2018]. http://pubs.rsc.org/marinlit/ Available from

[104] AntiBase A Data Base for the Identification of Natural Products [homepage on the Internet].Göttingen: Hartmut Laatsch, Department of Organic Chemistry, University of Göttingen [cited: 20th Apr 2018] Available from: http://wwwuser.gwdg.de/~hlaatsc/antibase.htm

[105] Whittle M, Willett P, Klaffke W, van Noort P. Evaluation of similarity measures for searching the dictionary of natural products database. J Chem Inf Comput Sci 2003; 43(2): 449-57.
[http://dx.doi.org/10.1021/ci025591m] [PMID: 12653508]

[106] Wang M, Carver JJ, Phelan VV, *et al.* Sharing and community curation of mass spectrometry data with global natural products social molecular networking. Nat Biotechnol 2016; 34(8): 828-37.
[http://dx.doi.org/10.1038/nbt.3597] [PMID: 27504778]

[107] Kurita KL, Linington RG. Connecting phenotype and chemotype: high-content discovery strategies for natural products research. J Nat Prod 2015; 78(3): 587-96.
[http://dx.doi.org/10.1021/acs.jnatprod.5b00017] [PMID: 25728167]

[108] Rochfort S. Metabolomics reviewed: a new "omics" platform technology for systems biology and implications for natural products research. J Nat Prod 2005; 68(12): 1813-20.
[http://dx.doi.org/10.1021/np050255w] [PMID: 16378385]

[109] Tawfike AF, Viegelmann C, Edrada-Ebel R. Metabolomics and dereplication strategies in natural products.Metabolomics tools for natural product discovery: Methods and Protocols. Totowa: Humana Press 2013; pp. 227-44.
[http://dx.doi.org/10.1007/978-1-62703-577-4_17]

[110] Macintyre L, Zhang T, Viegelmann C, *et al.* Metabolomic tools for secondary metabolite discovery from marine microbial symbionts. Mar Drugs 2014; 12(6): 3416-48. [http://dx.doi.org/10.3390/md12063416] [PMID: 24905482]

[111] Grootveld M. Introduction to the applications of chemometric techniques in _omics' research: Common pitfalls, misconceptions and _rights and wrongs'. Metabolic profiling: Disease and xenobiotics 2012; pp. 1-34.

[112] Hou Y, Braun DR, Michel CR, *et al.* Microbial strain prioritization using metabolomics tools for the discovery of natural products. Anal Chem 2012; 84(10): 4277-83. [http://dx.doi.org/10.1021/ac202623g] [PMID: 22519562]

[113] Zhang Y, Adnani N, Braun DR, *et al.* Micromonohalimanes A and B: Antibacterial Halimane-Type Diterpenoids from a Marine Micromonospora Species. J Nat Prod 2016; 79(11): 2968-72. [http://dx.doi.org/10.1021/acs.jnatprod.6b00555] [PMID: 27813411]

[114] Ito T, Odake T, Katoh H, Yamaguchi Y, Aoki M. High-throughput profiling of microbial extracts. J Nat Prod 2011; 74(5): 983-8. [http://dx.doi.org/10.1021/np100859a] [PMID: 21495658]

[115] Nielsen KF, Månsson M, Rank C, Frisvad JC, Larsen TO. Dereplication of microbial natural products by LC-DAD-TOFMS. J Nat Prod 2011; 74(11): 2338-48. [http://dx.doi.org/10.1021/np200254t] [PMID: 22026385]

[116] Bose U, Hewavitharana AK, Ng YK, Shaw PN, Fuerst JA, Hodson MP. LC-MS-based metabolomics study of marine bacterial secondary metabolite and antibiotic production in Salinispora arenicola. Mar Drugs 2015; 13(1): 249-66. [http://dx.doi.org/10.3390/md13010249] [PMID: 25574739]

[117] Cheng C, MacIntyre L, Abdelmohsen UR, *et al.* Biodiversity, anti-trypanosomal activity screening, and metabolomic profiling of actinomycetes isolated from mediterranean sponges. PLoS One 2015; 10(9): e0138528. [http://dx.doi.org/10.1371/journal.pone.0138528] [PMID: 26407167]

[118] Watrous J, Roach P, Alexandrov T, *et al.* Mass spectral molecular networking of living microbial colonies. Proc Natl Acad Sci USA 2012; 109(26): E1743-52. [http://dx.doi.org/10.1073/pnas.1203689109] [PMID: 22586093]

[119] Yang JY, Sanchez LM, Rath CM, *et al.* Molecular networking as a dereplication strategy. J Nat Prod 2013; 76(9): 1686-99. [http://dx.doi.org/10.1021/np400413s] [PMID: 24025162]

[120] Halouska S, Zhang B, Gaupp R, *et al.* Revisiting protocols for the NMR analysis of bacterial metabolomes. J Integr OMICS 2013; 3(2): 120-37. [PMID: 26078915]

[121] Schroeder FC, Gibson DM, Churchill AC, *et al.* Differential analysis of 2D NMR spectra: new natural products from a pilot-scale fungal extract library. Angew Chem Int Ed Engl 2007; 46(6): 901-4. [http://dx.doi.org/10.1002/anie.200603821] [PMID: 17183517]

[122] Schaal KP, Yassin AF, Stackebrandt E. The family actinomycetaceae: The genera actinomyces, actinobaculum, arcanobacterium, varibaculum and mobiluncus The prokaryotes. Springer 2006; pp. 430-537.

[123] Waksman SA. The actinomycetes: Their natura, occurance, activities and importance. Waltham: Chronica Botanica Company 1950; pp. 1-246.

[124] Li Q, Chen X, Jiang Y, Jiang C. Morphological identification of actinobacteria.Dhanasekaran D, Jiang Y. Actinobacteria - Basics and Biotechnological Applications Rijeka: In Tech Open 2016; pp. 59-86. [http://dx.doi.org/10.5772/61461]

[125] Santhanam R, Rong X, Huang Y, Andrews BA, Asenjo JA, Goodfellow M. Streptomyces bullii sp.

nov., isolated from a hyper-arid atacama desert soil. Antonie van leeuwenhoek 2013; 103(2): 367-73. [http://dx.doi.org/10.1007/s10482-012-9816-x] [PMID: 23011007]

[126] Luo Y, Xiao J, Wang Y, Xu J, Xie S, Xu J. Streptomyces indicus sp. nov., an actinomycete isolated from deep-sea sediment. Int J Syst Evol Microbiol 2011; 61(Pt 11): 2712-6. [http://dx.doi.org/10.1099/ijs.0.029389-0] [PMID: 21169463]

[127] Passari AK, Mishra VK, Saikia R, Gupta VK, Singh BP. Isolation, abundance and phylogenetic affiliation of endophytic actinomycetes associated with medicinal plants and screening for their *in vitro* antimicrobial biosynthetic potential. Front Microbiol 2015; 6: 273. [http://dx.doi.org/10.3389/fmicb.2015.00273] [PMID: 25904906]

[128] Abdelmohsen UR, Pimentel-Elardo SM, Hanora A, *et al.* Isolation, phylogenetic analysis and anti-infective activity screening of marine sponge-associated actinomycetes. Mar Drugs 2010; 8(3): 399-412. [http://dx.doi.org/10.3390/md8030399] [PMID: 20411105]

[129] Mahmoud HM, Kalendar AA. Coral-associated Actinobacteria: diversity, abundance, and biotechnological potentials. Front Microbiol 2016; 7(204): 204. [PMID: 26973601]

[130] Bérdy J. Bioactive microbial metabolites. J Antibiot (Tokyo) 2005; 58(1): 1-26. [http://dx.doi.org/10.1038/ja.2005.1] [PMID: 15813176]

[131] Subramani R, Aalbersberg W. Culturable rare Actinomycetes: diversity, isolation and marine natural product discovery. Appl Microbiol Biotechnol 2013; 97(21): 9291-321. [http://dx.doi.org/10.1007/s00253-013-5229-7] [PMID: 24057404]

[132] Fenical W, Jensen PR. Developing a new resource for drug discovery: marine actinomycete bacteria. Nat Chem Biol 2006; 2(12): 666-73. [http://dx.doi.org/10.1038/nchembio841] [PMID: 17108984]

[133] Abdelmohsen UR, Bayer K, Hentschel U. Diversity, abundance and natural products of marine sponge-associated actinomycetes. Nat Prod Rep 2014; 31(3): 381-99. [http://dx.doi.org/10.1039/C3NP70111E] [PMID: 24496105]

[134] Ward AC, Bora N. Diversity and biogeography of marine actinobacteria. Curr Opin Microbiol 2006; 9(3): 279-86. [http://dx.doi.org/10.1016/j.mib.2006.04.004] [PMID: 16675292]

[135] Nett M, König GM. The chemistry of gliding bacteria. Nat Prod Rep 2007; 24(6): 1245-61. [http://dx.doi.org/10.1039/b612668p] [PMID: 18033578]

[136] Udwary DW, Zeigler L, Asolkar RN, *et al.* Genome sequencing reveals complex secondary metabolome in the marine actinomycete Salinispora tropica. Proc Natl Acad Sci USA 2007; 104(25): 10376-81. [http://dx.doi.org/10.1073/pnas.0700962104] [PMID: 17563368]

[137] Demain AL. Importance of microbial natural products and the need to revitalize their discovery. J Ind Microbiol Biotechnol 2014; 41(2): 185-201. [http://dx.doi.org/10.1007/s10295-013-1325-z] [PMID: 23990168]

[138] Williams PG, Asolkar RN, Kondratyuk T, Pezzuto JM, Jensen PR, Fenical W. Saliniketals A and B, bicyclic polyketides from the marine actinomycete Salinispora arenicola. J Nat Prod 2007; 70(1): 83-8. [http://dx.doi.org/10.1021/np0604580] [PMID: 17253854]

[139] Asolkar RN, Freel KC, Jensen PR, *et al.* Arenamides A-C, cytotoxic NFkappaB inhibitors from the marine actinomycete Salinispora arenicola. J Nat Prod 2009; 72(3): 396-402. [http://dx.doi.org/10.1021/np800617a] [PMID: 19117399]

[140] Wu M-C, Law B, Wilkinson B, Micklefield J. Bioengineering natural product biosynthetic pathways for therapeutic applications. Curr Opin Biotechnol 2012; 23(6): 931-40. [http://dx.doi.org/10.1016/j.copbio.2012.03.008] [PMID: 22487048]

[141] Zotchev SB. Marine actinomycetes as an emerging resource for the drug development pipelines. J Biotechnol 2012; 158(4): 168-75.
[http://dx.doi.org/10.1016/j.jbiotec.2011.06.002] [PMID: 21683100]

[142] Olano C, Méndez C, Salas JA. Antitumor compounds from marine actinomycetes. Mar Drugs 2009; 7(2): 210-48.
[http://dx.doi.org/10.3390/md7020210] [PMID: 19597582]

[143] Kwon HC, Kauffman CA, Jensen PR, Fenical W. Marinomycins A-D, antitumor-antibiotics of a new structure class from a marine actinomycete of the recently discovered genus "marinispora". J Am Chem Soc 2006; 128(5): 1622-32.
[http://dx.doi.org/10.1021/ja0558948] [PMID: 16448135]

[144] Asolkar RN, Jensen PR, Kauffman CA, Fenical W. Daryamides A-C, weakly cytotoxic polyketides from a marine-derived actinomycete of the genus Streptomyces strain CNQ-085. J Nat Prod 2006; 69(12): 1756-9.
[http://dx.doi.org/10.1021/np0603828] [PMID: 17190455]

[145] Durán N, Justo GZ, Ferreira CV, Melo PS, Cordi L, Martins D. Violacein: properties and biological activities. Biotechnol Appl Biochem 2007; 48(Pt 3): 127-33.
[http://dx.doi.org/10.1042/BA20070115] [PMID: 17927569]

[146] Feling RH, Buchanan GO, Mincer TJ, Kauffman CA, Jensen PR, Fenical W. Salinosporamide A: a highly cytotoxic proteasome inhibitor from a novel microbial source, a marine bacterium of the new genus salinospora. Angew Chem Int Ed Engl 2003; 42(3): 355-7.
[http://dx.doi.org/10.1002/anie.200390115] [PMID: 12548698]

[147] Fenical W, Jensen PR, Palladino MA, Lam KS, Lloyd GK, Potts BC. Discovery and development of the anticancer agent salinosporamide A (NPI-0052). Bioorg Med Chem 2009; 17(6): 2175-80.
[http://dx.doi.org/10.1016/j.bmc.2008.10.075] [PMID: 19022674]

[148] Valliappan K, Sun W, Li Z. Marine actinobacteria associated with marine organisms and their potentials in producing pharmaceutical natural products. Appl Microbiol Biotechnol 2014; 98(17): 7365-77.
[http://dx.doi.org/10.1007/s00253-014-5954-6] [PMID: 25064352]

[149] clinicaltrials.gov. [homepage on the Internet]. Bethesda: The U.S. National Institute of Health [cited:20th April 2018]. https://clinicaltrials.gov/ Available from

[150] Kohonen T. The self-organizing map. Neurocomputing 1998; 21(1): 1-6.
[http://dx.doi.org/10.1016/S0925-2312(98)00030-7]

[151] Blunt JW, Carroll AR, Copp BR, Davis RA, Keyzers RA, Prinsep MR. Metronidazole resistance in *Clostridium difficile* is heterogeneous J Clin Microbiol 2008; 46(9): 3028-2.
[http://dx.doi.org/10.1039/C7NP00052A]

[152] Blunt JW, Copp BR, Keyzers RA, Munro MH, Prinsep MR. Marine natural products. Nat Prod Rep 2016; 33(3): 382-431.
[http://dx.doi.org/10.1039/C5NP00156K] [PMID: 26837534]

[153] Blunt JW, Copp BR, Keyzers RA, Munro MHG, Prinsep MR. Marine natural products. Nat Prod Rep 2017; 34(3): 235-94.
[http://dx.doi.org/10.1039/C6NP00124F] [PMID: 28290569]

[154] Tuteja R. Malaria - an overview. FEBS J 2007; 274(18): 4670-9.
[http://dx.doi.org/10.1111/j.1742-4658.2007.05997.x] [PMID: 17824953]

[155] World Health Organization. World malaria report 2016 Licence No: CC BY-NC-SA 30 IGO. Geneva: World Health Organization 2016.

[156] Shahinas D, Folefoc A, Pillai DR. Targeting Plasmodium falciparum Hsp90: towards reversing antimalarial resistance. Pathogens 2013; 2(1): 33-54.

[http://dx.doi.org/10.3390/pathogens2010033] [PMID: 25436880]

[157] Michalakis Y, Renaud F. Malaria: Evolution in vector control. Nature 2009; 462(7271): 298-300. [http://dx.doi.org/10.1038/462298a] [PMID: 19924207]

[158] Abdelmohsen UR, Balasubramanian S, Oelschlaeger TA, *et al.* Potential of marine natural products against drug-resistant fungal, viral, and parasitic infections. Lancet Infect Dis 2017; 17(2): e30-41. [http://dx.doi.org/10.1016/S1473-3099(16)30323-1] [PMID: 27979695]

[159] Lubell Y, Dondorp A, Guérin PJ, *et al.* Artemisinin resistance--modelling the potential human and economic costs. Malar J 2014; 13(1): 452. [http://dx.doi.org/10.1186/1475-2875-13-452] [PMID: 25418416]

[160] Prudhomme J, McDaniel E, Ponts N, *et al.* Marine actinomycetes: a new source of compounds against the human malaria parasite. PLoS One 2008; 3(6): e2335. [http://dx.doi.org/10.1371/journal.pone.0002335] [PMID: 18523554]

[161] Schulze CJ, Navarro G, Ebert D, DeRisi J, Linington RG. Salinipostins A-K, long-chain bicyclic phosphotriesters as a potent and selective antimalarial chemotype. J Org Chem 2015; 80(3): 1312-20. [http://dx.doi.org/10.1021/jo5024409] [PMID: 25584395]

[162] Na M, Meujo DA, Kevin D, Hamann MT, Anderson M, Hill RT. A new antimalarial polyether from a marine Streptomyces sp. H668. Tetrahedron Lett 2008; 49(44): 6282-5. [http://dx.doi.org/10.1016/j.tetlet.2008.08.052] [PMID: 19865468]

[163] Buedenbender L, Grkovic T, Duffy S, Kurtböke DI, Avery VM, Carroll AR. Naseseazine C, a new anti-plasmodial dimeric diketopiperazine from a marine sediment derived Streptomyces sp. Tetrahedron Lett 2016; 57(52): 5893-5. [http://dx.doi.org/10.1016/j.tetlet.2016.11.071]

[164] Lipinski CA, Lombardo F, Dominy BW, Feeney PJ. Experimental and computational approaches to estimate solubility and permeability in drug discovery and development settings. Adv Drug Deliv Rev 2001; 46(1-3): 3-26. [http://dx.doi.org/10.1016/S0169-409X(00)00129-0] [PMID: 11259830]

[165] Camp D, Davis RA, Campitelli M, Ebdon J, Quinn RJ. Drug-like properties: guiding principles for the design of natural product libraries. J Nat Prod 2012; 75(1): 72-81. [http://dx.doi.org/10.1021/np200687v] [PMID: 22204643]

[166] Quinn RJ, Carroll AR, Pham NB, *et al.* Developing a drug-like natural product library. J Nat Prod 2008; 71(3): 464-8. [http://dx.doi.org/10.1021/np070526y] [PMID: 18257534]

[167] Schmidt EW, Donia MS, McIntosh JA, Fricke WF, Ravel J. Origin and variation of tunicate secondary metabolites. J Nat Prod 2012; 75(2): 295-304. [http://dx.doi.org/10.1021/np200665k] [PMID: 22233390]

[168] Crawford JM, Clardy J. Bacterial symbionts and natural products. Chem Commun (Camb) 2011; 47(27): 7559-66. [http://dx.doi.org/10.1039/c1cc11574j] [PMID: 21594283]

[169] Chen L, Fu C, Wang G. Microbial diversity associated with ascidians: a review of research methods and application. Symbiosis 2016; •••: 1-8.

[170] Schreiber L, Kjeldsen KU, Funch P, *et al.* Endozoicomonas are specific, facultative symbionts of sea squirts. Front Microbiol 2016; 7: 1042. [http://dx.doi.org/10.3389/fmicb.2016.01042] [PMID: 27462299]

[171] Behrendt L, Larkum AW, Trampe E, Norman A, Sørensen SJ, Kühl M. Microbial diversity of biofilm communities in microniches associated with the didemnid ascidian Lissoclinum patella. ISME J 2012; 6(6): 1222-37. [http://dx.doi.org/10.1038/ismej.2011.181] [PMID: 22134643]

[172] Tianero MDB, Kwan JC, Wyche TP, *et al.* Species specificity of symbiosis and secondary metabolism in ascidians. ISME J 2015; 9(3): 615-28. [http://dx.doi.org/10.1038/ismej.2014.152] [PMID: 25171330]

[173] Steinert G, Taylor MW, Schupp PJ. Diversity of actinobacteria associated with the marine ascidian Eudistoma toealensis. Mar Biotechnol (NY) 2015; 17(4): 377-85. [http://dx.doi.org/10.1007/s10126-015-9622-3] [PMID: 25678260]

[174] Meziti A, Kormas KA, Pancucci-Papadopoulou M-A, Thessalou-Legaki M. Bacterial phylotypes associated with the digestive tract of the sea urchin Paracentrotus lividus and the ascidian Microcosmus sp. Russ J Mar Biol 2007; 33(2): 84-91. [http://dx.doi.org/10.1134/S1063074007020022]

[175] Breznak JA, Brune A. Role of microorganisms in the digestion of lignocellulose by termites. Annu Rev Entomol 1994; 39(1): 453-87. [http://dx.doi.org/10.1146/annurev.en.39.010194.002321]

[176] McDonald LA, Capson TL, Krishnamurthy G, *et al.* Namenamicin, a new enediyne antitumor antibiotic from the marine ascidian Polysyncraton lithostrotum. J Am Chem Soc 1996; 118(44): 10898-9. [http://dx.doi.org/10.1021/ja961122n]

[177] Lee MD, Ellestad GA, Borders DB. Calicheamicins: discovery, structure, chemistry, and interaction with DNA. Acc Chem Res 1991; 24(8): 235-43. [http://dx.doi.org/10.1021/ar00008a003]

[178] Golik J, Clardy J, Dubay G, *et al.* Esperamicins, a novel class of potent antitumor antibiotics. 2. Structure of esperamicin X. J Am Chem Soc 1987; 109(11): 3461-2. [http://dx.doi.org/10.1021/ja00245a048]

[179] Horton PA, Longley RE, McConnell OJ, Ballas LM. Staurosporine aglycone (K252-c) and arcyriaflavin A from the marine ascidian, Eudistoma sp. Experientia 1994; 50(9): 843-5. [http://dx.doi.org/10.1007/BF01956468] [PMID: 7925852]

[180] Andréo MA, Jimenez PC, Siebra JB, *et al.* Systematic UPLC-ESI-MS/MS study on the occurrence of staurosporine and derivatives in associated marine microorganisms from Eudistoma vannamei. J Braz Chem Soc 2012; 23(2): 335-43. [http://dx.doi.org/10.1590/S0103-50532012000200021]

[181] He H, Ding W-D, Bernan VS, *et al.* Lomaiviticins A and B, potent antitumor antibiotics from Micromonospora lomaivitiensis. J Am Chem Soc 2001; 123(22): 5362-3. [http://dx.doi.org/10.1021/ja010129o] [PMID: 11457405]

[182] Jimenez PC, Ferreira EG, Araújo LA, *et al.* Cytotoxicity of actinomycetes associated with the ascidian Eudistoma vannamei (Millar, 1977), endemic of northeastern coast of Brazil/Citotoxicidad de actinomicetos asociada a la ascidia Eudistoma vannamei (Millar, 1977), endémica de la costa noreste de Brasil. Lat Am J Aquat Res 2013; 41(2): 335.

[183] Li A, Piel J. A gene cluster from a marine Streptomyces encoding the biosynthesis of the aromatic spiroketal polyketide griseorhodin A. Chem Biol 2002; 9(9): 1017-26. [http://dx.doi.org/10.1016/S1074-5521(02)00223-5] [PMID: 12323376]

[184] Hayakawa Y, Shirasaki S, Shiba S, *et al.* Piericidins C7 and C8, new cytotoxic antibiotics produced by a marine Streptomyces sp. J Antibiot (Tokyo) 2007; 60(3): 196-200. [http://dx.doi.org/10.1038/ja.2007.22] [PMID: 17446692]

[185] Charan RD, Schlingmann G, Janso J, Bernan V, Feng X, Carter GT. Diazepinomicin, a new antimicrobial alkaloid from a marine Micromonospora sp. J Nat Prod 2004; 67(8): 1431-3. [http://dx.doi.org/10.1021/np040042r] [PMID: 15332871]

[186] Wyche TP, Hou Y, Vazquez-Rivera E, Braun D, Bugni TS. Peptidolipins B-F, antibacterial lipopeptides from an ascidian-derived Nocardia sp. J Nat Prod 2012; 75(4): 735-40.

[http://dx.doi.org/10.1021/np300016r] [PMID: 22482367]

[187] Buedenbender L, Carroll AR, Ekins M, Kurtböke Dİ. Taxonomic and Metabolite Diversity of Actinomycetes Associated with Three Australian Ascidians. Diversity (Basel) 2017; 9(4): 53. [http://dx.doi.org/10.3390/d9040053]

[188] Buedenbender L, Habener LJ, Grkovic T, *et al.* HSQC–TOCSY fingerprinting for prioritization of polyketide- and peptide-producing microbial isolates. J Nat Prod 2018; 81(4): 957-65. [http://dx.doi.org/10.1021/acs.jnatprod.7b01063] [PMID: 29498849]

[189] Buedenbender L. Integrated approaches to marine actinomycete biodiscovery Griffith University 2017.

[190] Supong K, Sripreechasak P, Tanasupawat S, Danwisetkanjana K, Rachtawee P, Pittayakhajonwut P. Investigation on antimicrobial agents of the terrestrial Streptomyces sp. BCC71188. Appl Microbiol Biotechnol 2017; 101(2): 533-43. [http://dx.doi.org/10.1007/s00253-016-7804-1] [PMID: 27554496]

[191] Grem JL, Morrison G, Guo X-D, *et al.* Phase I and pharmacologic study of 17-(allylamino)-17-demethoxygeldanamycin in adult patients with solid tumors. J Clin Oncol 2005; 23(9): 1885-93. [http://dx.doi.org/10.1200/JCO.2005.12.085] [PMID: 15774780]

[192] Kurtböke DI, Okazaki T, Vobis G. Actinobacteria in marine environments: from terrigenous origin to adapted functional diversity. Encyclopedia of Marine Biotechnology Kim S-K, (ed). 2018. in press, ISBN: 978-1-119-14377-2.

[193] Baltz RH. Renaissance in antibacterial discovery from actinomycetes. Curr Opin Pharmacol 2008; 8(5): 557-63. [http://dx.doi.org/10.1016/j.coph.2008.04.008] [PMID: 18524678]

CHAPTER 2

Therapeutic Use of Commensal Microbes: Fecal/Gut Microbiota Transplantation

Patricia A. Vanny[1,2], Aline Machiavelli[2,3], Catherine W. Wippel[1], Jéssica de Andrade[1], Caetana P. Zamparette[4], Daniela C. Tartari[4], Thaís C.M. Sincero[4], Aguinaldo R. Pinto[3] and **Carlos R. Zárate-Bladés[2,*]**

[1] *Hospital Infection Control Service, SCIH, University Hospital "Polydoro Ernani de Sao Thiago", Federal University of Santa Catarina, UFSC, Florianopolis, 88040-900, SC, Brazil*

[2] *Laboratory of Immunoregulation, iREG, Department of Microbiology, Immunology and Parasitology, Federal University of Santa Catarina, UFSC, Florianopolis, 88040-900, SC, Brazil*

[3] *Laboratory of Applied Immunology, LIA, Department of Microbiology, Immunology and Parasitology, Federal University of Santa Catarina, UFSC, Florianopolis, 88040-900, SC, Brazil*

[4] *Laboratory of Applied Molecular Microbiology, MIMA, Department of Clinical Analysis; Federal University of Santa Catarina, UFSC, Florianopolis, 88040-900, SC, Brazil*

Abstract: Recognition of the microbial component of human physiology has gained attention in recent years. Several discoveries in different medical disciplines about the participation of commensal microorganisms, broadly known as the microbiota, have guided both basic research and the development of novel therapeutic strategies. In parallel, the broad use of antibiotics has directly impacted microbiota composition with collateral effects, such as the emergence of antibiotic-associated diarrheas (AADs), with the best example being intestinal infection with *Clostridium difficile*. An ancient Chinese methodology called "fecal microbiota transplantation (FMT)" – referred to here as "gut microbiota transplantation (GMT)" – is currently being used around the world with impressive success against recurrent *C. difficile* infections (CDIs). This success has inspired several investigators to test this microbiota-based therapy in other conditions, including non-infectious diseases. We introduce this chapter with a review on the main microbiota effects on human health and the immune system. Then, a comprehensive discussion of GMT is provided including the pathogenesis of CDI, GMT history, methodology and ethical aspects, followed by a review of clinical and experimental studies on the therapeutic effects and mechanisms of GMT against recurrent CDI. Finally, GMT testing in other conditions is discussed.

[*] **Corresponding author Carlos R. Zárate-Bladés:** Laboratory of Immunoregulation, iREG, Federal University of Santa Catarina, UFSC, Florianopolis, SC, 88040-900, Brazil; Tel: 55-48-3721-5210; Fax: 55-48-3721-9672; E-mail: zarate.blades@ufsc.br

Atta-ur-Rahman (Ed.)

Keywords: Antibiotic, Antibiotic-Associated Diarrhea, Clinical Trial, *Clostridium difficile*, Colonoscopy, Diarrhea, Experimental Infection, Fecal Pills, Fecal Transfer, Gut, Immune Response, Infection, Infusion, Microbiome, Microbiota, Microbiota Transplantation, Mucosal Immunity, Nasogastric Tube, Recurrent, Pseudomembranous Colitis, Sequencing, Stool Donor, Transplant, Toxin.

INTRODUCTION

It has been calculated that the human microbial ecosystem consists of anywhere from equal numbers of microbial and human cells to ten times more microbial cells than human cells [1, 2]. The intestinal microbiota is the largest in quantity and complexity, including bacteria, archaea, fungi, protists and viruses, with approximately 10^{11} to 10^{12} microbial cells/ml of intestinal content. The human intestinal microbiota consists of 1,200 different species, and the corresponding microbiome (a term that refers to the genes present in the commensal bacteria genomes) is composed of 3.3 million unique genes, many of which encode proteins with functions absent in the human genome [3, 4]. This explains why the contribution of microbial commensals to human physiology is largely associated with the ability of microbial commensals to execute numerous metabolic processes that exceed human capacity.

The association of the microbiota with its host has gained attention in recent years, leading to the emergence of the term the "human holobiont". This is a reflection of the astonishing quantity of scientific data describing the participation of commensal microbiota in several human systems throughout a lifetime and therefore validates the view of commensal microbiota as a key component of human physiology and the interdependence of their relationship. Currently, it is well accepted that the commensal microbiota participates in i) metabolism by providing metabolic capabilities encoded in the microbiome that are absent in the human genome; ii) enhancement of the barrier function of skin and mucosa by avoiding colonization of the host by pathogens or inhibiting pathogen expansion; and iii) stimulation of the immune system by providing signals important for the development and maintenance of innate and adaptive immune cells both locally and with central effects, especially in the bone marrow (BM) [5]. Moreover, recent data suggest the influence of the microbiota in several other phenomena, including cognitive functions [6, 7]. Conversely, alteration of the composition of the microbiota, referred to as "dysbiosis" [8], has been linked to a wide range of pathologies, including infective, metabolic, autoimmune, tumoural and behavioral pathologies [9 - 15]. Therefore, the microbiota has become a vital part of human physiology, influencing several aspects of our homeostasis [16].

Although there are some suggestions that offspring colonization by the microbiota could start during pregnancy [17], it is well known that the method of delivery (vaginal or Cesarean section) is a key element that defines the composition of the newborn's microbiota [18]. While newborns will start to form their own microbiota for their skin and mucosa, which will differ in composition enormously, much of our understanding refers to the intestinal microbiota. Therefore, for the purpose of this chapter, the term microbiota will be used to refer to the intestinal microbiota unless otherwise stated.

Thus, the microbiota is initially formed from vaginal- and intestinal-associated species (*Prevotella*, *Sneathia*, *Lactobacillus*) or skin- and oral-associated species (*Propionibacterium*, *Corynebacterium*, *Streptococcus*) [18 - 21], depending on the mode of delivery. In addition, other key elements that participate in this initial colonization are feeding, the environment and contact with other persons. The mode of feeding is the most important of these elements, with clear differences in colonization between breast milk and formula feeding.

Over time, the microbiota will be inevitably affected by other circumstances, such as environmental factors and antibiotic use. These features, together with sex, host genetics and the function of the individual's immune system, will actively influence the microbiome composition into a configuration that will ensure a homeostatic relationship and consequently establish the uniqueness of each individual's microbiota [5, 16].

The adult human intestinal microbiota is composed of more than a thousand species of bacteria in four major phyla: Actinobacteria, Bacteroidetes, Firmicutes and Proteobacteria. Nonetheless, there are important differences in composition along the gastrointestinal (GI) tract. The stomach has the lowest microbiota density ($<10^3$-10^4 cells/ml), and the predominant genera are *Streptococcus*, *Lactobacillus* and *Propionibacterium* [22 - 24]. The small intestine presents important microbiota that increase in number and complexity in its three parts (duodenum, jejunum and ileum); *Streptococcus* and *Veillonella* are the main genera. The large intestine, composed of the cecum and colon, presents the densest microbiota, with *Akkermansia* representing one of the most abundant genera [22, 25].

Although is well recognized that the characterization of healthy microbiota composition is of utmost importance, the microbiota is still far from being fully understood. Our current knowledge in relation to specific functions/effects of the microbiota is restricted to a few specific commensal members or a group of them. Nonetheless, the diversity and richness of the microbiota are features that are strongly associated with health [26 - 28]. Pathogenic bacteria can act as a cause or an effect of dysbiosis, depending on several factors associated with the invading pathogen, such as the type of pathogen and the pathogen's virulence and capacity for adaptation to the intestinal environment; dysbiosis can also be caused by non-pathogen-related factors, such as host genetics, diet and drug use, with antibiotic use being the most important type of drug use. In addition, dysbiosis can also promote the proliferation of some members of the commensal microbiota, which in turn can cause pathology and become a part of the dysbiotic state of the microbiota. These potential pathogenic microbiota members are known as "pathobionts" [29]. Pathobionts differ from common pathogens because they do not cause disease unless at least one of two factors occurs: dysbiosis or alteration of the immune system [29, 30].

Clostridium difficile, in our opinion, is the best example of an intestinal pathobiont whose pathogenic activity is strongly linked to antibiotic use and dysbiosis [31, 32]. Most importantly, the recurrent form of this infection is currently a unique disease that can be corrected with high success using fecal microbiota transplantation (FMT) [33, 34]. FMT, as will be discussed in this chapter, is a very simple treatment method to correct dysbiosis and consists of the transfer of intestinal commensal microbiota obtained from the feces of a healthy donor [35]. Other designations for FMT have also been proposed to facilitate its acceptance as a therapeutic option, such as "bacteriotherapy" [36, 37], "human probiotic infusion" [38, 39], "intestinal microbiota transplantation" [33] and "gut microbiome transplantation" [40]. This chapter will use the designation "gut microbiota transplantation" (GMT) because, although close to the last two previous designations, in our opinion, the designation proposed here is simpler and more closely related to what is really being used in the transplantation, *i.e.*, different types of microbes, not only bacteria, and entire microbes, not only their genes; this terminology also mentions the procedure being performed.

The impressive success of this methodology to treat recurrent forms of *C. difficile* infections (CDIs) has motivated several investigators to test it in other intestinal infectious and non-infectious inflammatory diseases and try to determine the underlying mechanisms. Thus, the following sections will discuss the main aspects of microbiota composition regarding human health and the immune system; microbiota disruption, which leads to CDI; and the pathogenic mechanisms of the disease. Subsequently, GMT history, methodology, main observations in humans and possible mechanisms extracted from animal models are presented. Finally, the applications of GMT in other diseases and future perspectives on this issue are also discussed.

MICROBIOTA BARRIER FUNCTION AND CROSSTALK WITH THE IMMUNE SYSTEM

Immunology is one of the main scientific fields that has made enormous gains with the manipulation of microbiota and the accessibility of metagenomic technology in the last decade. Although much of the knowledge in this regard comes from murine microbiota analysis, the substantial data obtained in mice models have served as fundamental starting points for human studies. The particular complexity of the human GI tract, which is complex not only in terms of its size but also in terms of the difficulties in its manipulation *ex vivo*, is certainly reason for the broad use of murine models in this area, besides the obvious impossibility of manipulating the human microbiota *in vivo*. Interestingly, mice and humans share similar intestinal bacterial numbers and phylum abundance, but they specifically differ in Firmicutes species [41]. As will be presented, data on immune physiology observed in mice have an impressive correlation with the human intestinal immune system in several cases. The main observations presented in this section are summarized in Fig. (**1**).

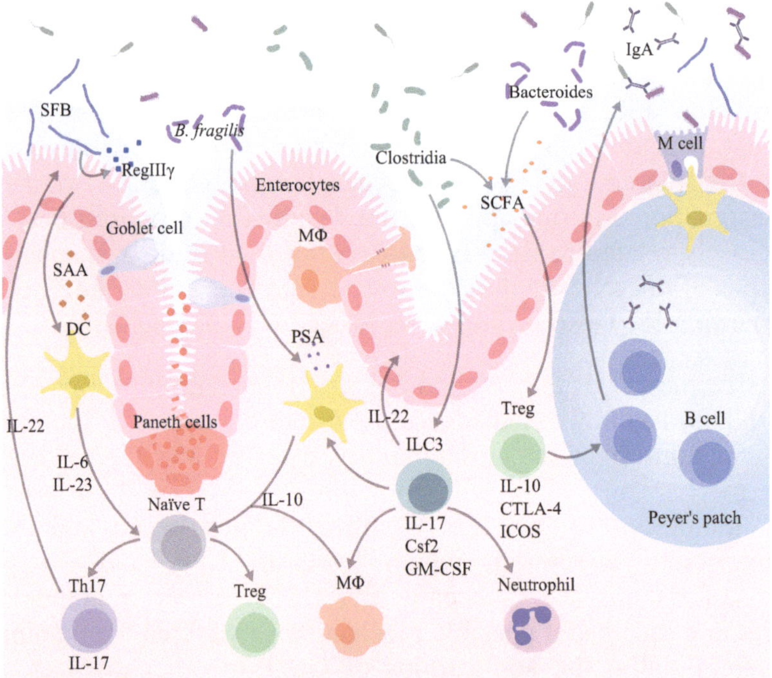

Fig. (1). Host-microbiota interactions at the steady state. The microbiota has a significant influence on cells present in the epithelial layer and in the lamina propria of the intestines. A constant production of defensins (by enterocytes and Paneth cells) and mucus (by goblet cells) in response to microbiota provides the first line of defense in the GI tract. Segmented filamentous bacteria (SFB) are in close contact with epithelial cells and stimulate production of serum amyloid A (SAA), which induces DCs to secrete IL-6 and IL-23. These cytokines promote the differentiation of naive T cells into Th17 cells. Enterocytes also produce the antimicrobial peptide RegIIIγ in response to SFB stimulation and in response to IL-22 secreted by Th17 cells and ILC3 cells. *Bacteroides* species, particularly *B. fragilis*, produce PSA, which stimulates production of IL-10 by lamina propria DCs. In turn, IL-10 stimulates Th0 cells to become Tregs. MΦs are capable of sampling luminal contents by dendritic projections called "trans-epithelial dendrites"; these cells present in the lamina propria are also able to secrete IL-10 at the steady state. Clostridia members induce ILC3 to secrete IL-22, an important cytokine for the enhancement of the epithelial barrier including mucus production. ILC3 also secrete IL-17, Csf2 and GM-CSF and participate in the translation of gut microbiota signals to maintain the homeostatic function of MΦs, DCs, and neutrophils. *Bacteroides* and Clostridia species also produce SCFAs from the fermentation of dietary fibers and stimulate Tregs to produce regulatory molecules (including IL-10, CTLA-4 and ICOS), as well increase their capacity to influence B cells to secrete IgA, which is important to restrict the microbiota presence mainly into the intestinal lumen.

Innate Immunity

Non-immune Cells

The first line of defense in the GI tract is the barrier function provided by epithelial cells, mucus and the microbiota. These layers essentially separate the internal connective tissue, the lamina propria, from the external lumen. The main epithelial cell types (that are not considered immune system cells) include enterocytes, enteroendocrine cells, tuft cells, M cells, Paneth cells and goblet cells. Although enterocytes participate in nutrient absorption, M cells transport antigens from the lumen to the lamina propria, Paneth cells produce defensins, and goblet cells produce mucus, they all express several innate immune receptors that allow them to sense and process signals from the microbiota. The microbiota component present in the intestine completes the barrier component against pathogens. The intestinal microbiota increases in quantity and complexity from an almost sterile duodenum to the most densely colonized cecum [3]. This protective function provided by the microbiota against pathogens is called "colonization resistance" [42]. Currently, colonization resistance includes the well-known microbiota ability to modulate the host immune system, which in turn combats pathogens. Non-immune cells and cells involved in innate immunity are able to contact the microorganisms through pattern recognition receptors (PRRs) and are therefore crucial for microbiota sensing and for processing this information for the adaptive immune system. The receptors that mediate this contact include several different families, with Toll-like receptors (TLRs) and NOD-like receptors (NLRs) being two of the main types. Several other microbe-sensing molecules and their downstream signaling mediators have essentially comparable effects, such as the NLRs NLRC4, NLRP3 and NLRP6; NFkB transcription factor; and AIM2 DNA sensor protein [43, 44]. These effects reflect the production of several antimicrobial products by epithelial cells, including antimicrobial peptides RegIIIγ, RegIIIβ, Ang4, and Itln1; cytokines such as IL-18, thymic stromal lymphopoietin (TSLP) and transforming growth factor-β (TGF-β); and chemokines such as CCL-2. All of these products act in concert to keep the microbiota in its appropriate niche and to form a link to immune system cells [16, 45]. Additionally, epithelial cells also respond to microbiota-derived metabolites, including spermine, taurine, indole and the short-chain fatty acids (SFCAs) acetate, butyrate and propionate. These metabolites enhance the barrier function of the epithelia by providing energy to the cells (SCFAs), fortifying tight junctions (indole) and activating the inflammasome (taurine, spermine) [16].

Immune Cells

Regarding innate immune cells, macrophages (MΦs) and dendritic cells (DCs) are present throughout the lamina propria. As the main antigen-presenting cells (APCs), they are in constant contact with microbial-derived signals and must decide whether to tolerate food and microbiota-derived signals or to trigger the inflammatory alarm against pathogens. These APCs have distinct but complementary roles in the immune system [46]. Although MΦs and DCs have distinct functional characteristics, each population is far from corresponding to a single well-defined cell type within the lamina propria; on the contrary, they are organized in different subpopulations, which seem to reflect functional specialization [47].

<u>Macrophages (MΦs)</u>

MΦs are the most common monocyte-derived population in the intestinal lamina propria in the steady state. In mice, two of the most important MΦ subpopulations in the lamina propria are CX3CR1+CD11b+CD11c+ and CX3CR1+CD11b+ CD11c- [47]. Initially identified as DCs, these subpopulations were later defined as MΦs and are distinct from other subpopulations of MΦs located near intestinal crypts and *muscularis mucosae*. They present the ability to sample luminal contents by dendritic projections called "trans-epithelial dendrites" that are located in between epithelial cells and are equipped with tight junction proteins [48, 49]. Interestingly, lamina propria-resident MΦs in the steady state present a phenotype prone to induce tolerance to antigens from food and microbiota, which is reflected in the low expression of TLRs, lack of production of inflammatory cytokines (TNF-α, IL-1, IL-6, IL-12 and IL-23) and constitutive production of IL-10. This phenotype is also important for Foxp3+ regulatory T cell (Treg) induction in the lamina propria [46, 50 - 52].

During immune responses, local MΦs are outnumbered by newly derived MΦs, which can develop a pro-inflammatory program. In humans, lamina propria MΦs can be identified by the expression of CD68 and CD13, but their biological activity regarding intestinal luminal sampling is not yet defined, nor is their ability to induce lamina propria Tregs. Nonetheless, human MΦs secrete CCL-20, which signals through CCR6 present on human Tregs that produce IL-10. Moreover, human lamina propria presents monocyte-derived MΦs during inflammatory responses [53].

The microbiota critically influences resident MΦs, since microbiota colonization of the intestine after birth is critical for the introduction and expansion of Ly6C+ monocyte-derived MΦs in the lamina propria, which will be present throughout life with constant renewal. In addition, antibiotic-induced dysbiosis in the intestine results in MΦ phenotypic changes, causing the migration of MΦs to mesenteric lymph nodes, the priming of T cell responses, and the induction of B cells to produce IgA [54]. Moreover, the presence of the microbiota is important for lamina propria-resident MΦs to produce IL-10, since in its absence, they produce TNF-α, IL-6, and other pro-inflammatory-related factors [55]. Interestingly, both the MΦs and DCs present in the lamina propria are derived from circulating monocytes, which in turn are derived from BM progenitors, particularly the granulocyte-MΦ progenitor (GMP). Commensal microbiota signals reach the BM and are important for the development of GMP and the subsequent production of the monocytes, MΦs and DCs that populate the intestinal lamina propria [56]. Compatible with this, an increase in microbiota complexity without infection also leads to a sustained increase in granulocytes by GMPs [57]. Therefore, commensal microbiota products are constantly stimulating progenitor cells in the BM to produce the innate cells that populate the intestinal lamina propria, including the main populations of APCs.

Dendritic Cells (DCs)

As mentioned recently, intestinal DCs derive from the same MΦ precursors but under different differentiation stimuli. They belong to two distinct subpopulations of CD103+CD11b+ and CD103+CD11b- DCs, which both constitutively express CCR7 and therefore have constant migratory properties from the lamina propria to mesenteric lymph nodes. As MΦs, DCs are programmed to induce tolerogenic responses in the steady state to food and microbiota antigens [58], which are due in part to the secretion of TSLP. This tolerogenic activity is also supported by CD103+ intestinal DCs, which stimulate naïve T (Th0) cells to differentiate into Tregs. This mechanism is dependent on TGF-β and retinoic acid secretion by DCs and is observed mainly in the colonic lamina propria [10, 59 - 61]. Interestingly, CD103+ DCs can also stimulate Th0 cells to differentiate into Th17 cells by a mechanism dependent on IL-6 and IL-23 secretion and interferon regulatory factor 4 (IRF4) expression on DCs [62, 63]. However, Th17 induction activity is mainly observed in the small intestinal lamina propria and, more specifically, in the ileum [10]. This anatomical segregation of Th17 and Treg induction by DCs, as will be discussed in next section, is dependent on different microbiota members

[10, 64 - 67].

During inflammatory responses, DCs are directly influenced by MΦs, secreting mainly IL-1, IL-6, IL-23 and TNF-α. Thus, under these stimuli, DCs change from a tolerogenic-prone to a pro-inflammatory program in which they migrate to lymph nodes, repress their retinoic acid production and activate Th0 cells, inducing them to differentiate into Th1 and Th17 cells [68, 69]. Although all these characteristics greatly resemble those in lamina propria-resident MΦs, DCs seem to be less dependent on microbiota signals to enter into this location [70]. Furthermore, MΦs and DCs are influenced in the steady state and during inflammatory responses by epithelial cells (enterocytes), including goblet cells and M cells, which can secrete many of the aforementioned cytokines and factors to induce tolerance and can also be the starting source of inflammatory signals, inducing the activation of both APCs.

Other Innate Immune Cells

As can be inferred, practically all cells in innate immunity (in addition to MΦs and DCs in the steady state and during dysbiosis) are influenced by the microbiota in the lamina propria and at distant sites. In this sense, neutrophils and type 3 innate lymphoid cells (ILC3) deserve special mention. Neutrophils are greatly mobilized in response to inflammation. They sense microbiota mainly by TLR-4 and MYD88 signaling, and the absence of microbiota in germ-free (GF) animals or antibiotic exposition has profoundly negative effects on the number of circulating neutrophils [71, 72]. Finally, intestinal ILC3 participate in the translation of microbiota signals secreting IL-17, IL-22 and Csf2, which are important for maintaining the homeostatic functions of MΦs, DCs and neutrophils [73, 74].

In conclusion, epithelial cells and innate cells in the intestine are closely interconnected. They capture and respond to food and microbiota antigens with the secretion of different mediators to induce tolerogenic responses in the steady state, which rapidly change to activate responses during dysbiosis or infection. The strategic localization of all these cells and their subpopulations provides coverage of the entire intestinal mucosa. The constant circulation of innate immune cells within the lamina propria and from the lamina propria to mesenteric lymph nodes allows these cells to interact with each other and to contact T and B lymphocytes to influence adaptive immunity.

Adaptive Immunity

Humoral and cellular immunity compose the adaptive immune system. The humoral immune response is mediated by B lymphocytes, which produce antibodies [75]. Cellular immunity is mediated by T cells, which are helper (Th) CD4+ cells or cytolytic (CTL) CD8+ cells. CD4+ Th cells are central to a fully functioning mammalian immune system because their differentiation into specialized subsets is necessary to maintain homeostasis; avoid tumoural, self-reactive and microbiota-reactive diseases; and develop a full response against pathogens [76]. To accomplish these functions, Th cells interact with all other cells of the immune system by supporting and/or regulating their function [76]. All forms of Th subsets (Th1, Th2, Th17), T follicular helper (Tfh) cells and Tregs are influenced by the microbiota, particularly Th17 cells and Tregs [77]. However, in contrast to MΦs and DCs, the absence of commensal microbes does not affect lymphoid progenitors [5, 77]. Thus, the effects of the microbiota on T cells are restricted to the differentiation of Th subsets rather than to thymic T cell progenitors.

Regulatory T Cells

Although conventional Tregs are identified by the presence of the transcription factor forkhead box P3 (FOXP3+), there are also several other FOXP3- regulatory cells in the intestine, which suggests that these cells may have some type of specialization in their suppressive activities and that they may work in cooperation to maintain tissue homeostasis [78]. The most robust information in Treg function and biology is associated with classical FOXP3+ Tregs. These cells are present in the lamina propria of the intestine at different frequencies. In the small intestine, they represent approximately 20% of all CD4+ T cells, whereas in the colon, they reach 30% representation. These numbers represent two and three times more FOXP3+ Tregs in the small and large intestines, respectively, than in any other part of the body, where they represent approximately 10% of CD4+ T cells [64, 65, 79 - 82]. Intestinal FOXP3+ Tregs are derived from the thymus or differentiate in the periphery from Th0 cells. The importance of the microbiota to both FOXP3+ Tregs cells is clearly observed in GF mice and in antibiotic-treated mice, which present severe reductions of these cells [64, 79 - 82]. Thymus-derived Tregs recognize self-antigens, whereas peripheral-derived Tregs recognize microbiota and food antigens, with the majority of colonic Tregs being derived from the thymus [83].

As mentioned, the microbiota exerts a strong influence on the FOXP3+ Tregs present in the colonic lamina propria. The main microbiota members that possess such an influence belong to the class Clostridia and to the genus *Bacteroides*. While 46 different strains of Clostridia present in the large intestine of mice have been associated with Treg induction, at least 17 strains present in healthy human stool were able to induce the same effect in mice and rats [82]. The effects on Tregs include the induction of regulatory molecules, such as IL-10, CTLA-4 and

ICOS [82, 84], and an increase in their capacity to induce IgA secretion by B cells, which in turn is important to restraining microbiota in the intestinal lumen and maintaining its diversity, especially for Clostridia species [82]. At the molecular level, the main molecules responsible for the aforementioned effects on Tregs appear to be the SCFAs produced from the fermentation of dietary fibers by Clostridia and *Bacteroides* species [85 - 87]. Among *Bacteroides*, *B. fragilis* is an especially important mammalian microbiota member due to its effects on Treg differentiation. Bacterial polysaccharide A (PSA) has been described as the molecule engaging TLR-2 in lamina propria DCs, which respond by producing IL-10, stimulating T cells to become IL-10-producing Tregs [65].

In conclusion, Clostridia and *Bacteroides* are the two main commensal members that are responsible for the induction and maintenance of Tregs in the colon; they function to maintain tolerance towards self-antigens as well as food and microbiota antigens in a steady state, and they participate in the regulation of immune responses during inflammation.

Th17 Cells

CD4+ Th17 cells are able to produce IL-17A, IL-17F and IL-22 and have critical functions in mucosal defense [76]. They are particularly abundant in the intestinal lamina propria, especially in the ileum of the small intestine, and are greatly reduced by antibiotic treatment or the absence of microbiota, such as in GF animals. Lamina propria Th17 cells derive from the differentiation of Th0 cells under the influence of IL-6 and IL-23, which are secreted by DCs in response to serum amyloid A (SAA). In turn, SAA is secreted by epithelial cells in response to segmented filamentous bacteria (SFB) [10], members of the murine intestinal microbiota that are present in human intestines in children under three years of age but are absent in adults [88, 89]. SFB, formally known as *Candidatus arthromitus*, is a commensal that does not translocate to the lamina propria but rather is located in the inner part of the mucus layer and has close contact with the outer plasma membrane of enterocytes [10, 90]. In addition to SAA, epithelial cells produce antimicrobial peptides, such as RegIIIγ, in response to SFB or in response to IL-22, which is secreted by Th17 cells. Thus, this immune loop is important to strengthen the mucosal barrier to keep pathogens far from the lamina propria, as has been observed with *Citrobacter rodentium* infection [10]. Interestingly, the Th17 cells induced by SFB are in fact specific for SFB peptides, meaning that the TCR repertoire includes the recognition of commensal members of the microbiota [91]. Consequently, SFB became the first identified commensal bacterium that induces an effector (not regulatory) T cell subtype [10, 92]. All these observations expand the classical paradigm that the adaptive immune system is generated to recognize offending microorganisms to initiate the response, and in accordance, two intriguing questions (which have not been answered yet) arise: 1) How are these SFB-specific Th17 cells important to the immune system against pathogens? 2) What mechanisms maintain these Th17 cells without causing constant inflammatory responses against SFB for life? As mentioned, SFB are absent in the adult human intestine; therefore, identifying the bacteria in adult humans equivalent to rodent SFB is a major objective in this field. Recent studies indicate that *Prevotella copri*, a non-spore-forming Gram-negative human commensal member that possess the ability to stimulate the production of IL-6 and IL-23 in BM-derived DCs, is able to skew T cells to the Th17 phenotype in mice and seems to be associated with the development of rheumatoid arthritis in humans [93, 94].

B Lymphocytes

B cells in the intestines are present in high numbers in the lamina propria, especially in the areas called GALT (gut-associated lymphoid tissues), which are best represented by Peyer's patches in the ileum. In these areas, B cells receive signals to fully mature into antibody-secreting plasma cells and become memory B cells [95]. Plasma B cells in the intestine produce IgA, IgG and IgM antibodies, with IgA as the main subtype in this location. The microbiota is considered to be targeted by B lymphocytes, since approximately 60% of commensal bacteria are coated by IgA antibodies [96, 97] and the levels of IgA in the lumen and the number of plasma cells in the lamina propria correlate with the presence of microbiota [98, 99]. Enterobacteriaceae (phylum Proteobacteria) seem to be preferentially targeted by IgA over other groups of bacteria, such as *Bacteroides, Prevotella* (phylum Bacteroidetes) and *Lactobacillus* (phylum Firmicutes) [100]. Nonetheless, SFB, which is part of the phylum Firmicutes, is one of the best-characterized commensals with the ability to induce production of high IgA levels by B cells in the lamina propria. Interestingly, the secretion of IgA in response to SFB does not preferentially occur in Peyer's patches but in isolated lymphoid follicles, and it induces tertiary lymphoid tissues in the lamina propria [101]. Moreover, as demonstrated for Th17 cells [91], only part of the SFB-induced IgA is actually specific against SFB antigens [101].

Altogether, these data indicate that commensals are able to induce specific and non-specific types of effector and regulatory cells of humoral and adaptive immunity in the lamina propria that are important to maintain commensals in their appropriate niche and trigger the appropriate effector immune response against pathogens Fig. (**1**).

ANTIBIOTIC-ASSOCIATED DIARRHEA AND *C. DIFFICILE* INFECTION

As discussed previously, the human intestinal microbiota has important physiological functions and consequently plays an important role in host health [102 - 104]. The use of antibiotics (properly or erroneously prescribed) can cause imbalances in the microbiota, favoring the appearance of opportunistic pathogens and causing diseases such as antibiotic-associated diarrhea (AAD), especially the form caused by *C. difficile* [32, 105, 106]. AAD, which is one of the most common medication side effects (35% of treated patients), is defined as diarrhea without a clear etiology that could be related to previous antibiotic treatment [107]. The effect of antibiotics on microbiota is different for each drug, and consequently, the incidence of AAD varies according to the antibiotic and individual patient characteristics. In addition, the effects of some antibiotics can remain in the human body for extended periods. The most common antibiotics related to AAD are amoxicillin-clavulanate and ampicillin (with rates of AAD of 10-25%), whereas fluoroquinolones, macrolides, tetracyclines, and cephalosporins induce AAD less often [107 - 109]. In this context, one of the most common and concerning conditions related to AAD is CDI.

C. difficile Infection (CDI)

C. difficile is an anaerobic Gram-positive sporulated bacillus first described in 1935 by Hall and O'Toole [110]. It can colonize both the lumen of the bowel and the epithelial brush-border surface of the large intestines of adults (1-15%) and children (15-20%, reaching 80% in newborns) [111]. Nonetheless, *C. difficile* overgrowth leads to the development of intestinal inflammation, which can cause a range of outcomes, from mild diarrhea to potentially lethal diseases, such as pseudomembranous colitis [112]. CDI is associated with antimicrobial therapy used to treat any kind of infection; antimicrobials usually affect the intestinal microbiota, allowing *C. difficile* overgrowth [113, 114]. Even patients who have not been colonized by *C. difficile* but have dysbiosis can develop CDI by acquiring spores of toxigenic strains from healthcare facilities, invasive procedures or contact with professionals or other patients [115]. Consequently, *C. difficile* is considered one of the most common pathogens of healthcare infections (HAIs) in the United States and is the most commonly recognized cause of infectious diarrhea in healthcare settings [116 - 118]. Approximately 20% of patients become colonized by *C. difficile* during hospitalization, and almost 30% of these patients develop CDI [119]. In a report published by the CDC (Centers for Disease Control and Prevention) in 2013 about antibiotic resistance, *C. difficile* was classified as an "Urgent Threat"; more than 500,000 infected patients per year were reported in the US, with high mortality rates resulting from this infection, which occasionally elevated healthcare costs [120, 121]. Additional risk factors related to CDI are advanced age, medical comorbidities, longer hospitalization, chemotherapy, proton-pump inhibitor use and chronic kidney disease [111, 122 - 124]. Among all risk factors, exposure to antimicrobials is the most important for CDI development (particularly clindamycin, penicillins, cephalosporins and fluoroquinolones) [114, 125].

Among hospitalized patients, more than half receive a broad-spectrum antibiotic during hospitalization, and studies show that 30-50% of prescribed antimicrobials are incorrect or unnecessary [122]. Even a short course of antibiotic therapy, such as when used for pre-surgical prophylaxis, may result in dysbiosis and subsequent CDI development [122].

Antimicrobial therapy can also induce *C. difficile* to sporulate and become infectious to other patients. Spores excreted into the environment from *C. difficile*-infected patients can survive against commonly employed disinfectants and can lead to an increase in the percentage of patients colonized by *C. difficile*. Once the spores are part of the microbial community, it is believed that a healthy microbiota allows spores to remain quiescent by converting cholate derivatives (primary bile salt) into deoxycholate (secondary bile salt), which prevents the vegetative growth of spores. On the other hand, microbiota disruption causes a relative increase in the concentration of primary bile salts (more cholate/taurocholate and less deoxycholate/chenodeoxycholate), leading to germination of the *C. difficile* spores in the intestines, outgrowth of vegetative cells, and the eventual production of toxins. This mechanism could be useful for CDI treatment; for example, deoxycholate could be added to the antibiotic regimens of infected patients [115, 126, 127].

The mechanism by which *C. difficile* causes disease is not well established. However, it is well known that the bacterium is not an invasive microorganism and that its toxins cause damage; these toxins are the main virulence factor in the *Clostridium* genus, since non-toxigenic strains are not capable of causing diarrhea [128].

Three toxins are produced by *C. difficile*. Toxin A, or enterotoxin (308 kDa), and toxin B, or cytotoxin (270 kDa), are large clostridial cytotoxins, and the third one, is a clostridial binary toxin (CDT) with ADP ribosyltransferase activity. Despite the fact that CDT clearly contributes to virulence in a limited number of strains, toxins A and B are still recognized as the main virulence factors [129]. The genes coding for toxin A (*tcdA*) and toxin B (*tcdB*) are located in a pathogenicity locus (PaLoc) of the *C. difficile* chromosome that is approximately 19.6 kb in size. The PaLoc region contains three other regulatory genes: *tcdC*, a negative regulator of toxin A and B expression; *tcdR*, a positive regulator; and the *tcdE* gene that putatively encodes proteins called holins, which allow toxins A and B to exit the cell [115, 129, 130]. On the other hand, CDT toxin is encoded by the *cdtA* and *cdtB* genes, which are located outside the PaLoc region [130].

The mechanisms of action of toxins A and B in human intestinal cells are not entirely known. However, the most important event for the pathogenicity of toxins A and B is translocation of the catalytic domain, which occurs when the C-terminal region of the toxins interacts with cell-surface carbohydrates, before both enter into the cell through receptor-mediated endocytosis. These receptors are in low quantity or not present in children younger than 2 years old, which is why this population group does not usually develop the disease despite its high rate of *C. difficile* toxigenic colonization. Once inside the cells, these toxins irreversibly inactivate the Rho, Rac and Cdc-42 GTPases through glycosylation, resulting in an interruption of vital cellular mechanisms, such as cytoskeletal actin regulation, cell junction, motility and cell migration, in addition to promoting inflammatory processes, such as the production of cytokines and chemokines or the infiltration of neutrophils in the intestinal lumen. All these mechanisms induced by toxins A and B lead to lysis of the intestinal cell and typical symptoms of CDI [115, 130, 131]. In addition, the action of toxins could be affected in different ways by both pH and calcium levels [115]. In the end, toxin production disrupts epithelial integrity, resulting in the activation of inflammatory mediators. As a result, an exchange of fluids occurs, causing diarrhea, the formation of a pseudomembrane and even epithelial necrosis [123, 132].

The variability in the *tcdA* and *tcdB* genes is of significant practical importance, as toxins A and B are targets for laboratory diagnostic tests and vaccine development. This variability can be assessed by a PCR-restriction fragment length polymorphism (RFLP)-based typing method that distributes strains into toxinotypes. *C. difficile* toxinotypes are a group of strains that have changes in the toxin A and B coding regions of the PaLoc compared to a reference strain. All strains with changes in the *tcdA* and *tcdB* genes are defined as variant strains and are distributed into 34 toxinotypes designated with Roman numerals (I to XXXIV). This method was first described in 1998 by Rupnik [133] and was reviewed a decade later [134]. Recently, the same group described some recent changes in the toxinotyping scheme and provided a summary of currently known *C. difficile* toxinotypes [129].

The immune responses to *C. difficile* are not completely understood, but as an extracellular bacterium, these responses are mediated and supported by IL-17-secreting cells, although IFN-γ-mediated responses are also important [135]. The response starts with microbial sensing by epithelial cells. In response, these cells are able to secrete inflammatory cytokines (*i.e.*, IL-1 and IL-8) to attract neutrophils. In parallel, ILC1 cells secrete IFN-γ to stimulate Th1 responses. DCs, which capture *C. difficile* antigens, are able to secrete IL-6 and IL-23 and will therefore attract more neutrophils and activate CD4+ T cells, polarizing them to Th17 cells [136, 137]. Subsequently, Th17 lymphocytes secrete large quantities of IL-17 and IL-22, supporting neutrophilic infiltration. Nonetheless, the remarkable resistance of *C. difficile* and its location outside the epithelia will aggravate the inflammatory response, which will affect the integrity of the epithelial barrier [138 - 140]. The lesions will subsequently expose immune cells to a higher number of intestinal antigens, attracting more leucocytes and enhancing the response. The prominent cellular infiltration in these areas will allow the formation of characteristic lesions called "pseudomembranes", which are a common finding during colonoscopy [34]. Pseudomembranes occur across the colonic wall just outside the epithelial level. Therefore, pseudomembranous colitis due to CDI is dependent not only on the damage caused by the bacteria but also on the exacerbation of damage due to the intensity of the immune response. The lack of complete effectiveness in bacterial clearance by antibiotic treatment, together with the dysbiosis induced by the treatment itself, is responsible for the chronification of the response, with a subsequent increase in tissue damage and recurrence of the disease [105, 135, 141].

Clinically, CDI is characterized by diarrhea associated with fecal tests positive for the presence of toxigenic *C. difficile* or its toxins or with colonoscopy and/or histopathology showing characteristics of the infection (pseudomembranous colitis) [122]. The severity of CDI is generally categorized into mild/moderate, severe, severe/complicated and recurrent [34, 122]. More recently, recurrent CDI was differentiated into recurrent non-severe CDI of one episode and recurrent non-severe CDI of two or more episodes [122, 142]. Approximately 15% of patients diagnosed with CDI are readmitted with recurrent CDI within 60 days after the first episode [118]. It is important to note that only 30% of patients with AAD develop CDI; thus, empiric therapy without CDI diagnosis is considered inappropriate [108]. Treatment options are specifically recommended for each category, with GMT as a therapeutic option for recurrent CDI [34, 142]; in fact, the treatment and management guidelines for CDI were recently reviewed [142 - 146]. Fig. (**2**) presents a summary of antibiotic treatment for CDI according to clinical categories.

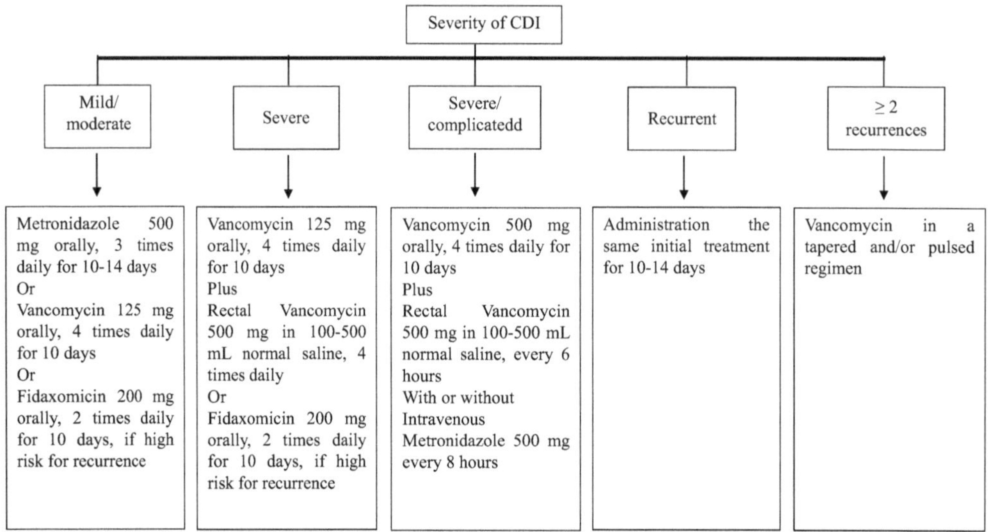

Fig. (2). Summary of conventional antibiotic treatment recommendations for CDI [34, 122, 142, 146].

Conventional Antibiotic Treatment of CDI

Metronidazole and Vancomycin

According to some guidelines, the first choice for treating mild-to-moderate CDI is oral metronidazole [34, 122, 142, 147], a nitronidazole that is active against a broad spectrum of anaerobic bacteria and parasites [148]. It has good clinical efficacy and is also inexpensive and unrelated to vancomycin-resistant enterococci (VRE) selection risk [147]. Metronidazole has high systemic absorption, which means that its concentration in the intestinal lumen is reduced, bordering the minimum inhibitory concentration (MIC). Therefore, it is possible that a decrease in treatment efficacy in moderate or severe CDIs is associated with a low drug concentration at the site of infection, which could lead to the development of bacillus resistance [114, 148, 149]. Although *C. difficile* isolates resistant to metronidazole are uncommon, there are some reports showing strains with high MIC values (> 32 µg/ml) [150, 151]. MIC values > 32 µg/ml observed *in vitro* are probably different from those present in the colon, since the metronidazole concentration found in patient stools with CDI has ranged from 0.8-24.2 µg/ml [152]. Moreover, it has recently been shown that sub-inhibitory concentrations of metronidazole may support biofilm formation in strains with susceptibility or low susceptibility to this antimicrobial, suggesting a possible role of biofilms in *C. difficile* resistance to metronidazole [153].

Vancomycin is a glycopeptide that inhibits cell wall synthesis in Gram-positive bacteria by inhibiting the synthesis of peptidoglycan, an essential component of bacterial cell walls [154], and is the drug of choice for treating moderate and severe CDI [147]. It has reduced absorption in the GI tract, resulting in high concentrations in the colon [155]. However, for patients in which vancomycin cannot reach colon segments by ileostomy, Hartman's pouch or colostomy, vancomycin therapy delivered *via* enema should be added to treatment until the patient improves. In these cases, oral metronidazole is not a good alternative as it is rapidly absorbed by the small intestine, and only low concentrations of the drug reach the colon [34].

A meta-analysis study reported that vancomycin was slightly superior to metronidazole as a symptomatic cure of CDI (79% *vs* 72%, respectively) [156]. In addition, vancomycin is recommended as an antimicrobial alternative in patients with mild-to-moderate CDI who are pregnant or breastfeeding or who are allergic or intolerant to metronidazole [34]. The chance of recurrence of CDI for both antibiotics is approximately 10-20% in the first infection; however, this rate increases to 40-65% if it is not the first CDI episode [113, 157].

There is great concern that vancomycin use may facilitate VRE selection. Indeed, oral vancomycin and metronidazole use has already been associated with excessive VRE growth in fecal samples during treatment for CDI, increasing the transmission risk of this pathobiont [147, 158]. Resistance to vancomycin is more frequent in *Enterococci* and *Staphylococci* but is rarely observed in *C. difficile*. Nevertheless, *C. difficile* strains with reduced susceptibility to vancomycin (MICs > 216 mg/l) have been described [159 - 161].

The vancomycin resistance mechanism in *C. difficile* has not been completely elucidated. Some studies have linked resistance to transposon Tn1549, which is also present in vancomycin-resistant *E. faecalis*, but in *C. difficile*, there is no VanB operon sequence [162, 163]. Notably, *C. difficile* biofilm producer strains are more resistant to high concentrations of vancomycin, and biofilm formation seems to be induced in the presence of sub-inhibitory and inhibitory concentrations of the antibiotic [164]. Although the treatment of CDI with metronidazole and vancomycin can be effective, both are broad-spectrum antimicrobial agents that cause dysbiosis, facilitating the colonization of other pathogens associated with HAIs, such as VRE and *Candida spp* [147].

Fidaxomicin

Fidaxomicin is the last antimicrobial that has been added to the guidelines for the treatment of CDI; it was approved in 2011 by the Food and Drug Administration (FDA) of the United States of America (USA) [165]. Fidaxomicin is a macrolide that selectively inhibits RNA synthesis by the RNA polymerase of *C. difficile*, and it is able to inhibit the bacillus sporulation, avoiding recurrent infection. It has been suggested to cause minimal disruption of the intestinal microbiota [148, 166, 167]. A meta-analysis found that the symptomatic cure rate for CDI was higher with fidaxomicin (71%) than with vancomycin (61%) [156]. Nonetheless, compared with vancomycin, fidaxomicin is expensive [123, 147, 168, 169], and it has low oral absorption and high availability in the intestine [170]. A reduction in the susceptibility to fidaxomicin was reported due to the substitution of only one amino acid in the target protein, which elevated the MIC significantly (> 256 μg/ml) [171, 172].

In conclusion, although new antibiotics, such as fidaxomicin, represent promising options for CDI treatment, it is important to note that they all affect the microbiota in some way, even with minimal effects. Therefore, it is reasonable to hypothesize that the use of these new antibiotics for the treatment of CDI also has the potential to induce or contribute to dysbiosis. Consequently, the new antibiotics will ultimately not be as effective as desired.

GUT MICROBIOTA TRANSPLANTATION (GMT)

Historical Use of GMT

The usage of fecal flora as treatment for some diseases, especially for recurrent CDI, has gained interest in recent times, opening a new therapeutic option to several critical situations. Interestingly, GMT is an ancient practice; the first descriptions emerged in the 4th century during the Dong-jin Dynasty in China. As reviewed by Zhang and coworkers [173], the most likely first report of GMT was written in the Chinese handbook of emergency medicine called "Handy Therapy for Emergencies", or "Zhou Hou Bei Ji Fang" in Chinese. Patients who suffered from severe diarrhea or food poisoning had human fecal suspensions prescribed by mouth by a well-known traditional Chinese medical doctor. Patients frequently recovered from a very severe ill state, and this practice was long considered a "medical miracle". Later, during the Ming Dynasty in the 16th century, another description was made in "Ben Cao Gang Mu", the "Compendium of Materia Medica", which is the most well-known book of traditional Chinese medicine. As described in the compendium, patients with abdominal diseases with severe diarrhea, fever, pain, vomiting, and/or constipation were treated with infant feces, fresh fecal suspension, fermented fecal solution or dry feces. The treatment was very successful in terms of time parameters and aesthetics, and the doctors used to label the fecal suspension as "yellow soup" or another invigorating name [173].

After this description, the use of GMT was neglected until the description of a case series that occurred in the 21st century. In 1958, Ben Eiseman, an American surgeon, reported the use of fecal enema as a supplementary treatment in cases of pseudomembranous enterocolitis caused by *Staphylococci* [35]. In this publication, Eiseman describes the cases of four patients and explains the motivation to experiment with the reestablishment of commensal flora based on the observations of Prohaska [174], who had stated that *Staphylococci* could not grow in the presence of normal flora. For this reason, Eiseman made an attempt to re-establish the balance of natural flora with the administration of normal feces into the colon of his patients. After the therapy, *Staphylococci* were no longer found in the stool of the patients, and the symptoms of the illness disappeared [35]. Later, in 1983, a new case of a GMT in a 65-year-old woman was reported. The patient was diagnosed with CDI, had a long history of intestinal diseases and was enduring treatment with a large number of antibiotics. After the diagnosis of CDI, treatment with vancomycin was attempted several times, but it always resulted in recurrence. GMT treatment was tested using normal feces from her husband. After two enemas, no adverse reactions presented, there was a normalization of bowel function and the physical characteristics of the patient's stools and weight gain, thus the treatment was considered a success [175].

Six years later, in 1989, a patient who had had ulcerative colitis (UC) for 7 years was treated with normal feces infusion by enema. Six months after the infusion, the patient was without symptoms, and again, the experiment was considered a success [176].

In May of the same year, the results obtained with 1 patient who received normal feces and 5 patients who received a mixture of ten different bacteria against recurrent *C. difficile* diarrhea were reported. The treatments were successful, and remarkably, the report mentioned the detection of *Bacteroides* sp. as a substitute for *C. difficile* in all patients. Moreover, the authors suggested that the absence of *Bacteroides* sp. could result in chronic and recurrent CDI [37]. Together with extensive data on the role of *Bacteroides* sp. in Treg induction in the colon to maintain tissue homeostasis, this result is probably the first indirect evidence of the importance of *Bacteroides* in this role.

In 1997, six patients in Australia were treated for chronic inflammatory bowel disease (IBD) with a mixture of 13 different bacteria by oral and rectal administration. As a result, two patients presented full clinical remission during their follow-up at 18 and 32 months [177].

A case series published in 2003 described the first use of nasogastric (NG) tube to administer GMT [178]. Also in 2003, Borody reported the treatment of six patients with UC using a "human probiotic infusion" [39].

In 2010, a case series of self-administered fecal enema was reported in which seven patients with CDI treated their infections. With only some equipment and instructions, the patients were able to self-administer the enema or receive it from a family member. After a 14 months follow-up, none of the patients presented a relapse of CDI [179]. Nonetheless, self-administered GMT is strongly discouraged.

One year later, the first systematic review of GMT in the treatment of recurrent CDI and pseudomembranous colitis was published. It analyzed 27 case series and reports, corresponding to a total of 317 patients. The conclusion was a 92% of success rate of GMT [33].

Since then, a massive number of articles regarding GMT have been published, most of which have been case reports and case series. In 2013, the American College of Gastroenterology published a guideline for "Diagnosis, Treatment, and Prevention of *Clostridium difficile* Infections". In this document, GMT is formally considered a treatment option for recurrent CDI [34]. In parallel, the European Society of Clinical Microbiology and Infectious Diseases published an "Update of the treatment guidance document for *Clostridium difficile* infection" in the same year. Here, in accordance with US guidelines, GMT is a strongly recommended option for multiple recurrent CDIs [147]. In 2015, Ganc and coworkers published the first results of GMT against recurrent CDI in South America with 90% success [180]. This was the first time that oral push enteroscopy was used to infuse the GMT [180].

Finally, the European Consensus Conference on Fecal Microbiota Transplantation in Clinical Practice held in Rome in 2016 elaborated detailed recommendations on the entire GMT procedure, including indications, preparation and clinical management [181]. Fig. (**3**) presents the main events cited in the history of GMT.

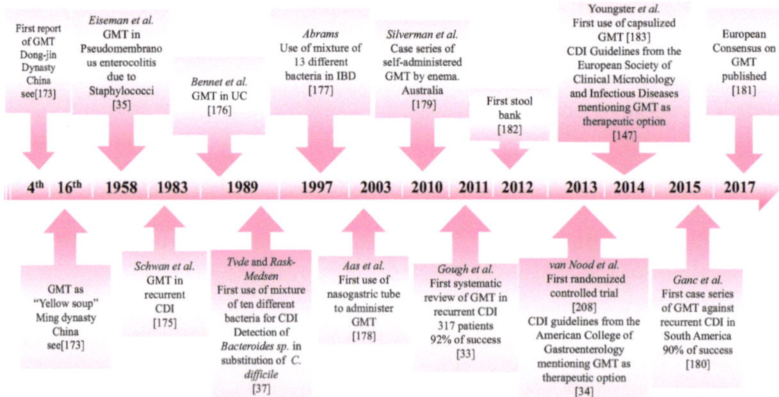

Fig. (3). History of Gut Microbiota Transplantation (GMT). In the boxes are main relevant events in chronological order.
CDI: *Clostridium difficile* infection, UC: ulcerative colitis; IBD: inflammatory bowel disease

New trends in GMT are the creation of stool banks and fecal pills. The first stool bank was launched in 2012, and it provides frozen, prescreened, ready-to-use preparations of human feces for distribution in hospitals and clinics for the treatment of CDI [182]. Finally, in 2014, Youngster reported the use of capsulized fecal material for the treatment of patients with recurrent CDI [183].

GMT Methods

The key elements in performing a GMT procedure are donor selection, stool preparation and route of administration. This information will be presented next and is summarized in Fig. (**4**), along with considerations on adverse effects and ethical aspects.

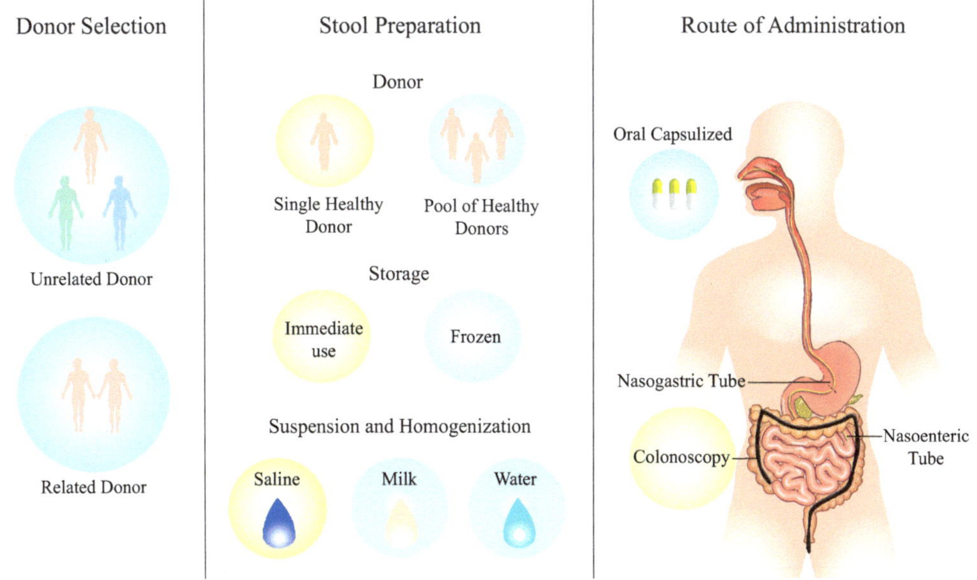

Fig. (4). GMT Methodology. The figure summarizes the main steps for GMT application. The images presented inside a yellow circle are the most common and/or the most recommended procedures used in each step.

Donor Selection

Immunologic matching between the donor and recipient is not required for GMT, which makes donor selection simpler for GMT than for other organs. However, maternal-line first-degree relatives can theoretically be more tolerant to the microbiota received from their donors [184]. GMT donors can be anyone over the age of 18 who is related to the recipient (a spouse or a close relative) or unrelated to the recipient and willing to be a donor. Many patients prefer an anonymous donor to eliminate the potentially awkward conversation about requesting a stool sample from a loved one. With mild variation, donor screening protocols largely focus on infectious screening to prevent the transmission of infectious organisms that may be harbored in donor stool. Donors are also excluded if they have recently received antibiotics. Potential donors should be questioned about their travel history, sexual behavior, previous operations, blood transfusions, and other factors that increase the risk of transmissible disease. Once a donor is selected, blood and fecal samples must be screened for pathogens. These and additional exclusion criteria can be found in the American and European guidelines [34, 181].

The patient's clinical condition should always be considered before performing the questionnaire and the test for donor exclusion. If the condition of the patient is precarious at that time, the risk of a transmissible disease from donor stool can be trumped by mutual agreement between the donor and recipient to proceed with GMT [184].

Stool Preparation

Because of how GMT is believed to work, correct stool preparation and conservation are key points for procedural success. However, there is still a lack of standardization in these processes. Stool contains undigested food elements, chemical and cellular compounds from the donor GI tract and a great number of microorganisms – which is the key point of GMT. Even though most studies present processing methods, including suspension, homogenization and filtration of the stool [185], GMT can be performed using whole homogenized stool; however, this option does not seem to satisfy patients as much, given its aesthetic features [186]. The substance used to homogenize stool is usually sterile saline, even though published data do not show a significantly different success rate when the donor stool is mixed in water, milk or saline [187]. Nevertheless, saline is presumed to be less likely to affect the microbiota of the donor, as suggested [188].

The interval between obtaining donor stool and its administration also varies between studies, from immediate administration to keeping the stool frozen. One of the biggest advantages of using frozen material is that it can be stored for long periods of time, which is useful since stool tests include pathogen cultures, PCR, and tests for viral pathogens, which can take significant periods of time. In addition, some diseases, such as hepatitis and HIV infections, can take several weeks before seroconversion occurs in donors who recently acquired the disease, a period known as the immunological window [189]. Thus, the opportunity to store the microbiota material before use enables the performance of rigorous testing and the elimination of uncertainty that may be associated with fresh material. Moreover, different approaches to stool sources have also been explored, such as synthetic stool preparation, in which stool is made from purified intestinal bacterial cultures derived from a single healthy donor [190], or the pooling of stool samples from several donors to create a standardized transplant material [191].

Routes of Administration

The administration of a fecal microbiota solution is performed *via* the lower (enema and colonoscopy) or upper (NG or nasoenteric, NE, tubes) GI tract. More recently, oral capsules have started to be tested. These methods are briefly discussed next.

Lower GI Tract Delivery Methods: Enema and Colonoscopy

An enema is the insertion of a liquid into the rectum or colon, usually up to the splenic flexure, by way of the anus. The efficacy of this method may depend upon the technique used to cleanse the colon before the administration of a fecal enema [192]. The cleansing process can reduce the number of *C. difficile* cells as well as inactive spores, which could potentially convert into vegetative forms, reducing the effectiveness of GMT [188]. Colonoscopy enables visual examination of the large intestine and even the distal part of the small intestine. The biggest difference in the use of a colonoscope, therapeutically speaking, is that, different from enemas, it allows the administration of GMT bacteria in the area where *C. difficile* spores reside. It should be emphasized, however, that this procedure should be conducted extremely carefully, given the risk of perforation in a likely inflamed bowel wall, especially in case of severe colitis [181]. One way to clean the intestine before GMT is giving the patient balanced electrolyte solutions that contain polyethylene glycol (PEG). These solutions are isosmotic, which minimizes fluid exchange along the colonic membrane and consequently reduces the potential for systemic electrolyte disturbances. In addition, PEG is a non-absorbable high-molecular-weight polymer that can exert an osmotic effect on the electrolyte solution, causing it to persist in the colon for a longer time than other solutions [188].

In the protocol suggested by Borody and Ramrakha [188], the patient first receives 500 mg of oral vancomycin twice daily for seven days. Subsequently, oral colon lavage with PEG (3 to 4 liters) is performed. Given that this procedure may cause electrolyte disbalance and dehydration, patients who are considered too ill to tolerate it can forego the lavage, but they should always receive the vancomycin pretreatment. Eight grams of loperamide (2 g every 2 hours) can also be included in pretreatment since it may help with retention of the enema. Additionally, the use of bowel preparation can theoretically assist with expulsion of the dysbiotic microbiota and facilitate implantation of the donor microbiota.

The treatment itself consists of a suspension of 200 to 300 g of donor stool homogenized in 200 to 300 ml of sterile normal saline administered within 10 minutes of preparation on two consecutive days. The first procedure should be performed through a colonoscope, and the second *via* an enema. A two-day procedure appears to be enough for patients with infection only due to *C. difficile*, but it is recommended to repeat the procedure over five days in patients who have Crohn's disease or colitis as a complication of CDI. For better results, patients should be encouraged to retain the enema for at least 6 hours, but an inability to accomplish this task does not have a great impact on procedural success. Another recommended option is adhesion to a high-fiber diet during the treatment period in order to "feed" the new colonic microbiota [188].

The advantages of using the lower GI tract as a route, other than those that have already been mentioned, are that enema administration is effective, cheap, and safe and that it carries lower procedural or institutional admission costs [188]. Colonoscopy offers the advantage of the direct assessment and evaluation of colonic mucosa, which can support the determination of disease severity, exclude concomitant pathology and more easily "implant" the microbiota being transferred.

Upper GI Tract Delivery Methods: Nasogastric (NG) / Nasoenteric (NE) Tubes / Endoscopy

In patients with contraindications to colonoscopy, an upper GI tract approach can be used. Contraindications include anatomical issues (recent colonic surgery, abdominal and/or pelvic surgeries, abdominal hernias, presence of a colostomy); the presence of an implanted cardiac defibrillator or some pacemakers, which may require special management during the application of electrocautery; comorbidities that may increase the risks of sedation; and the use of medication or drugs that may increase tolerance to sedation (benzodiazepines, narcotics) [188, 193]. The high risk of colonic perforation should also be considered by the transplant team.

Patient preparation is basically the same as for the lower GI tract procedure. The only added step is that in order to decrease gastric acid, a proton-pump inhibitor drug should be administered on the evening before the procedure [194]. The procedure itself consists of placing a NE (preferably) or NG tube into the upper GI tract. NE tubes (in the duodenum or jejunum) should be prioritized over NG tubes since they will be placed directly in the small intestine and may have reduced microbiota degradation by gastric acid by avoiding passage through the stomach. Tube position is confirmed radiographically before proceeding with the infusion.

Interestingly, there is evidence that 25 to 30 g of stool suspended in 50 ml of saline is normally sufficient for disease resolution; however, it is recommended to maintain the probe for a period of 3 to 5 days, since multiple instillations are often needed before the patient develops a normal bowel pattern [194]. Using the upper GI tract approach has some advantages. A lower risk of complications, easier preparation of the patient and stool sample (since smaller volumes are used) and lower costs are some of the factors that should be considered when choosing which technique will be used for GMT. On the other hand, vomiting as an occasional complication of oral or NG probing is a disadvantage to fecal material administration *via* the upper GI route [193]. In addition, the infusion of colonic microbiota *via* the upper GI tract seems non-physiological and could disturb the local and infused microbiota. Nonetheless, the use of oral endoscopy to the jejunum to infuse GMT has experienced a high success rate [180].

Combined Jejunal and Colonic Approach

A study from Garg *et al.* used a combined approach for the treatment of recurrent CDI [195]. The protocol was modified to include 180 ml of fecal filtrate administered through endoscopy and 270 ml administered through colonoscopy. The primary end point was the resolution of diarrhea and a negative test for the *C. difficile* toxin, whereas the secondary end points included changes in body weight, white blood cell count and abdominal pain. The study included 27 patients, all of whom reached the primary end point. Among the secondary end points, resolution of abdominal pain was seen in 15 out of 17 patients (88.2%), and resolution of bloating was seen in 9 of 9 patients after GMT. The advantages of this procedure include the delivery of a larger volume of fecal filtrate into the GI tract of the recipient without rapid expulsion from the colon and reduced problems with oral regurgitation of the fecal filtrate. Patient acceptance of this combined approach seems to be satisfactory since the delivery is made at once and the patient does not have to be hospitalized for a longer time. On the other hand, the higher cost of two endoscopic procedures (rather than one) is a potential disadvantage of this method.

Oral Capsulized GMT

Few studies have evaluated oral capsules for GMT, but they have produced very encouraging results [183, 196, 197]. Differences in the stool preparation and encapsulation procedure need to be standardized in order to make GMT more accessible and less awkward for the patient through the use of pills. Nonetheless, these studies reflect that the use of microbiota pills will certainly became an option in the near future [183, 196, 197].

Adverse Effects

GMT is considered a safe therapy with minimal adverse effects. Overall, any adverse events are mild and include abdominal discomfort, transient fever, diarrhea, constipation, nausea, vomiting, bloating, and flatulence. These symptoms are often transient and resolve within a few hours after the procedure [198, 199]. However, it is often difficult to know if these symptoms are related to the underlying disease that triggered the need for GMT, the sedative medications received during endoscopy, or the use of air or CO_2 during colonoscopy or if they are a true result of the infused donor stool [198]. In a study by Kelly and coworkers [200], one patient died of aspiration during sedation for GMT administered *via* a colonoscopy. As donor screening testing protocols are quite extensive, the development of a new infection is rare, and the overall risk of pathogen exposure is thought to be low. Another theoretical risk is that GMT could modify the microbiota of the recipient after transplant and make the recipient susceptible to chronic conditions such as obesity, asthma, diabetes mellitus type 2, IBD, atherosclerosis, nonalcoholic fatty liver disease, irritable bowel syndrome (IBS), and autism [187, 201]. Nevertheless, although more prospective studies are required to identify long-term concerns related to the safety and potential risks of GMT, the benefit/risk ratio is clearly in favor of GMT.

Ethical Aspects

GMT has emerged as the most effective single therapy for the management of patients with recurrent CDI, but it is still considered investigational. The regulatory status of GMT has been controversial in some countries. In the USA and France, authorities have classified human feces used for medical purposes as a drug, in contrast to the Denmark, the UK and the Netherlands, which have classified it as a human tissue [202, 203]. The regulatory status of GMT within the FDA remains uncertain. Despite FDA oversight, physicians may perform GMT outside of a clinical trial (without an investigational new drug-IND application or approval) for CDI that has not been responsive to standard antibiotic therapies [204]. This compassionate exception proposed by the FDA allows many people suffering from CDI to benefit from the use of GMT while the agency develops appropriate policies for the use of GMT as an IND. However, an IND must be submitted to the FDA for non-recurrent CDI applications. Although GMT appears to be safe and without serious adverse effects, from an ethical point of view, it is necessary to consider several aspects of this innovative therapeutic intervention, such as the donor's recruitment, safety concerns and informed consent [205].

The FDA recommends that the treating physician obtain adequate informed consent from the patient or his or her legally authorized representative for the use of GMT. The informed consent process represents a key element expressing fundamental ethical and medico-legal value, and it should be based on an understanding of the objectives, risks, process and possible benefits of GMT. Patients must be informed that they are receiving stool donated from a healthy individual, as this kind of biological sample may cause anxiety in the recipient. The risk of transmission of an infectious agent or diseases/conditions that were not prescreened in the donor must be stated in the informed consent [202]. Moreover, some risks are specific to the method of delivery. For example, with colonoscopy, there is a risk of discomfort during the procedure, reaction to anesthetic agents, GI bleeding, and perforation. Furthermore, greater care is needed when information about possible future risks is not yet available, as in the case of GMT [198]. Information should also be clearly provided to the donor. This information must describe the procedures that he/she must undergo before becoming a donor (their stool and blood will be screened for ova, parasites, HIV, HCV and others). Moreover, donors must be informed that microbiome screening can reveal susceptibility to disease and that they have the right to choose "to know" or "not know" about these susceptibilities [205].

The procedures of GMT must be overseen by Research Ethics Committees (RECs). RECs play an important role in GMT by formulating more specific operating procedures and are also invested in the responsibility of guaranteeing the safety and protection of individuals enrolled in GMT procedures [206]. Although GMT represents a new challenge in both its scientific-clinical characteristics and ethical aspects, RECs are helpful in formulating more sharable procedures for this innovative therapeutic resource. Furthermore, although there are positive expectations for GMT, RECs should demand long-term follow-up to evaluate its safety. There is high pressure from patients and clinical doctors performing GMT for necessary regulations. The lack of such regulations leads patients to use a "do-it-yourself" method of GMT, provided by social media, which may influence future patient outcomes but certainly is strongly discouraged [187, 201].

THERAPEUTIC MECHANISMS OF GMT

Human Studies

As mentioned previously, the first report of GMT was in 1958 by Eiseman and coworkers [35], who described the use of fecal enemas for patients with severe or fulminant forms of pseudomembranous colitis. Since then, there have been numerous case reports and case series with > 90% efficacy in the treatment of recurrent CDI without any serious adverse events. However, despite the huge amount of research conducted on GMT, there has been a paucity of robust methodological studies evaluating this procedure, such as randomized clinical trials. Some of the key trials recently conducted on GMT for recurrent CDI are listed in Table **1** and are described below.

Table 1. Most relevant trials of GMT for recurrent CDI.

Year	Design	N	Route	Donor considerations	Follow-up (weeks)	Outcome	Authors, References
2012	Case series in patients with and without IBD	43	Colonoscopy	- Individual donor indicated by patients - Standard donor (fresh or frozen material)	8	Cure rate of 86% regarding recurrent CDI	Hamilton *et al.* [207]
2013	Open-label randomized controlled trial	41	Nasoduodenal tube	Donor pool	10	Overall cure rate of 94%	Van Nood *et al.* [208]
2014	Open-label randomized controlled trial	20	Nasogastric tube or colonoscopy	All unrelated donors	8	Overall cure rate of 90% Nasogastric administration appeared to be as effective as colonoscopic administration.	Youngster *et al.* [209]
2014	Open-label single-group	20	Oral capsules	All unrelated donors	8	Overall rate of clinical resolution of 90%	Youngster *et al.* [183]
2014	Retrospective multicenter case series of immunocompromised patients with CDI	80	Varied methods, but mainly colonoscopy	Not informed	12	Overall cure rate of 89%	Kelly *et al.* [200]
2016	Prospective study	180	Oral capsules	All unrelated donors	8	Cure rate of 91%	Youngster *et al.* [197]
2016	Double-blind randomized controlled trial	43	Colonoscopy	- Donor indicated by patient or unrelated - Autologous	8	Cure rate of 62.5% by autologous GMT Cure rate of 90.9% by donor GMT	Kelly *et al.* [210]

(Table 1) cont.....

Year	Design	N	Route	Donor considerations	Follow-up (weeks)	Outcome	Authors, References
2016	Randomized, double-blind frozen vs fresh FMT dose	232	Retention enema	Unrelated donor (fresh or frozen material)	13	Cure rate of 83.5% by frozen GMT Cure of 85.1% by fresh GMT	Lee *et al.* [189]

In 2012, Hamilton and coworkers [207] reported clinical experience with 43 patients with CDI who were treated with GMT at the University of Minnesota Fairview Medical Center, Minneapolis, USA. In this work, donor identification and screening were moved from patient-identified individual donors to standard volunteer donors. Material preparation shifted from the endoscopy suite to a standardized process in the laboratory and then ultimately to banked frozen processed fecal material that was ready to use when needed. The overall rate of infection clearance was 86% in response to a single infusion of donor fecal material, as evidenced by symptom resolution and negative PCR testing for *C. difficile* toxin B after 2 months of follow-up. In all, 30% of patients who received GMT using material from patient-identified individual donors had a recurrence of CDI. Two standard donors were employed for the remaining 33 cases in this series, but the majority of material was prepared from a single donor. In all, 3/33 patients who received GMT from a standard donor (fresh or frozen) had a recurrence of CDI. The difference in donor source, patient-identified *vs*. standard, was not significant. There was also no significant difference in clearing the infection with fresh or frozen donor material. The standardization of material preparation significantly simplified the practical aspects of GMT without a loss of apparent efficacy in clearing recurrent CDI. Approximately 30% of the patients had underlying IBD, and GMT was equally effective in this group. Moreover, no serious adverse events were noted following GMT in any of the patients with either fresh or frozen materials.

The first randomized controlled trial for GMT in recurrent CDI following vancomycin therapy was performed in the Netherlands by van Nood and coworkers [208]. Patients were randomly assigned to 3 groups. Group one received an initial vancomycin regimen followed by bowel lavage and subsequent GMT through a nasoduodenal tube. The second group received a standard vancomycin regimen. The third group received a standard vancomycin regimen with bowel lavage. The primary end point measured was cessation of diarrhea associated with CDI and no relapse after 10 weeks with 3 consecutive negative stool tests for *C. difficile* toxin. There was an 81% resolution of CDI-associated diarrhea following the first infusion and 94% resolution following a second infusion in those unresponsive to the first treatment. The vancomycin group had

31% resolution, and the vancomycin and bowel lavage group had 23% resolution. The overall cure rate of microbiota transplantation was statistically superior compared to vancomycin alone or to vancomycin with bowel lavage. The trial was stopped early due to the significant efficacy of GMT relative to controls. There was no significant difference in adverse events among the 3 groups, except for belching, mild diarrhea, and abdominal cramping on the day of GMT in the infusion group. After donor-feces infusion, patients showed increased fecal bacterial diversity, similar to that in healthy donors, with an increase in Bacteroidetes species and *Clostridium* clusters IV and XIVa and a decrease in Proteobacteria species.

Youngster and coworkers reported in 2014 [209] a trial comparing NG *versus* colonoscopic GMT delivery using a frozen fecal suspension from a healthy, screened, and unrelated donor. A total of 20 patients were enrolled, 10 in each treatment arm. Although this study was underpowered to detect a meaningful difference between treatment arms, a single GMT administered by colonoscopy yielded an 80% (8 of 10 patients) clinical cure rate, whereas the CDI resolution rate with NG administration was 60% (6 of 10 patients). Five patients who did not achieve a clinical cure were retreated by NG administration, with 4 obtaining a cure, resulting in an overall cure rate of 90%. No patients relapsed clinically at the 8-week follow-up, and no serious or unexpected adverse events occurred. Fecal microbiota was analyzed in 14 stool samples from 4 donors and 65 samples from 19 recipients (21 pre-GMT and 44 at different time points after GMT). The Shannon diversity index of fecal microbiota obtained from recipients evaluated prior to GMT was consistently low, and it increased after GMT to a diversity level comparable to that of the donors. This level persisted over time, and there was no significant difference between the diversity index in the stool samples obtained in the first week after the procedure and those obtained up to 6 months later. The route of administration made no difference in the mean Shannon diversity index obtained after GMT.

In 2014, GMT in immunocompromised (IC) patients with CDI [200] was reported for the first time. Eighty patients from 16 centers were included (75 adult, 5 pediatric). The reasons for the patients being IC included HIV/AIDS (n=3), solid organ transplant (n=19), oncologic condition (n=7), immunosuppressive therapy for IBD (n=36), and other medical conditions/medications (n=15). The CDI cure rate after a single GMT was 78%, with 62 patients suffering no recurrence at least 12 weeks post-GMT. Twelve patients underwent repeated GMT, of whom eight had no further CDI. Thus, the overall cure rate was 89%. Twelve (15%) had serious adverse effects within 12 weeks post-GMT, of which 10 were hospitalizations. Two deaths occurred within 12 weeks of GMT, one of which was the result of aspiration during sedation for GMT administered *via* colonoscopy; the other was unrelated to GMT. No patients suffered infections definitely related to GMT, but two patients developed unrelated infections, and five had self-limited diarrheal illness in which no causal organism was identified. One patient had a superficial mucosal tear caused by the colonoscopy performed in GMT, and three patients reported mild, self-limited abdominal discomfort post-GMT. Patients with IBD may experience disease flare following GMT, although whether this is precipitated by the CDI, GMT itself, or progression of the underlying disease state is not known. Three UC patients underwent colectomy related to course of UC >100 days after GMT. This report concludes that GMT appears to be effective and safe in IC patients.

In 2016, a study of colonoscopic GMT for the treatment of recurrent CDI performed in two academic medical centers in the USA, Montefiore Medical Center in the Bronx, New York (NY), and The Miriam Hospital in Providence, Rhode Island (RI), was reported [210]. Cure rates and adverse events were compared between donor GMT and autologous GMT (given as a "placebo") in patients who had at least 3 CDI recurrences. In the intention-to-treat (ITT) analysis, 20 out of 22 patients (90.9%) in the donor GMT group achieved a clinical cure compared with 15 out of 24 (62.5%) in the autologous GMT group. Donor GMT was statistically superior to autologous GMT, but efficacy varied by site. In RI, the cure rate with donor GMT was 90.0% *versus* 42.9% with autologous GMT, whereas in NY, 91.7% of patients achieved clinical cure after donor GMT compared with 90.0% after autologous GMT. Resolution after autologous GMT differed by site. All 9 patients who developed recurrent CDI after autologous GMT were free of further CDI after subsequent donor GMT. There were no serious adverse effects related to GMT. In this study, the authors deeply examined the gut microbiome through 16S ribosomal RNA gene amplification and sequencing. Donor GMT restored gut bacterial community diversity and composition to a state that resembled that of healthy donors. Before GMT, all patient samples showed marked dysbiosis with lower alpha diversity, more Gammaproteobacteria and Betaproteobacteria, and fewer Firmicutes and Bacteroidetes than those of the donor samples. Donor GMT was associated with the normalization of fecal microbiota and the restoration of alpha diversity. Similar patterns were seen in microbiota from post-GMT samples after donor GMT was performed as a rescue procedure after an initial failure. In contrast, dysbiosis persisted in autologous GMT patients who did not have relapse and was characterized by decreased Proteobacteria and a nonsignificant change in Bacteroidetes. *Klebsiella*, *Enterobacter*, *Escherichia*, *Lactobacillus*, and *Veillonella* were abundant in pre-GMT samples and were restricted in donors and donor GMT patients. Relative abundances of *Bacteroides*, *Blautia*, *Faecalibacterium*, and *Roseburia* were greater in donor fecal samples and in patients undergoing donor GMT. The fecal bacterial community composition of both donor and pre-

GMT samples differed between sites. Samples from RI patients contained more Gammaproteobacteria and Betaproteobacteria, whereas pre-GMT samples from NY patients commonly showed more Verrucomicrobia. There was a trend toward more Bacteroidetes in the NY samples, but there was no difference in Bacteroidetes between patients with or without a history of fidaxomicin treatment. Fidaxomicin use correlated with more Verrucomicrobia and fewer Gammaproteobacteria before GMT. After autologous GMT, NY patients had greater abundances of *Clostridium* species. Feces from the NY patient who underwent donor GMT and relapsed after the 2-week follow-up still had a high proportion of Gammaproteobacteria (12%) despite an expansion of Bacteroidetes (27%) and increased alpha diversity. In contrast, the RI patient who had a relapse at approximately 8 weeks after autologous GMT showed a community that taxonomically resembled that of donor samples through the end of the study; the expansion of Bacteroidetes and Firmicutes was observed after autologous GMT, but dysbiosis was still apparent because alpha diversity did not recover.

In 2016, one study reported the use of oral administration of frozen encapsulated fecal material from unrelated donors to treat recurrent CDI. This study was conducted at Massachusetts General Hospital, Boston, USA, in a large cohort with structured follow-up. Patients were treated using stools obtained from seven donors; GMT capsules were generated and stored at −80°C. The primary endpoint was defined as clinical resolution 8 weeks after last capsule ingestion while off antibiotics for CDI. Safety was defined as any FMT-related adverse event grade 2 or above. Overall, 180 patients aged 7-95 years with a minimal follow-up of 8 weeks were included in the analysis. Of the 180 patients who reached 8 weeks, 147 were cured of CDI after the first administration of fecal capsules (82%). Twenty-six individuals relapsed within 8 weeks and were re-treated, with 17 responding, resulting in an overall cure rate of 91% with one or two treatments. Six individuals declined retreatment. Three patients were cured after a third administration but were considered “non-responders” as per protocol definition. One patient received three treatments, relapsed, and was advised to continue suppressive vancomycin. Five patients relapsed in the 2- to 6-month window, one due to antibiotic treatment, one due to chemotherapy, and three spontaneously. Four patients were retreated, of whom one was lost to follow-up. Of the four remaining retreated patients, all durably resolved, but one died of recurrent cancer. Three adverse events Grade 2 or above, deemed related or possibly related to FMT, were observed. This report confirms the effectiveness and safety of oral administration of frozen encapsulated fecal material for treating recurrent CDI. This approach can make GMT accessible to a wider population of patients, making the procedure safer and significantly reducing its cost, and is now offered as standard care for recurrent or refractory CDI at Massachusetts General Hospital [197].

The largest randomized clinical trial of GMT to date was conducted in Canada; it involved 6 academic medical centers and was published in 2016 by Lee and colleagues [189]. Patients with CDI received frozen or fresh GMT *via* enema. A total of 219 patients (n=108 in the frozen GMT group and n=111 in the fresh GMT group) were included in the modified ITT (mITT) population, and 178 (frozen GMT: n=91, fresh GMT: n=87) in the per-protocol population. In the per-protocol population, the proportion of patients with clinical resolution was 83.5% for the frozen GMT group and 85.1% for the fresh GMT group. In the mITT population, the clinical resolution was 75.0% for the frozen GMT group and 70.3% for the fresh GMT group. There were no differences in the proportion of adverse or serious adverse events between the treatment groups. Among adults with recurrent or refractory CDI, the use of frozen compared with fresh GMT did not result in a worse proportion of clinical resolution of diarrhea. Given the potential advantages of providing frozen GMT, its use is a reasonable option in this setting.

The intestinal tract of healthy humans is dominated by bacteria from the phyla Bacteroidetes and Firmicutes [211]. However, there is an evident decrease in microbial diversity in cases of recurrent CDI [212], and the main goal of GMT is to restore the "normal" population of bacteria in this dysbiotic colonic environment. Nevertheless, few studies in humans have examined the gut microbiome through gene sequencing approaches in patients with CDI before and after GMT. Khoruts [213] and Hamilton [214] have shown that before GMT, the patient microbiota was deficient in members of the bacterial divisions of Firmicutes and Bacteroidetes. GMT increases the quantity of Firmicutes and Bacteroidetes and decreases that of Proteobacteria and Actinobacteria, suggesting rapid donor engraftment. There is clearly a shift in the bacterial populations in the post-GMT gut that mirrors the donor stool. It is important to note that Firmicutes and Bacteroidetes phyla are thought to play a large role in gut homeostasis. *Bacteroides* spp. are thought to inhibit *C. difficile* proliferation in addition to the previously mentioned role in Treg induction. In accordance, Firmicutes produce butyrate, an important SCFA, which plays a critical role in maintaining the integrity of the colonic epithelium and regulating mucosal immune responses. Manipulating a patient's fecal microbiota can be a microbiological therapy, but more investigation is needed to clarify the exact microbial and nonmicrobial components of the donor material that leads to *C. difficile* eradication [213].

Experimental Studies

Studies using isolated cells *in vitro* as well as those using animal models, such as specific pathogen-free (SPF), GF and/or genetically modified animals, allow the development of assays that are not possible in humans. Unlike in animal studies, evaluation of the mechanisms in patients receiving GMT, despite being much more valuable, is difficult to perform, subjected to high variation and offers very limited manipulation of the variables [215]. Thus, experimental studies are being used to help to shed some light on the mechanisms involved in GMT treatment.

Initial studies related to the mechanism involved in GMT success tried to identify which specific bacteria present in the intestinal microbiota could have an effect on CDI. In 1981, Wilson, Silva and Fekety [216] used a hamster model to show that *C. difficile* inhibition was related to the presence of Gram-negative anaerobes or other Clostridia species. Subsequently, Corthier and coworkers [217] tested the impact of several bacterial strains isolated from human feces on *C. difficile* growth and toxin production in mice. The authors observed that gnotobiotic mice that received strains of *E. coli* or *Bifidobacterium bifidum* prior to CDI were protected against *C. difficile*-induced pathology. Although the *C. difficile* population did not decrease considerably in these mice, toxin production was lower, which led to decreased mortality. The increased growth of *C. difficile* was also associated with a decreased number of *Lactobacilli*, an increased number of Bacteroidaceae and a decreased frequency of other Clostridia species, indicating that alterations in the abundance of these bacteria in the intestine may be related to CDI susceptibility, which could also be important for its treatment through GMT [218]. However, when gnotobiotic mice were inoculated with *Lactobacilli*, *Bacteroides* or *Clostridium* strains isolated from the microbiota of healthy mice and then infected with *C. difficile*, none of these isolates was able to ameliorate the infection. Nonetheless, when gnotobiotic mice received whole feces or feces treated with chloroform from conventional mice or whole feces from mice with a microbiota composed mainly of Clostridia, the infection was inhibited [218], which indicates that the administration of a single bacterial species is not sufficient to inhibit CDI and that a more complex group of microorganisms is therefore necessary.

More recently, Lawley [219] used a CDI murine model to determine a group of bacteria isolated from mice feces that was capable of recovering the microbiota after CDI. Mice infected with *C. difficile* were treated with combinations of different bacteria until a mixture capable of suppressing the infection was found. This mixture contained *Lactobacillus reuteri, Staphylococcus warneri, Enterococcus hirae* and previously undescribed species of Bacteroidetes, Anaerostipes and Enterorhabdus. Once again, in this study, autoclaved feces, fecal filtrates or individual bacterial strains were not capable of reproducing these results, indicating that in order to achieve CDI resolution, a group of phylogenetically diverse bacteria with distinct physiology is required [219].

The presence of bacteria from the family Lachnospiraceae in feces has also been related to inhibition of CDI. Reeves [220] tested this by colonizing GF mice with either Lachnospiraceae or *E. coli* isolated from healthy mice cecum and then challenging these mice with *C. difficile*. Mice colonized with Lachnospiraceae exhibited fewer signs of colonic histopathology, had lower cytotoxin levels in the intestine and had significantly decreased *C. difficile* colonization, whereas all the mice pre-colonized with *E. coli* and the GF mice died within 48 h of CDI, indicating that Lachnospiraceae is associated with resistance against *C. difficile*. However, it is important to note that colonization with Lachnospiraceae was merely able to partially restore resistance against *C. difficile*, whereas only the transplantation of feces from a wild-type mouse to GF mice was able to restore *C. difficile* resistance completely. This corroborates the hypothesis that interactions between groups of bacteria are necessary, rather than a single bacterium being responsible for the resolution of CDI.

Further studies tried to reveal the molecular mechanisms responsible for the good GMT treatment outcomes for CDI. In general, two main categories of mechanistic effects for microbiota transplantation have been proposed against CDI: a) competition between the donor microbiota and *C. difficile* for nutrients in the host's intestine and compounds produced by this newly introduced microbiota that act against *C. difficile* and b) interaction between the donor and host gut microbiota, with subsequent effects on host physiology and crosstalk with the immune system [221]. In this sense, when new bacteria from a healthy donor are introduced through GMT, they could outcompete *C. difficile*. Accordingly, there is evidence that pre-colonization with a non-pathogenic *C. difficile* strain may prevent infection with a pathogenic one by competition [222, 223]. Thus, one of the mechanisms that may be involved in microbiota-dependent CDI resistance is the competitive exclusion of toxigenic *C. difficile* by a non-toxigenic strain [224].

Nonetheless, nutrient competition is believed to be one of the main mechanisms of competitive exclusion [225]. Wilson and Perini [226] evaluated the impact of competition for single nutrients in the control of *C. difficile* using an anaerobic continuous-flow culture inoculated with contents from mice cecum, which was very similar to the ecosystem found in live mice. The authors observed that *C. difficile* could not successfully compete for carbon sources with this isolated microbiota. When inoculated in a medium containing these cultures from mouse gut, *C. difficile* had a lower growth rate. However, if glucose, N-acetylglucosamine, or N-acetylneuraminic acid was added to this medium, *C. difficile* showed higher growth rates, which according to the authors suggests that bacteria present in mice intestine compete more efficiently than does *C. difficile* for monomeric glucose, N-acetylglucosamine, and sialic acids found in colonic contents [226].

In relation to bacterial factors produced by microbiota, bacteriocins may also be responsible for the inhibition of *C. difficile*. An example of a component produced by intestinal bacteria that affects *C. difficile* is thuricin CD, an antimicrobial produced by *Bacillus thuringiensis* present in human fecal samples [227]. Thuricin CD is a protein with two distinct peptides that have activity against *C. difficile* but no impact on most commensal bacteria [227]. This fact indicates that this antimicrobial may be an important component in the efficiency of GMT treatment against CDI. Nisin is another polypeptide with activity against multiple pathogens, including *C. difficile.* This bacteriocin is produced by several *Lactobacillus lactis* strains, some of which have already been isolated from human feces [228]. Nisin not only is able to inhibit *C. difficile* vegetative cells but also has an activity against spore germination *in vitro* [228]. Another mechanism proposed to explain the protection ensured by GMT against CDI could be related to SCFAs, since these metabolites showed *C. difficile* growth inhibition *in vitro* [229] but had opposing results *in vivo* [230, 231].

The transformation of bile acids by microbiota could also be involved in CDI resistance [224]. Bile acids have an important impact on the germination of *C. difficile* spores upon infection. Antibiotic use may reduce the population of bacteria in the intestine that are able to process primary bile acids into secondary bile acids, increasing primary bile acids in the intestine and promoting the germination of *C. difficile* spores [232]. Accordingly, GMT may transfer bacteria that can process bile acids, increasing secondary bile acids in the intestine instead of primary ones, which compromises the germination of *C. difficile* spores. Therefore, receiving bacteria that can metabolize bile acids may protect against *C. difficile* [215]. To test this hypothesis, Buffie [215] mathematically modulated microbiota dynamics using a systems biology approach to identify bacteria related to *C. difficile* resistance and tested the potential of a bacterial consortium composed of four representative isolates from the gut microbiota associated with *C. difficile* resistance in a murine model. Since the mathematical model suggested that *C. scindens* had the strongest correlation with CDI resistance, this bacterium was tested alone and was also included in the consortium. Administration of this mixture or *C. scindens* alone in mice infected with *C. difficile* mitigated the infection and significantly reduced weight loss and mortality when compared with the control. Individual administration of other bacteria present in this mixture did not have an effect. According to these results, a single bacterium could also be able to inhibit CDI, which contradicts previous studies on CDI treatment through GMT [219, 220]. The authors further characterized this bacteria and related *C. scindens* protection to its production of metabolites derived from host bile salts. Further analyses with the functional gene prediction software PICRUSt showed that *C. difficile* resistance correlated to a microbiome with an increased number of copies of genes related to the biosynthesis of secondary bile acids. Recovery from

CDI was also linked to the presence of the baiCD gene in the microbiota, a gene that encodes a critical enzyme for the biosynthesis of secondary bile acids [215].

Bacteria present in healthy gut microbiota may also stimulate the host immune system, helping fight *C. difficile*. One way this may occur is through TLR signaling [224]. For instance, Myd88−/− mice seem to be more susceptible to CDI, as these mice showed *C. difficile*-associated disease symptoms when treated with clindamycin and colonized by *C. difficile*, whereas wild-type mice did not develop disease when exposed to the same treatment, suggesting that signaling through TLRs functions to protect mice [233]. It is possible that GMT could reconstitute dysbiotic gut microbiota, introducing bacteria that activate TLR pathways, stimulating the immune system in a way that helps suppress CDI. In this scenario, the dysbiosis caused by prolonged exposure to antibiotics could be lacking bacteria that promote TLR signaling, rendering the individual more susceptible to CDI in a similar way to the dysbiosis in Myd88−/− mice. Another indication that TLR signaling could be related to CDI resistance is the fact that administration of flagellin from *Salmonella* in mice infected with *C. difficile* seems to repress the infection, inhibiting *C. difficile* growth, reducing toxin concentrations in the intestine, reducing damage to the intestinal epithelial barrier, and consequently preventing *C. difficile*-induced death. Moreover, these effects were not reproducible in TLR5-/- mice, indicating that the protection mediated by flagellin is dependent on TLR5 signaling and, therefore, that the activation of TLR5 by *Salmonella* flagellin has a protective effect against CDI [234].

Microbiota depletion by antibiotics has an effect not only on gut colonization but also on local and systemic immunity [235]. Thus, restoration with GMT may also have a broad effect on the immune system. Ekmekciu and coworkers [235] evaluated these effects in a mouse model and showed that after microbiota depletion through administration of antibiotics, GMT leads to a recovery of the relative abundances and absolute numbers of Th lymphocytes, cytotoxic T cells, Tregs and B lymphocytes in the lamina propria of the small and large intestines and in the numbers of monocytes, MΦs, Foxp3+ Tregs, and B and T lymphocytes in the colon. In addition to these local effects, Treg and B cell populations in the mesenteric lymph nodes that decreased following antibiotic treatment were also recovered after GMT. In contrast, the numbers of CD4+, CD8+ and B cells in the spleen increased after antibiotic treatment and decreased after GMT, indicating that when the intestinal microbiota is depleted, there is an accumulation of these immune cells in the spleen. The authors also assessed apoptosis and cell proliferation and observed that GMT restored numbers of proliferating cells in the epithelia of ileum and colon. The expression of surface markers and cytokine profiles was also evaluated, revealing that CD4+ and CD8+ cells expressing CD44 were less abundant in all lymphoid tissues evaluated after antibiotic treatment and that the production of IFN-γ, IL-17, IL-22 and IL-10 in the small and large intestines, mesenteric lymph nodes and spleen was reduced. Importantly, all these effects could be restored to pre-antibiotic state after GMT [235].

The most relevant studies cited above are listed in Table **2**.

Table 2. Most relevant experimental studies on GMT for CDI.

Animal model	**Design**	**Main conclusions**	**Authors, References**
Hamster	Transfer of cecal contents from healthy animals incubated with antibiotics and subsequent CDI	*C. difficile* inhibition related to the presence of Gram-negative anaerobes or other Clostridia species	Wilson, Silva and Fekety [216]
Gnotobiotic mice	Animals received distinct bacterial strains prior to CDI and the impact on *C. difficile* growth and toxin production was assessed	Lower toxin production and decreased mortality in mice that received *E. coli* or *Bifidobacterium bifidum* strains prior to CDI	Corthier *et al.* [217]
Conventional and gnotobiotic mice	-Culture-based identification of bacteria present in feces after administration of ampicillin - Animals were inoculated with several Lactobacilli, Bacteroides or *Clostridium* strains isolated from healthy mice and then infected with *C. difficile*	- Increased growth of *C. difficile* was associated with a decreased number of Lactobacilli, an increased number of Bacteroidaceae and a decreased frequency of other Clostridia species - Infection was inhibited in animals that received whole feces, feces treated with chloroform or whole feces from mice with a microbiota mainly composed of Clostridia	Itoh *et al.* [218]
Gnotobiotic mice	Transfer of microbiota from conventional hamster with different treatments to GF mice and evaluation of the volatile fatty acids in mice cecal contents	The presence of volatile fatty acids did not inhibit the growth of *C. difficile* in gnotobiotic mice	Su *et al.* [231].
Hamster	Animals were treated with antibiotics and colonized with toxigenic and non-toxigenic strains of *C. difficile*	Animals treated with clindamycin and colonized with non-toxigenic strains of *C. difficile* did not develop CDI-associated disease when infected with pathogenic *C. difficile*	Merrigan *et al.* [222]
Hamster	Animals were treated with clindamycin and infected with distinct non-toxigenic strains of *C. difficile* and then challenged with a toxigenic strain	Each non-toxigenic strain prevented disease in 87–97% of hamsters that were challenged with toxigenic strains	Sambol *et al.* [223]

(Table 2) cont.....

Animal model	Design	Main conclusions	Authors, References
Germ-free mice	Animals were colonized with either Lachnospiraceae or *E. coli* isolated from healthy mice cecum and then challenged with *C. difficile*	Mice colonized with Lachnospiraceae exhibited fewer signs of colonic histopathology, had lower cytotoxin levels in the intestine and had significantly decreased *C. difficile* colonization	Reeves *et al.* [220]
Mouse	- Transfer of feces from healthy mice that were serially passed in a nutrient broth to *C. difficile*-infected mice - *C. difficile*-infected animals were treated with combinations of different bacteria until a mixture capable of suppressing the infection was found	A combination of six phylogenetically distinct bacteria was able to suppress *C. difficile* infection	Lawley *et al.* [219]
Mouse	Administration of distinct consortiums of bacteria after CDI	A mixture composed of four representative isolates from the healthy gut microbiota or *C. scindens* alone mitigated the infection	Buffie *et al.* [215]
Myd88−/− mice	Myd88−/− and wild-type mice treated with clindamycin and subsequently infected with *C. difficile*	Myd88−/− developed CDI associated disease after treatment while wild-type mice did not	Lawlay *et al.* [233]
TLR-5−/− mice	TLR-5 deficient mice treated with antibiotic and infected with *C. difficile* received flagellin intraperitoneally	Flagellin administration inhibited *C. difficile* growth, reduced toxin concentrations in the intestine, and reduced intestinal epithelial barrier damage, and these effects were not reproducible in TLR5-/- mice	Jarchum *et al.* [234]
Mouse	Evaluation of the impact of antibiotic exposure followed by GMT in different immune cell subsets and pro- and anti-inflammatory cytokines	GMT lead to a recovery in relative abundances and absolute numbers of CTL, Th, Treg and B cells in the intestinal lamina propria; numbers of monocytes, macrophages, Th, Treg cells, and B cells in the colon; Treg and B cells in the MLN; numbers of proliferating cells in the epithelia in ileum and colon; and production of IFN-γ, IL-17, IL-22 and IL-10 in small and large intestines, MLN and spleen	Ekmekciu *et al.* [235]

In conclusion, data suggest that different mechanisms may be involved in the resolution of CDI with the use of GMT rather than a single mechanism Fig. (**5**). It is also possible that in different patients (differing in microbiota composition, dysbiotic states and not infected with the same subtype of *C. difficile*), distinct mechanisms may also contribute differently to resolution.

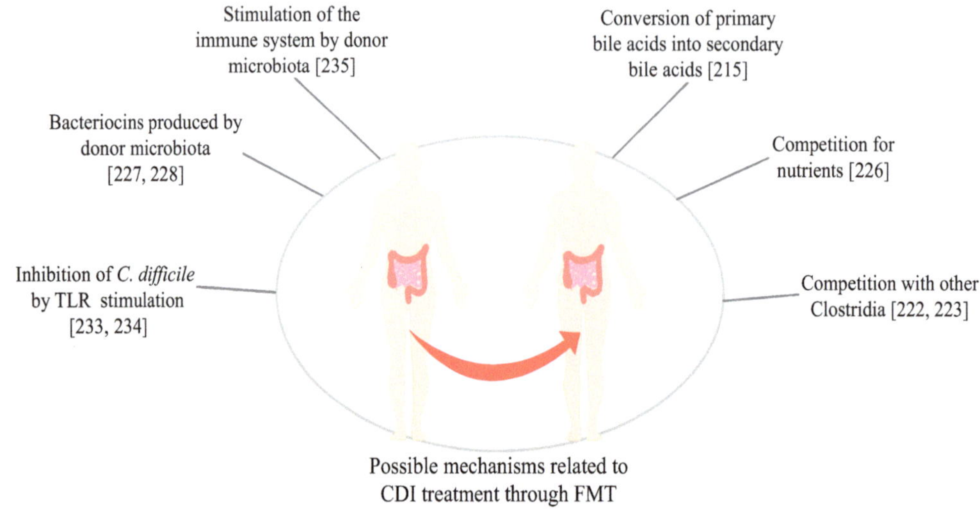

Fig. (5). Possible mechanisms for resolution of CDI by GMT.

OTHER POSSIBLE APPLICATIONS OF GMT

As stated before, the microbiota is essential for human health, and dysbiosis has been implicated in an increasing number of pathological conditions [187, 236]. The main example of dysbiosis as a cause of comorbidity is CDI, but other diseases of the GI tract, such as IBD, irritable colon syndrome and chronic constipation, and extra-intestinal pathologies, such as allergies, neurological diseases and metabolic syndrome, are also related to a loss of balance of the microbiota [236, 237]. Based on this evidence and on the success of GMT in the treatment of recurrent CDI, the possibility of applying this technique to the therapy of other diseases that may be associated with intestinal microbiota disorders should be considered. Nonetheless, and as will be discussed below, there is already substantial data regarding the application of GMT in other conditions but with no clear effectiveness as observed against recurrent CDI.

Inflammatory Bowel Disease (IBD)

IBD is a term used to denote diseases that are characterized by chronic non-infectious inflammation of the intestinal mucosa, are recurrent and manifest clinically due to diarrhea, abdominal pain, weight loss and nausea. Crohn's disease and non-specific UC are the main presentations of IBD. Although disease-related mortality is low, morbidity is considerable [238].

The pathogenesis of IBD is not yet fully understood, but its origin is considered to be multifactorial, with genetic, environmental, and immunological factors and factors related to microbiota involvement [239, 240]. Until now, research has indicated that reducing the diversity of the microbiota is related to the development of IBD; however, it is not yet determined whether this reduction is at the origin of the disease or if arises as a consequence, *i.e.*, whether IBD results from an altered immune system response to a commensal microbiota, a disorder in the microbiota induces an abnormal immune system response in a susceptible host, or if there are one or more microorganisms whose presence or absence plays a key role in triggering the disease [239 - 241]. Nevertheless, with recurrent evidence pointing to dysbiosis as a common feature among patients with IBD, the effects of GMT in these cases have been studied.

In 2014, a systematic review with meta-analysis was performed with the objective of evaluating the efficacy of GMT as a treatment for patients with IBD [242]. The review included 9 cohort studies, case studies, and the only randomized and controlled clinical trial conducted so far. The analysis showed clinical remission in 54 of 119 patients (45%). To minimize publication bias due to the inclusion of case studies, a meta-analysis including only cohort studies was performed, and clinical remission after GMT was detected in 36.2% of the patients. A subgroup of this meta-analysis, including only patients with UC, showed remission in 22% of the patients. The subgroup analysis also showed noticeably greater efficacy in younger populations (although heterogeneous results limit the conclusion) and better results in patients with Crohn's disease (60.5%) than in patients with UC (22%). Nonetheless, the subgroup of CD consisted of only 4 studies with moderate heterogeneity [242].

In 2015, two randomized, placebo-controlled, double-blind clinical studies were published on patients with UC. In the first, researchers compared a group of patients who received GMT from anonymous donors *via* enema in a 50-ml volume once a week for 6 weeks. The control group received water enema [243]. The study identified a statistically significant difference in the remission of colitis in 24% of patients in the GMT group and 5% in the placebo group. There was no difference in the occurrence of adverse events between groups. The study also identified a better response in patients with less than one year of disease progression and in those taking immunosuppressive therapy and suggested that the donor may interfere with the GMT result (of the nine patients who had a response, seven received feces from the same donor). In the second study, 23 patients were allocated in a group that received GMT from a healthy donor *via* NE probe in a 500-ml volume with previous intestinal lavage, and after 3 weeks, the whole procedure was repeated [244]. In this study, the control group, composed of 25 patients, received autologous GMT. There was no significant difference in clinical and endoscopic remission between the two groups, but the researchers identified distinct characteristics (an increase in species diversity) in the microbiota of responding patients when compared to non-responders. In the group that received GMT from a donor, the microbiota of the responders became similar to that of the donor. In the group that received autologous GMT, there was also a change in the microbiota of the responders, but the species with increased representation differed from those that increased in the group of responders who had received fecal material from donors. No serious adverse events were identified, and the procedure was considered safe, with the exception of the study having a short observation period of only 12 weeks. The authors considered that sample size, the route of administration of fecal material and the frequency of the procedure may have contributed to the absence of differences between groups in response to treatment. In fact, in relation to the route of administration in the UC cases, available evidence corroborates the authors' conclusion and indicates worse results and a higher incidence of adverse events when the transplantation is performed using a NE tube [243 - 245].

Despite the initial expectation, current data indicate that the success of GMT in CDI will not be present in IBD; the more complex pathogenesis of IBD is possibly an explanation for the difference in outcomes. There are still open questions regarding the route of administration, the number of infusions required, the ideal interval of the procedure, the subgroup of patients with the greatest benefit, the ideal donor microbiota and the long-term consequences [240 - 244]. However, with an increase in the understanding of the mechanisms of both IBD and GMT, it is expected that the GMT will certainly find its place among the therapeutic arsenal against IBD, although it is possible that only a portion of individuals affected by IBD will benefit from this treatment.

Pouchitis, an idiopathic inflammatory condition of the mucosa, is the most frequent complication among patients submitted to total proctocolectomy with ileo-pouch-anal anastomosis for UC treatment, occurring in approximately 50% of patients [246]. Acute conditions are managed with antibiotics, but in 10-20% of patients, there is no response to standard treatment, and the condition becomes chronic. Management of the chronic patient implies recurrent use of antibiotics, steroids, immunomodulators and probiotics [246]. Dysbiosis has been implicated in the pathogenesis of pouchitis [247, 248], but few studies have evaluated the benefit of GMT in these cases. The controversial results and small number of studies currently prevent definitive conclusions about the therapeutic potential of GMT for pouchitis [248, 249].

Functional Diseases of the Intestine

According to the Rome III classification system, functional bowel diseases include IBS, functional stuffiness, functional constipation, and non-specific functional bowel disease [250]. IBS is the most relevant. It is a chronic condition with recurrent character and is diagnosed based on symptomatology according to the following criteria: recurrent abdominal pain or discomfort for at least 3 days per month, occurring in the last three months and associated with at least two of the following: improvement with defecation, onset associated with change in stool frequency or onset associated with change in stool appearance. Moreover, the symptoms must have started at least 6 months before diagnosis [251]. A systematic review with meta-analysis, including eighty cross-sectional studies, found 11% overall prevalence for IBS, with significant variation according to geographical region and diagnostic criteria [252].

IBS is a complex entity, and its pathogenesis has not yet been fully elucidated; however, evidence indicates that it results from a combination of factors: genetic predisposition, environmental exposures, psychosocial factors, altered bowel mobility, visceral hypersensitivity, increased intestinal permeability, low degree of inflammation, alteration of the intestinal microbiota and deregulation of the central nervous system - enteric nervous system axis [250, 253]. The microbiota is essential in multiple functions of the intestine, from motility to proper functioning of the communication between the enteric and the central nervous system [254]. Evidence indicates an important role of dysbiosis in the pathogenesis of functional diseases of the GI tract, including the development of IBS in patients with gastroenteritis [255], and multiple studies have shown a difference in microbiota composition among IBS patients compared to healthy individuals [254, 256]. Thus, in view of these data, research in the field of modulation of the microbiota for treatment of this pathology has gained strength. Probiotics, prebiotics, symbiotics, antibiotics, and, more recently, GMT are being studied. Currently, limited data are available on the efficacy of GMT in IBS or in other functional diseases of the GI tract. Recent literature reviews have found only a few published case series [251, 254, 256].

Although the initial results are promising, their limitations and the numerous questions that remain to be answered prevent the indication of GMT for the treatment of functional diseases of the GI tract. This scenario should change over the next few years, considering that there are currently 10 ongoing clinical trials on the topic [257].

Bacterial Resistance Control

For patients who develop multi-drug-resistant (MDR) bacterial infection, therapeutic options are limited, often less effective, and may include more toxic antibiotics [258]. Since the intestine is the site of initial colonization and persistence of many of the known MDR bacteria, this organ is a potential source of microorganisms for subsequent infections [258] and therefore the main target for a decolonization strategy of the patient aiming to reduce the risk of infection. The use of antibiotics for this purpose, although the strategy is logical and can demonstrate success [259, 260], should be managed with caution considering that these drugs are part of the origin of the problem by inducing dysbiosis of the intestinal microbiota, allowing establishment of MDR strains [261]. Moreover, even when the decontamination program succeeds in reducing colonization, this effect is not maintained for a long time [262]. Thus, it is necessary to consider alternatives for the decolonization of the patients.

The data showing the high success rates of GMT in the treatment of recurrent CDI, the low incidence of complications, and the persistence of the result for long periods have led researchers to consider the application of this procedure in the decolonization of patients with MDR bacteria. In addition, evidence generated by CDI surveys has indicated the additional benefit of MDR eradication in patients undergoing GMT. It is known that CDI leads to repeated courses of antibiotics, and of course, an increase in the induction of bacterial resistance genes is expected. This association was evidenced in a cohort of 20 patients submitted to GMT for treatment of CDI, where the improvement of colitis symptoms was accompanied by a reduction in the number and diversity of antibiotic resistance genes previously identified in patients submitted to transplantation [263]. Additionally, since the risk factors for CDI acquisition and colonization by MDR bacteria overlap, some patients with CDI submitted to GMT demonstrated colonization by MDR, and the follow-up of these patients after treatment showed eradication of MDR colonization [264] or reduction in the number of MDR strains [265]. In 2016, a review on the subject was published [258] summarizing the report of eight cases described in the literature. All of them showed a response to the GMT, with eradication or reduction of colonization by different MDR bacteria in different scenarios, in addition to showing few adverse events, although without long-term monitoring data. Also based on a case report, the promising GMT safety profile for MDR bacteria eradication was evidenced in an IC patient [266].

A question always pertinent when considering the modulation of the microbiota for the treatment of a specific condition is whether a particular species can be associated with the expected result. In the field of bacterial resistance, the genus *Barnesiella* was related to the elimination of colonization by vancomycin-resistant *Enterococcus faecium* in rats and the protection against colonization by this microorganism in patients undergoing hematopoietic cell halogen transplantation

[267]. In *in vitro* experiments, *Lactobacillus* species exhibited inhibition properties against *K. pneumonia* [268].

The published evidence for the application of GMT in decolonization of patients with MDR bacteria is currently restricted to case reports and cohort studies [247], and many issues including safety procedures are open. However, as our understanding of the microbiota role in the prevention and/or treatment of infections due to opportunistic pathogens and by MDR bacteria increases, it is also possible to observe a light of hope in a battle that seemed lost.

Obesity and Metabolic Diseases

Obesity is related to numerous pathologies such as hypertension, type 2 diabetes, dyslipidemia, coronary, respiratory and gallbladder diseases, stroke, osteoarthritis and cancer [269]. The prevalence of obesity in adults, children and adolescents has increased in recent decades, in both developed and developing countries [270]. The approach to multifactorial diseases must also consider the various elements of their pathogenesis. Microbiota alterations, such as global reduction of the diversity and increase in Firmicutes with reduction of *Bacteroides*, have been demonstrated in obese individuals [255]. In another study, dysbiosis with reduction of bacterial diversity and predominance of the genera *Prevotella* and *Klebsiella* was related to arterial hypertension [271]. The authors attributed the results to a direct influence of the microbiota on blood pressure levels, since blood pressure elevation was observed in rats submitted to transplantation of stool samples from hypertensive human donors [271].

The mechanisms between the intestinal microbiota and the development of obesity and other metabolic diseases such as diabetes are complex and involve the production of energy for the host by digestion of soluble dietary fiber and formation of SCFAs, modulation of inflammatory response and influence of the microbiota on satiety across the cerebrum-intestine axis [272, 273].

Before considering GMT as a therapeutic option for obesity and metabolic syndrome, many gaps need to be filled regarding safety, long-term effect, and specificity of the bacteria to be transplanted, among other factors; the expectation is high considering the accumulation of evidence on the role of the intestinal microbiota in human metabolism. There are currently eight ongoing clinical trials to assess the impact of GMT on obesity, type II diabetes and metabolic syndrome [257].

Neurological and Psychiatric Disorders

An exciting current field of research is the two-way path known as the intestinal-microbiota-brain axis, through which the brain and intestine interact [274] Revisions published in recent years have reported the influence of the microbiota on numerous neuropsychiatric diseases, such as depression, schizophrenia, attention deficit hyperactivity disorder, autistic spectrum disorders, multiple sclerosis, myoclonic dystonia, chronic fatigue syndrome, and Parkinson disease. Among the evidence of this relationship is the presence of dysbiosis in sick individuals, improvement of symptoms with probiotics or antibiotics and relation to infectious episodes [274 - 278]. Data on the use of GMT for the treatment of these pathologies are scarce. In a review from 2013, Aroniadis and Brandt cite personal correspondence with a researcher who described improvement of autism symptoms in two children after GMT and in five children after daily administration of *Bacteroidetes* and Clostridia for several weeks [237]. Additionally, a cohort of 60 patients diagnosed with chronic fatigue syndrome were submitted to transcolonoscopic infusion of thirteen non-pathogenic enteric bacteria, and the treatment success rate was 70% [279].

Hepatic Diseases

The intestinal microbiota is related to liver diseases, and there is evidence of its influence in the pre-cirrhotic state and on non-alcoholic fatty liver disease (NAFLD) and alcoholic liver disease, as well as in cirrhosis itself [280]. NAFLD is studied in the context of the metabolic syndrome [281], and therefore, mechanisms related to microbiota have also been described in the pathogenesis of NAFLD [282]. When compared to patients with hepatic steatosis and healthy subjects, patients with non-alcoholic steatohepatitis had a reduced percentage of Bacteroidetes and increased percentage of *C. coccoides* [283]. Given that not all chronic alcohol users develop alcoholic liver disease, the presence of co-factors involved in alcohol-induced tissue injury has been considered. Alteration in the composition of the intestinal microbiota related to high serum levels of endotoxin in some but not all chronic alcohol users may possibly contribute to liver injury [284].

Important complications of cirrhosis, notably, spontaneous bacterial peritonitis and hepatic encephalopathy, have origins in bacterial translocation, and mechanisms related to intestinal microbiota are associated with such occurrence [285]. As in other pathologies, evidence of the relationship of dysbiosis with the pathogenesis of liver diseases has led to experiments modulating the microbiota to treat these conditions; however, only two studies using GMT have been reported. In the first study, a man with hepatic encephalopathy was submitted to stool transplantation weekly for five weeks. There was clinical improvement, but the benefits were transient [286]. The second study was a case-control study involving 18 patients diagnosed with chronic hepatitis B with positive serology for HBeAg after more than three years of entecavir or tenofovir therapy. Five patients underwent stool transplantation *via* gastroscopy every 4 weeks until clearance of HBeAg, without discontinuing prior therapy. The other 13 patients maintained only the antiviral therapy already in use. Participants received from one to seven stool transplants. Three participants in the GMT group had clearance of HBeAg, whereas this result was not achieved by any of the participants in the control group. Despite the small number of participants, a significant difference was observed [287].

Currently, an open, single-arm pilot study is underway to assess the impact of GMT on primary sclerosing cholangitis (see www.clinicaltrials.gov) [257]; NCT 02424175). Therefore, it is necessary to await results from undergoing clinical trials so that a possible indication of GMT can be evaluated in the therapy of liver diseases.

Allogeneic Transplants

A recent review [288] described the considerable amount of evidence regarding the relationship between changes in microbiota, immune system and incidence of complications following liver, kidney, and hematopoietic stem cell transplantation. The encouraging results of studies using probiotics, prebiotics and antibiotics to prevent these complications through the modulation of the intestinal microbiota are also noted in this review. In addition, the safety and efficacy of GMT for the treatment of graft *versus* acute host disease after stem cell transplantation were demonstrated in 4 patients [289].

Sepsis

Available data from the case report of a patient with sepsis and diarrhea treated with antibiotics, probiotics and supportive measures presented resolution of the condition after GMT, suggesting a potential role of this intervention as an alternative therapy in cases of sepsis [290].

Finally, there are many studies presenting data demonstrating a relationship between dysbiosis and different types of pathologies in which GMT could be considered an option to ameliorate the disease. These conditions include polycystic ovarian syndrome, some types of tumoural and autoimmune diseases, and allergic diseases such as eczema, asthma, rhinitis and food-triggered allergy (*i.e.*, [291 - 296]).

FUTURE PERSPECTIVES

From the data discussed so far, it is possible to foresee that microbiota-based therapies will become increasingly common in the near future. This will be possible through an increased understanding of some of the key aspects surrounding this issue, including the following.

There is relatively little understanding of the function and importance of some immune cell populations present in the intestine, including intraepithelial lymphocytes as gamma-delta T cells. Less is known regarding their behavior in relation to food and microbiota antigens. Thus, increasing understanding of the mucosal immune system and its relation to microbiota is imperative. In parallel, it is necessary to define the effects on human health of the different members of the microbiota. Moreover, microbiota members other than bacteria, such viruses and fungi, may contribute to restore homeostasis. In fact, ultra-deep sequencing has recently shown that higher abundance and lower diversity of some types of bacteriophages are associated with a better outcome of GMT against CDI. In this context, molecular high-throughput technologies play an important role in supporting the advancement of this exciting area, as they are especially needed due to the complexity of the microbiota.

Microbiota-related research continuously surprises with discoveries of the roles of commensal microbes in the most unexpected aspects of human physiology. Therefore, several disorders considered idiopathic are possibly associated with dysbiosis, and better therapeutic options will be increasingly needed, especially those that aim to maintain or recover the integrity of the microbiota.

CONSENT FOR PUBLICATION

Not applicable.

ACKNOWLEDGEMENTS

We thank Dr. Christina Surawicz from University of Washington School of Medicine for comments that greatly improved this chapter.

CONFLICT OF INTEREST

A.M., C.P.Z. and D.C.T. have financial support from CAPES (Coordenação de Aperfeiçoamento de Pessoal de Nível Superior) and A.R.P. from CNPq (Conselho Nacional de Desenvolvimento Científico e Tecnológico). All other authors confirm that they have no conflicts of interest to declare.

REFERENCES

[1] Sender R, Fuchs S, Milo R. Revised estimates for the Number of Human and bacteria cells in the body. PLoS Biol 2016; 14(8): e1002533. [http://dx.doi.org/10.1371/journal.pbio.1002533] [PMID: 27541692]

[2] Berg RD. The indigenous gastrointestinal microflora. Trends Microbiol 1996; 4(11): 430-5. [http://dx.doi.org/10.1016/0966-842X(96)10057-3] [PMID: 8950812]

[3] Eckburg PB, Bik EM, Bernstein CN, *et al.* Diversity of the human intestinal microbial flora. Science 2005; 308(5728): 1635-8. [http://dx.doi.org/10.1126/science.1110591] [PMID: 15831718]

[4] Proctor LM. The Human Microbiome Project in 2011 and beyond. Cell Host Microbe 2011; 10(4): 287-91. [http://dx.doi.org/10.1016/j.chom.2011.10.001] [PMID: 22018227]

[5] Gorjifard S, Goldszmid RS. Microbiota-myeloid cell crosstalk beyond the gut. J Leukoc Biol 2016; 100(5): 865-79. [http://dx.doi.org/10.1189/jlb.3RI0516-222R] [PMID: 27605211]

[6] Palm NW, de Zoete MR, Flavell RA. Immune-microbiota interactions in health and disease. Clin Immunol 2015; 159(2): 122-7. [http://dx.doi.org/10.1016/j.clim.2015.05.014] [PMID: 26141651]

[7] Burokas A, Moloney RD, Dinan TG, Cryan JF. Microbiota regulation of the Mammalian gut-brain axis. Adv Appl Microbiol 2015; 91: 1-62. [http://dx.doi.org/10.1016/bs.aambs.2015.02.001] [PMID: 25911232]

[8] Zoppi G. Probiotics and prebiotics in the treatment of infections due to vancomycin-dependent Enterococcus faecalis and of imbalance of the intestinal ecosystem (dysbiosis). Acta Paediatr 1997; 86(10): 1148-50. [http://dx.doi.org/10.1111/j.1651-2227.1997.tb14829.x] [PMID: 9350905]

[9] Turnbaugh PJ, Ley RE, Mahowald MA, Magrini V, Mardis ER, Gordon JI. An obesity-associated gut microbiome with increased capacity for energy harvest. Nature 2006; 444(7122): 1027-31. [http://dx.doi.org/10.1038/nature05414] [PMID: 17183312]

[10] Ivanov II, Atarashi K, Manel N, *et al.* Induction of intestinal Th17 cells by segmented filamentous bacteria. Cell 2009; 139(3): 485-98. [http://dx.doi.org/10.1016/j.cell.2009.09.033] [PMID: 19836068]

[11] Berer K, Mues M, Koutrolos M, *et al.* Commensal microbiota and myelin autoantigen cooperate to trigger autoimmune demyelination. Nature 2011; 479(7374): 538-41. [http://dx.doi.org/10.1038/nature10554] [PMID: 22031325]

[12] Grivennikov SI, Wang K, Mucida D, *et al.* Adenoma-linked barrier defects and microbial products drive IL-23/IL-17-mediated tumour growth. Nature 2012; 491(7423): 254-8. [http://dx.doi.org/10.1038/nature11465] [PMID: 23034650]

[13] Horai R, Zárate-Bladés CR, Dillenburg-Pilla P, *et al.* Microbiota-dependent activation of an autoreactive T cell receptor provokes autoimmunity in an immunologically privileged site. Immunity 2015; 43(2): 343-53. [http://dx.doi.org/10.1016/j.immuni.2015.07.014] [PMID: 26287682]

[14] Zárate-Bladés CR, Horai R, Mattapallil MJ, *et al.* Gut microbiota as a source of a surrogate antigen that triggers autoimmunity in an immune privileged site. Gut Microbes 2017; 8(1): 59-66. [http://dx.doi.org/10.1080/19490976.2016.1273996] [PMID: 28045579]

[15] Wang L, Christophersen CT, Sorich MJ, Gerber JP, Angley MT, Conlon MA. Increased abundance of Sutterella spp. and ruminococcus torques in feces of children with autism spectrum disorder. Mol Autism 2013; 4(1): 42. [http://dx.doi.org/10.1186/2040-2392-4-42] [PMID: 24188502]

[16] Thaiss CA, Zmora N, Levy M, Elinav E. The microbiome and innate immunity. Nature 2016; 535(7610): 65-74. [http://dx.doi.org/10.1038/nature18847] [PMID: 27383981]

[17] Jiménez E, Marín ML, Martín R, *et al.* Is meconium from healthy newborns actually sterile? Res Microbiol 2008; 159(3): 187-93. [http://dx.doi.org/10.1016/j.resmic.2007.12.007] [PMID: 18281199]

[18] Bäckhed F, Roswall J, Peng Y, *et al.* Dynamics and stabilization of the human gut microbiome during the first year of life. Cell host microbe 2015; 17(5): 690-703. [http://dx.doi.org/10.1016/j.chom.2015.04.004] [PMID: 25974306]

[19] Mackie RI, Sghir A, Gaskins HR. Developmental microbial ecology of the neonatal gastrointestinal tract. Am J Clin Nutr 1999; 69(5): 1035S-45S. [http://dx.doi.org/10.1093/ajcn/69.5.1035s] [PMID: 10232646]

[20] Dominguez-Bello MG, Costello EK, Contreras M, *et al.* Delivery mode shapes the acquisition and structure of the initial microbiota across multiple body habitats in newborns. Proc Natl Acad Sci USA 2010; 107(26): 11971-5. [http://dx.doi.org/10.1073/pnas.1002601107] [PMID: 20566857]

[21] MacIntyre DA, Chandiramani M, Lee YS, *et al.* The vaginal microbiome during pregnancy and the postpartum period in a European population. Sci Rep 2015; 5: 8988. [http://dx.doi.org/10.1038/srep08988] [PMID: 25758319]

[22] Proctor LM. The national institutes of health human microbiome project. Semin Fetal Neonatal Med 2016; 21(6): 368-72. [http://dx.doi.org/10.1016/j.siny.2016.05.002] [PMID: 27350143]

[23] Telesford KM, Yan W, Ochoa-Reparaz J, *et al.* A commensal symbiotic factor derived from Bacteroides fragilis promotes human CD39(+)Foxp3(+) T cells and treg function. Gut Microbes 2015; 6(4): 234-42. [http://dx.doi.org/10.1080/19490976.2015.1056973] [PMID: 26230152]

[24] Ianiro G, Molina-Infante J, Gasbarrini A. Gastric Microbiota. Helicobacter 2015; 20 (Suppl. 1): 68-71. [http://dx.doi.org/10.1111/hel.12260] [PMID: 26372828]

[25] El Aidy S, van den Bogert B, Kleerebezem M. The small intestine microbiota, nutritional modulation and relevance for health. Curr Opin Biotechnol 2015; 32: 14-20. [http://dx.doi.org/10.1016/j.copbio.2014.09.005] [PMID: 25308830]

[26] Gill SR, Pop M, Deboy RT, *et al.* Metagenomic analysis of the human distal gut microbiome. Science 2006; 312(5778): 1355-9. [http://dx.doi.org/10.1126/science.1124234] [PMID: 16741115]

[27] Wang Y, Hoenig JD, Malin KJ, *et al.* 16S rRNA gene-based analysis of fecal microbiota from preterm infants with and without necrotizing enterocolitis. ISME J 2009; 3(8): 944-54. [http://dx.doi.org/10.1038/ismej.2009.37] [PMID: 19369970]

[28] Claesson MJ, Jeffery IB, Conde S, *et al.* Gut microbiota composition correlates with diet and health in the elderly. Nature 2012; 488(7410): 178-84. [http://dx.doi.org/10.1038/nature11319] [PMID: 22797518]

[29] Chow J, Mazmanian SK. A pathobiont of the microbiota balances host colonization and intestinal inflammation. Cell Host Microbe 2010; 7(4): 265-76. [http://dx.doi.org/10.1016/j.chom.2010.03.004] [PMID: 20413095]

[30] Levy M, Kolodziejczyk AA, Thaiss CA, Elinav E. Dysbiosis and the immune system. Nat Rev Immunol 2017; 17(4): 219-32. [http://dx.doi.org/10.1038/nri.2017.7] [PMID: 28260787]

[31] Spigaglia P, Barbanti F, Mastrantonio P. European Study Group on Clostridium difficile (ESGCD). Multidrug resistance in European Clostridium difficile clinical isolates. J Antimicrob Chemother 2011; 66(10): 2227-34. [http://dx.doi.org/10.1093/jac/dkr292] [PMID: 21771851]

[32] Crowther GS, Wilcox MH. Antibiotic therapy and clostridium difficile infection - primum non nocere - first do no harm. Infect Drug Resist 2015; 8: 333-7.
[PMID: 26396535]

[33] Gough E, Shaikh H, Manges AR. Systematic review of intestinal microbiota transplantation (fecal bacteriotherapy) for recurrent Clostridium difficile infection. Clin Infect Dis 2011; 53(10): 994-1002.
[http://dx.doi.org/10.1093/cid/cir632] [PMID: 22002980]

[34] Surawicz CM, Brandt LJ, Binion DG, *et al.* Guidelines for diagnosis, treatment, and prevention of Clostridium difficile infections. Am J Gastroenterol 2013; 108(4): 478-98.
[http://dx.doi.org/10.1038/ajg.2013.4] [PMID: 23439232]

[35] Eiseman B, Silen W, Bascom GS, Kauvar AJ. Fecal enema as an adjunct in the treatment of pseudomembranous enterocolitis. Surgery 1958; 44(5): 854-9.
[PMID: 13592638]

[36] Zoppi G, Balsamo V, Deganello A, Iacono G, Saccomani F, Benoni G. Oral bacteriotherapy in clinical practice. II. The use of different preparations in the treatment of acute diarrhoea. Eur J Pediatr 1982; 139(1): 22-4.
[http://dx.doi.org/10.1007/BF00442073] [PMID: 6816602]

[37] Tvede M, Rask-Madsen J. Bacteriotherapy for chronic relapsing Clostridium difficile diarrhoea in six patients. Lancet 1989; 1(8648): 1156-60.
[http://dx.doi.org/10.1016/S0140-6736(89)92749-9] [PMID: 2566734]

[38] McCune VL, Struthers JK, Hawkey PM. Faecal transplantation for the treatment of Clostridium difficile infection: a review. Int J Antimicrob Agents 2014; 43(3): 201-6.
[http://dx.doi.org/10.1016/j.ijantimicag.2013.10.009] [PMID: 24636428]

[39] Borody TJ, Warren EF, Leis S, Surace R, Ashman O. Treatment of ulcerative colitis using fecal bacteriotherapy. J Clin Gastroenterol 2003; 37(1): 42-7.
[http://dx.doi.org/10.1097/00004836-200307000-00012] [PMID: 12811208]

[40] Jayasinghe TN, Chiavaroli V, Holland DJ, Cutfield WS, O'Sullivan JM. The New Era of Treatment for Obesity and metabolic Disorders: Evidence and expectations for Gut Microbiome Transplantation. Front Cell Infect Microbiol 2016; 6: 15.
[http://dx.doi.org/10.3389/fcimb.2016.00015] [PMID: 26925392]

[41] Chung H, Pamp SJ, Hill JA, *et al.* Gut immune maturation depends on colonization with a host-specific microbiota. Cell 2012; 149(7): 1578-93.
[http://dx.doi.org/10.1016/j.cell.2012.04.037] [PMID: 22726443]

[42] Bohnhoff M, Drake BL, Miller CP. The effect of an antibiotic on the susceptibility of the mouse's intestinal tract to Salmonella infection. Antibiot Annu 1955-1956; 3: 453-5.
[PMID: 13355309]

[43] Man SM, Zhu Q, Zhu L, *et al.* Critical role for the DNA sensor AIM2 in stem cell proliferation and cancer. Cell 2015; 162(1): 45-58.
[http://dx.doi.org/10.1016/j.cell.2015.06.001] [PMID: 26095253]

[44] Song-Zhao GX, Srinivasan N, Pott J, Baban D, Frankel G, Maloy KJ. Nlrp3 activation in the intestinal epithelium protects against a mucosal pathogen. Mucosal Immunol 2014; 7(4): 763-74.
[http://dx.doi.org/10.1038/mi.2013.94] [PMID: 24280937]

[45] Artis D. Epithelial-cell recognition of commensal bacteria and maintenance of immune homeostasis in the gut. Nat Rev Immunol 2008; 8(6): 411-20.
[http://dx.doi.org/10.1038/nri2316] [PMID: 18469830]

[46] Flannigan KL, Geem D, Harusato A, Denning TL. Intestinal antigen-presenting cells: Key regulators of immune homeostasis and inflammation. Am J Pathol 2015; 185(7): 1809-19.
[http://dx.doi.org/10.1016/j.ajpath.2015.02.024] [PMID: 25976247]

[47] Gross M, Salame TM, Jung S. Guardians of the gut - murine intestinal macrophages and dendritic cells. Front Immunol 2015; 6: 254.
[http://dx.doi.org/10.3389/fimmu.2015.00254] [PMID: 26082775]

[48] Rescigno M, Urbano M, Valzasina B, *et al.* Dendritic cells express tight junction proteins and penetrate gut epithelial monolayers to sample bacteria. Nat Immunol 2001; 2(4): 361-7.
[http://dx.doi.org/10.1038/86373] [PMID: 11276208]

[49] Vallon-Eberhard A, Landsman L, Yogev N, Verrier B, Jung S. Transepithelial pathogen uptake into the small intestinal lamina propria. J Immunol 2006; 176(4): 2465-9.
[http://dx.doi.org/10.4049/jimmunol.176.4.2465] [PMID: 16456006]

[50] Denning TL, Wang YC, Patel SR, Williams IR, Pulendran B. Lamina propria macrophages and dendritic cells differentially induce regulatory and interleukin 17-producing T cell responses. Nat Immunol 2007; 8(10): 1086-94.
[http://dx.doi.org/10.1038/ni1511] [PMID: 17873879]

[51] Smythies LE, Sellers M, Clements RH, *et al.* Human intestinal macrophages display profound inflammatory anergy despite avid phagocytic and bacteriocidal activity. J Clin Invest 2005; 115(1): 66-75.
[http://dx.doi.org/10.1172/JCI200519229] [PMID: 15630445]

[52] Kamada N, Hisamatsu T, Okamoto S, *et al.* Abnormally differentiated subsets of intestinal macrophage play a key role in Th1-dominant chronic colitis through excess production of IL-12 and IL-23 in response to bacteria. J Immunol 2005; 175(10): 6900-8.
[http://dx.doi.org/10.4049/jimmunol.175.10.6900] [PMID: 16272349]

[53] Kamada N, Hisamatsu T, Okamoto S, *et al.* Unique CD14 intestinal macrophages contribute to the pathogenesis of Crohn disease *via* IL-23/IFN-gamma axis. J Clin Invest 2008; 118(6): 2269-80.
[PMID: 18497880]

[54] Diehl GE, Longman RS, Zhang JX, *et al.* Microbiota restricts trafficking of bacteria to mesenteric lymph nodes by CX(3)CR1(hi) cells. Nature 2013; 494(7435): 116-20.
[http://dx.doi.org/10.1038/nature11809] [PMID: 23334413]

[55] Ueda Y, Kayama H, Jeon SG, *et al.* Commensal microbiota induce LPS hyporesponsiveness in colonic macrophages *via* the production of IL-10. Int Immunol 2010; 22(12): 953-62.
[http://dx.doi.org/10.1093/intimm/dxq449] [PMID: 21051439]

[56] Khosravi A, Yáñez A, Price JG, *et al.* Gut microbiota promote hematopoiesis to control bacterial infection. Cell Host Microbe 2014; 15(3): 374-81.
[http://dx.doi.org/10.1016/j.chom.2014.02.006] [PMID: 24629343]

[57] Balmer ML, Schürch CM, Saito Y, *et al.* Microbiota-derived compounds drive steady-state granulopoiesis *via* MyD88/TICAM signaling. J Immunol 2014; 193(10): 5273-83.
[http://dx.doi.org/10.4049/jimmunol.1400762] [PMID: 25305320]

[58] Worbs T, Bode U, Yan S, *et al.* Oral tolerance originates in the intestinal immune system and relies on antigen carriage by dendritic cells. J Exp Med 2006; 203(3): 519-27.
[http://dx.doi.org/10.1084/jem.20052016] [PMID: 16533884]

[59] Coombes JL, Siddiqui KR, Arancibia-Cárcamo CV, *et al.* A functionally specialized population of mucosal CD103+ DCs induces Foxp3+ regulatory T cells *via* a TGF-beta and retinoic acid-dependent mechanism. J Exp Med 2007; 204(8): 1757-64.
[http://dx.doi.org/10.1084/jem.20070590] [PMID: 17620361]

[60] Mucida D, Park Y, Kim G, *et al.* Reciprocal TH17 and regulatory T cell differentiation mediated by retinoic acid. Science 2007; 317(5835): 256-60.
[http://dx.doi.org/10.1126/science.1145697] [PMID: 17569825]

[61] Sun CM, Hall JA, Blank RB, *et al.* Small intestine lamina propria dendritic cells promote de novo generation of Foxp3 T reg cells *via* retinoic acid. J Exp Med 2007; 204(8): 1775-85.

[http://dx.doi.org/10.1084/jem.20070602] [PMID: 17620362]

[62] Persson EK, Uronen-Hansson H, Semmrich M, *et al.* IRF4 transcription-factor-dependent CD103(+)CD11b(+) dendritic cells drive mucosal T helper 17 cell differentiation. Immunity 2013; 38(5): 958-69. [http://dx.doi.org/10.1016/j.immuni.2013.03.009] [PMID: 23664832]

[63] Schlitzer A, McGovern N, Teo P, *et al.* IRF4 transcription factor-dependent CD11b+ dendritic cells in human and mouse control mucosal IL-17 cytokine responses. Immunity 2013; 38(5): 970-83. [http://dx.doi.org/10.1016/j.immuni.2013.04.011] [PMID: 23706669]

[64] Atarashi K, Tanoue T, Shima T, *et al.* Induction of colonic regulatory T cells by indigenous Clostridium species. Science 2011; 331(6015): 337-41. [http://dx.doi.org/10.1126/science.1198469] [PMID: 21205640]

[65] Round JL, Mazmanian SK. Inducible Foxp3+ regulatory T-cell development by a commensal bacterium of the intestinal microbiota. Proc Natl Acad Sci USA 2010; 107(27): 12204-9. [http://dx.doi.org/10.1073/pnas.0909122107] [PMID: 20566854]

[66] Dasgupta S, Erturk-Hasdemir D, Ochoa-Reparaz J, Reinecker HC, Kasper DL. Plasmacytoid dendritic cells mediate anti-inflammatory responses to a gut commensal molecule *via* both innate and adaptive mechanisms. Cell Host Microbe 2014; 15(4): 413-23. [http://dx.doi.org/10.1016/j.chom.2014.03.006] [PMID: 24721570]

[67] Mazmanian SK, Round JL, Kasper DL. A microbial symbiosis factor prevents intestinal inflammatory disease. Nature 2008; 453(7195): 620-5. [http://dx.doi.org/10.1038/nature07008] [PMID: 18509436]

[68] Laffont S, Siddiqui KR, Powrie F. Intestinal inflammation abrogates the tolerogenic properties of MLN CD103+ dendritic cells. Eur J Immunol 2010; 40(7): 1877-83. [http://dx.doi.org/10.1002/eji.200939957] [PMID: 20432234]

[69] Hart AL, Al-Hassi HO, Rigby RJ, *et al.* Characteristics of intestinal dendritic cells in inflammatory bowel diseases. Gastroenterology 2005; 129(1): 50-65. [http://dx.doi.org/10.1053/j.gastro.2005.05.013] [PMID: 16012934]

[70] Wilson NS, Young LJ, Kupresanin F, *et al.* Normal proportion and expression of maturation markers in migratory dendritic cells in the absence of germs or Toll-like receptor signaling. Immunol Cell Biol 2008; 86(2): 200-5. [http://dx.doi.org/10.1038/sj.icb.7100125] [PMID: 18026177]

[71] Deshmukh HS, Liu Y, Menkiti OR, *et al.* The microbiota regulates neutrophil homeostasis and host resistance to Escherichia coli K1 sepsis in neonatal mice. Nat Med 2014; 20(5): 524-30. [http://dx.doi.org/10.1038/nm.3542] [PMID: 24747744]

[72] Kristensen MB, Metzdorff SB, Bergström A, *et al.* Neonatal microbial colonization in mice promotes prolonged dominance of CD11b(+)Gr-1(+) cells and accelerated establishment of the CD4(+) T cell population in the spleen. Immun Inflamm Dis 2015; 3(3): 309-20. [http://dx.doi.org/10.1002/iid3.70] [PMID: 26417445]

[73] Longman RS, Diehl GE, Victorio DA, *et al.* CX_3CR1^+ mononuclear phagocytes support colitis-associated innate lymphoid cell production of IL-22. J Exp Med 2014; 211(8): 1571-83. [http://dx.doi.org/10.1084/jem.20140678] [PMID: 25024136]

[74] Mortha A, Chudnovskiy A, Hashimoto D, *et al.* Microbiota-dependent crosstalk between macrophages and ILC3 promotes intestinal homeostasis. Science 2014; 343(6178): 1249288. [http://dx.doi.org/10.1126/science.1249288] [PMID: 24625929]

[75] Howard M, Paul WE. Regulation of B-cell growth and differentiation by soluble factors. Annu Rev Immunol 1983; 1: 307-33. [http://dx.doi.org/10.1146/annurev.iy.01.040183.001515] [PMID: 6242466]

[76] O'Shea JJ, Paul WE. Mechanisms underlying lineage commitment and plasticity of helper CD4+ T

cells. Science 2010; 327(5969): 1098-102.
[http://dx.doi.org/10.1126/science.1178334] [PMID: 20185720]

[77] Ivanov II, Honda K. Intestinal commensal microbes as immune modulators. Cell Host Microbe 2012; 12(4): 496-508.
[http://dx.doi.org/10.1016/j.chom.2012.09.009] [PMID: 23084918]

[78] Tanoue T, Atarashi K, Honda K. Development and maintenance of intestinal regulatory T cells. Nat Rev Immunol 2016; 16(5): 295-309.
[http://dx.doi.org/10.1038/nri.2016.36] [PMID: 27087661]

[79] Geuking MB, Cahenzli J, Lawson MA, *et al.* Intestinal bacterial colonization induces mutualistic regulatory T cell responses. Immunity 2011; 34(5): 794-806.
[http://dx.doi.org/10.1016/j.immuni.2011.03.021] [PMID: 21596591]

[80] Weiss JM, Bilate AM, Gobert M, Ding Y, Curotto de Lafaille MA, Parkhurst CN, *et al.* Neuropilin 1 is expressed on thymus-derived natural regulatory T cells, but not mucosa-generated induced Foxp3+ T reg cells. J Exp Med 2012; 209(10): 1723-42, S1.
[http://dx.doi.org/10.1084/jem.20120914] [PMID: 22966001]

[81] Stefka AT, Feehley T, Tripathi P, *et al.* Commensal bacteria protect against food allergen sensitization. Proc Natl Acad Sci USA 2014; 111(36): 13145-50.
[http://dx.doi.org/10.1073/pnas.1412008111] [PMID: 25157157]

[82] Atarashi K, Tanoue T, Oshima K, *et al.* Treg induction by a rationally selected mixture of Clostridia strains from the human microbiota. Nature 2013; 500(7461): 232-6.
[http://dx.doi.org/10.1038/nature12331] [PMID: 23842501]

[83] Korn LL, Hubbeling HG, Porrett PM, Yang Q, Barnett LG, Laufer TM. Regulatory T cells occupy an isolated niche in the intestine that is antigen independent. Cell Reports 2014; 9(5): 1567-73.
[http://dx.doi.org/10.1016/j.celrep.2014.11.006] [PMID: 25482559]

[84] Kawamoto S, Maruya M, Kato LM, *et al.* Foxp3(+) T cells regulate immunoglobulin a selection and facilitate diversification of bacterial species responsible for immune homeostasis. Immunity 2014; 41(1): 152-65.
[http://dx.doi.org/10.1016/j.immuni.2014.05.016] [PMID: 25017466]

[85] Smith PM, Howitt MR, Panikov N, *et al.* The microbial metabolites, short-chain fatty acids, regulate colonic Treg cell homeostasis. Science 2013; 341(6145): 569-73.
[http://dx.doi.org/10.1126/science.1241165] [PMID: 23828891]

[86] Furusawa Y, Obata Y, Fukuda S, *et al.* Commensal microbe-derived butyrate induces the differentiation of colonic regulatory T cells. Nature 2013; 504(7480): 446-50.
[http://dx.doi.org/10.1038/nature12721] [PMID: 24226770]

[87] Arpaia N, Campbell C, Fan X, *et al.* Metabolites produced by commensal bacteria promote peripheral regulatory T-cell generation. Nature 2013; 504(7480): 451-5.
[http://dx.doi.org/10.1038/nature12726] [PMID: 24226773]

[88] Yin Y, Wang Y, Zhu L, *et al.* Comparative analysis of the distribution of segmented filamentous bacteria in humans, mice and chickens. ISME J 2013; 7(3): 615-21.
[http://dx.doi.org/10.1038/ismej.2012.128] [PMID: 23151642]

[89] Ericsson AC, Hagan CE, Davis DJ, Franklin CL. Segmented filamentous bacteria: commensal microbes with potential effects on research. Comp Med 2014; 64(2): 90-8.
[PMID: 24674582]

[90] Jepson MA, Clark MA, Simmons NL, Hirst BH. Actin accumulation at sites of attachment of indigenous apathogenic segmented filamentous bacteria to mouse ileal epithelial cells. Infect Immun 1993; 61(9): 4001-4.
[PMID: 8359925]

[91] Yang Y, Torchinsky MB, Gobert M, *et al.* Focused specificity of intestinal TH17 cells towards

commensal bacterial antigens. Nature 2014; 510(7503): 152-6.
[http://dx.doi.org/10.1038/nature13279] [PMID: 24739972]

[92] Ivanov II, Littman DR. Segmented filamentous bacteria take the stage. Mucosal Immunol 2010; 3(3): 209-12.
[http://dx.doi.org/10.1038/mi.2010.3] [PMID: 20147894]

[93] Maeda Y, Kurakawa T, Umemoto E, *et al.* Dysbiosis Contributes to Arthritis Development *via* Activation of Autoreactive T Cells in the Intestine. Arthritis Rheumatol 2016; 68(11): 2646-61.
[http://dx.doi.org/10.1002/art.39783] [PMID: 27333153]

[94] Scher JU, Sczesnak A, Longman RS, *et al.* Expansion of intestinal Prevotella copri correlates with enhanced susceptibility to arthritis. eLife 2013; 2: e01202.
[http://dx.doi.org/10.7554/eLife.01202] [PMID: 24192039]

[95] Kubinak JL, Round JL. Do antibodies select a healthy microbiota? Nat Rev Immunol 2016; 16(12): 767-74.
[http://dx.doi.org/10.1038/nri.2016.114] [PMID: 27818504]

[96] Bunker JJ, Flynn TM, Koval JC, *et al.* Innate and Adaptive Humoral Responses Coat Distinct Commensal Bacteria with Immunoglobulin A. Immunity 2015; 43(3): 541-53.
[http://dx.doi.org/10.1016/j.immuni.2015.08.007] [PMID: 26320660]

[97] Planer JD, Peng Y, Kau AL, *et al.* Development of the gut microbiota and mucosal IgA responses in twins and gnotobiotic mice. Nature 2016; 534(7606): 263-6.
[http://dx.doi.org/10.1038/nature17940] [PMID: 27279225]

[98] Macpherson AJ, Gatto D, Sainsbury E, Harriman GR, Hengartner H, Zinkernagel RM. A primitive T cell-independent mechanism of intestinal mucosal IgA responses to commensal bacteria. Science 2000; 288(5474): 2222-6.
[http://dx.doi.org/10.1126/science.288.5474.2222] [PMID: 10864873]

[99] Shroff KE, Meslin K, Cebra JJ. Commensal enteric bacteria engender a self-limiting humoral mucosal immune response while permanently colonizing the gut. Infect Immun 1995; 63(10): 3904-13.
[PMID: 7558298]

[100] Tsuruta T, Inoue R, Iwanaga T, Hara H, Yajima T. Development of a method for the identification of S-IgA-coated bacterial composition in mouse and human feces. Biosci Biotechnol Biochem 2010; 74(5): 968-73.
[http://dx.doi.org/10.1271/bbb.90801] [PMID: 20460734]

[101] Lécuyer E, Rakotobe S, Lengliné-Garnier H, *et al.* Segmented filamentous bacterium uses secondary and tertiary lymphoid tissues to induce gut IgA and specific T helper 17 cell responses. Immunity 2014; 40(4): 608-20.
[http://dx.doi.org/10.1016/j.immuni.2014.03.009] [PMID: 24745335]

[102] Guinane CM, Cotter PD. Role of the gut microbiota in health and chronic gastrointestinal disease: understanding a hidden metabolic organ. Therap Adv Gastroenterol 2013; 6(4): 295-308.
[http://dx.doi.org/10.1177/1756283X13482996] [PMID: 23814609]

[103] Kamada N, Seo SU, Chen GY, Núñez G. Role of the gut microbiota in immunity and inflammatory disease. Nat Rev Immunol 2013; 13(5): 321-35.
[http://dx.doi.org/10.1038/nri3430] [PMID: 23618829]

[104] West CE, Renz H, Jenmalm MC, *et al.* in-FLAME Microbiome Interest Group. The gut microbiota and inflammatory noncommunicable diseases: associations and potentials for gut microbiota therapies. J Allergy Clin Immunol 2015; 135(1): 3-13.
[http://dx.doi.org/10.1016/j.jaci.2014.11.012] [PMID: 25567038]

[105] Britton RA, Young VB. Role of the intestinal microbiota in resistance to colonization by Clostridium difficile. Gastroenterology 2014; 146(6): 1547-53.
[http://dx.doi.org/10.1053/j.gastro.2014.01.059] [PMID: 24503131]

[106] Kamada N, Núñez G. Role of the gut microbiota in the development and function of lymphoid cells. J Immunol 2013; 190(4): 1389-95.
[http://dx.doi.org/10.4049/jimmunol.1203100] [PMID: 23378581]

[107] Silverman MA, Konnikova L, Gerber JS. Impact of Antibiotics on Necrotizing Enterocolitis and Antibiotic-Associated Diarrhea. Gastroenterol Clin North Am 2017; 46(1): 61-76.
[http://dx.doi.org/10.1016/j.gtc.2016.09.010] [PMID: 28164853]

[108] Bartlett JG. Clinical practice. Antibiotic-associated diarrhea. N Engl J Med 2002; 346(5): 334-9.
[http://dx.doi.org/10.1056/NEJMcp011603] [PMID: 11821511]

[109] Jernberg C, Löfmark S, Edlund C, Jansson JK. Long-term impacts of antibiotic exposure on the human intestinal microbiota. Microbiology 2010; 156(Pt 11): 3216-23.
[http://dx.doi.org/10.1099/mic.0.040618-0] [PMID: 20705661]

[110] Hall IC, O'Toole E. Intestinal flora in new-born infants with a description of a new pathogenic anaerobe, *Bacillus difficilis*. Am J Dis Child 1935; 49(2): 12.
[http://dx.doi.org/10.1001/archpedi.1935.01970020105010]

[111] Napolitano LM, Edmiston CE Jr. Clostridium difficile disease: Diagnosis, pathogenesis, and treatment update. Surgery 2017; 162(2): 325-48.
[http://dx.doi.org/10.1016/j.surg.2017.01.018] [PMID: 28267992]

[112] Bauer MP, Notermans DW, van Benthem BH, *et al.* ECDIS Study Group. Clostridium difficile infection in Europe: a hospital-based survey. Lancet 2011; 377(9759): 63-73.
[http://dx.doi.org/10.1016/S0140-6736(10)61266-4] [PMID: 21084111]

[113] Aljarallah KM. Conventional and alternative treatment approaches for*Clostridium difficile*infection. Int J Health Sci (Qassim) 2017; 11(1): 1-10.
[PMID: 28293151]

[114] Spigaglia P. Recent advances in the understanding of antibiotic resistance in Clostridium difficile infection. Ther Adv Infect Dis 2016; 3(1): 23-42.
[http://dx.doi.org/10.1177/2049936115622891] [PMID: 26862400]

[115] Rineh A, Kelso MJ, Vatansever F, Tegos GP, Hamblin MR. Clostridium difficile infection: molecular pathogenesis and novel therapeutics. Expert Rev Anti Infect Ther 2014; 12(1): 131-50.
[http://dx.doi.org/10.1586/14787210.2014.866515] [PMID: 24410618]

[116] Gerding DN, Olson MM, Peterson LR, *et al.* Clostridium difficile-associated diarrhea and colitis in adults. A prospective case-controlled epidemiologic study. Arch Intern Med 1986; 146(1): 95-100.
[http://dx.doi.org/10.1001/archinte.1986.00360130117016] [PMID: 3942469]

[117] Bartlett JG. Antibiotic-associated colitis. Dis Mon 1984; 30(15): 1-54.
[PMID: 6391879]

[118] Kwon JH, Olsen MA, Dubberke ER. The morbidity, mortality, and costs associated with Clostridium difficile infection. Infect Dis Clin North Am 2015; 29(1): 123-34.
[http://dx.doi.org/10.1016/j.idc.2014.11.003] [PMID: 25677706]

[119] Ananthakrishnan AN. Clostridium difficile infection: epidemiology, risk factors and management. Nat Rev Gastroenterol Hepatol 2011; 8(1): 17-26.
[http://dx.doi.org/10.1038/nrgastro.2010.190] [PMID: 21119612]

[120] Dubberke ER, Olsen MA. Burden of Clostridium difficile on the healthcare system. Clin Infect Dis 2012; 55 (Suppl. 2): S88-92.
[http://dx.doi.org/10.1093/cid/cis335] [PMID: 22752870]

[121] CDC Center for Disease Control and Prevention. Antibiotic Resistance Threats in the United States. U.S. Department of Health and Human Services 2013.

[122] Cohen SH, Gerding DN, Johnson S, *et al.* Society for Healthcare Epidemiology of America; Infectious Diseases Society of America. Clinical practice guidelines for Clostridium difficile infection in adults:

2010 update by the society for healthcare epidemiology of America (SHEA) and the infectious diseases society of America (IDSA). Infect Control Hosp Epidemiol 2010; 31(5): 431-55. [http://dx.doi.org/10.1086/651706] [PMID: 20307191]

[123] Bagdasarian N, Rao K, Malani PN. Diagnosis and treatment of Clostridium difficile in adults: a systematic review. JAMA 2015; 313(4): 398-408. [http://dx.doi.org/10.1001/jama.2014.17103] [PMID: 25626036]

[124] Centers for Disease Control and Prevention [homepage on the Internet]. Atlanta: Nearly half a million Americans suffered from Clostridium difficile infections in a single year c2015-17. [updated March 22nd 2017; cited 8th July 2017]. [about 2 screens]. Available from: http://www.cdc.gov/media/releases/2015/p0225-clostridium-difficile.html/

[125] Owens RC Jr, Donskey CJ, Gaynes RP, Loo VG, Muto CA. Antimicrobial-associated risk factors for Clostridium difficile infection. Clin Infect Dis 2008; 46 (Suppl. 1): S19-31. [http://dx.doi.org/10.1086/521859] [PMID: 18177218]

[126] Rolfe RD, Helebian S, Finegold SM. Bacterial interference between Clostridium difficile and normal fecal flora. J Infect Dis 1981; 143(3): 470-5. [http://dx.doi.org/10.1093/infdis/143.3.470] [PMID: 6785366]

[127] Ridlon JM, Kang DJ, Hylemon PB. Bile salt biotransformations by human intestinal bacteria. J Lipid Res 2006; 47(2): 241-59. [http://dx.doi.org/10.1194/jlr.R500013-JLR200] [PMID: 16299351]

[128] Lyerly DM, Krivan HC, Wilkins TD. Clostridium difficile: its disease and toxins. Clin Microbiol Rev 1988; 1(1): 1-18. [http://dx.doi.org/10.1128/CMR.1.1.1] [PMID: 3144429]

[129] Rupnik M, Janezic S. An update on clostridium difficile toxinotyping. J Clin Microbiol 2016; 54(1): 13-8. [http://dx.doi.org/10.1128/JCM.02083-15] [PMID: 26511734]

[130] Alcalá Hernández L, Reigadas Ramírez E, Bouza Santiago E. Clostridium difficile infection. Med Clin (Barc) 2017; 148(10): 456-63. [http://dx.doi.org/10.1016/j.medcli.2017.01.033] [PMID: 28396132]

[131] Voth DE, Ballard JD. Clostridium difficile toxins: mechanism of action and role in disease. Clin Microbiol Rev 2005; 18(2): 247-63. [http://dx.doi.org/10.1128/CMR.18.2.247-263.2005] [PMID: 15831824]

[132] Zanella Terrier MC, Simonet ML, Bichard P, Frossard JL. Recurrent Clostridium difficile infections: the importance of the intestinal microbiota. World J Gastroenterol 2014; 20(23): 7416-23. [http://dx.doi.org/10.3748/wjg.v20.i23.7416] [PMID: 24966611]

[133] Rupnik M, Avesani V, Janc M, von Eichel-Streiber C, Delmée M. A novel toxinotyping scheme and correlation of toxinotypes with serogroups of Clostridium difficile isolates. J Clin Microbiol 1998; 36(8): 2240-7. [PMID: 9665999]

[134] Rupnik M. Heterogeneity of large clostridial toxins: importance of Clostridium difficile toxinotypes. FEMS Microbiol Rev 2008; 32(3): 541-55. [http://dx.doi.org/10.1111/j.1574-6976.2008.00110.x] [PMID: 18397287]

[135] Buonomo EL, Petri WA Jr. The microbiota and immune response during Clostridium difficile infection. Anaerobe 2016; 41: 79-84. [http://dx.doi.org/10.1016/j.anaerobe.2016.05.009] [PMID: 27212111]

[136] Jarchum I, Liu M, Shi C, Equinda M, Pamer EG. Critical role for MyD88-mediated neutrophil recruitment during Clostridium difficile colitis. Infect Immun 2012; 80(9): 2989-96. [http://dx.doi.org/10.1128/IAI.00448-12] [PMID: 22689818]

[137] Hasegawa M, Yamazaki T, Kamada N, *et al.* Nucleotide-binding oligomerization domain 1 mediates

recognition of Clostridium difficile and induces neutrophil recruitment and protection against the pathogen. J Immunol 2011; 186(8): 4872-80.
[http://dx.doi.org/10.4049/jimmunol.1003761] [PMID: 21411735]

[138] Bulusu M, Narayan S, Shetler K, Triadafilopoulos G. Leukocytosis as a harbinger and surrogate marker of Clostridium difficile infection in hospitalized patients with diarrhea. Am J Gastroenterol 2000; 95(11): 3137-41.
[http://dx.doi.org/10.1111/j.1572-0241.2000.03284.x] [PMID: 11095331]

[139] El Feghaly RE, Stauber JL, Tarr PI, Haslam DB. Intestinal inflammatory biomarkers and outcome in pediatric Clostridium difficile infections. J Pediatr 2013; 163(6): 1697-1704.e2.
[http://dx.doi.org/10.1016/j.jpeds.2013.07.029] [PMID: 24011765]

[140] El Feghaly RE, Stauber JL, Deych E, Gonzalez C, Tarr PI, Haslam DB. Markers of intestinal inflammation, not bacterial burden, correlate with clinical outcomes in Clostridium difficile infection. Clin Infect Dis 2013; 56(12): 1713-21.
[http://dx.doi.org/10.1093/cid/cit147] [PMID: 23487367]

[141] Buffie CG, Pamer EG. Microbiota-mediated colonization resistance against intestinal pathogens. Nat Rev Immunol 2013; 13(11): 790-801.
[http://dx.doi.org/10.1038/nri3535] [PMID: 24096337]

[142] Taylor KN, McHale MT, Saenz CC, Plaxe SC. Diagnosis and treatment of Clostridium difficile (C. diff) colitis: Review of the literature and a perspective in gynecologic oncology. Gynecol Oncol 2017; 144(2): 428-37.
[http://dx.doi.org/10.1016/j.ygyno.2016.11.024] [PMID: 27876339]

[143] Tariq R, Khanna S. Clostridium difficile infection: Updates in management. Indian J Gastroenterol 2017; 36(1): 3-10.
[http://dx.doi.org/10.1007/s12664-016-0719-z] [PMID: 27995486]

[144] Al-Jashaami LS, DuPont HL. Management of *Clostridium difficile* Infection. Gastroenterol Hepatol (N Y) 2016; 12(10): 609-16.
[PMID: 27917075]

[145] Fehér C, Mensa J. A comparison of current guidelines of five international societies on clostridium difficile infection management. Infect Dis Ther 2016; 5(3): 207-30.
[http://dx.doi.org/10.1007/s40121-016-0122-1] [PMID: 27470257]

[146] Kobayashi K, Sekiya N, Ainoda Y, Kurai H, Imamura A. Adherence to clinical practice guidelines for the management of Clostridium difficile infection in Japan: a multicenter retrospective study. Eur J Clin Microbiol Infect Dis 2017; 36(10): 1947-53.
[http://dx.doi.org/10.1007/s10096-017-3018-4] [PMID: 28577158]

[147] Debast SB, Bauer MP, Kuijper EJ. European society of clinical microbiology and infectious diseases. European society of clinical microbiology and infectious diseases: Update of the treatment guidance document for Clostridium difficile infection. Clin Microbiol Infect 2014; 20 (Suppl. 2): 1-26.
[http://dx.doi.org/10.1111/1469-0691.12418] [PMID: 24118601]

[148] Jarrad AM, Karoli T, Blaskovich MA, Lyras D, Cooper MA. Clostridium difficile drug pipeline: challenges in discovery and development of new agents. J Med Chem 2015; 58(13): 5164-85.
[http://dx.doi.org/10.1021/jm5016846] [PMID: 25760275]

[149] Pepin J, Alary ME, Valiquette L, *et al.* Increasing risk of relapse after treatment of Clostridium difficile colitis in Quebec, Canada. Clin Infect Dis 2005; 40(11): 1591-7.
[http://dx.doi.org/10.1086/430315] [PMID: 15889355]

[150] Huang H, Weintraub A, Fang H, Nord CE. Antimicrobial resistance in Clostridium difficile. Int J Antimicrob Agents 2009; 34(6): 516-22.
[http://dx.doi.org/10.1016/j.ijantimicag.2009.09.012] [PMID: 19828299]

[151] Peláez T, Cercenado E, Alcalá L, *et al.* Metronidazole resistance in Clostridium difficile is

heterogeneous. J Clin Microbiol 2008; 46(9): 3028-32. [http://dx.doi.org/10.1128/JCM.00524-08] [PMID: 18650353]

[152] Bolton RP, Culshaw MA. Faecal metronidazole concentrations during oral and intravenous therapy for antibiotic associated colitis due to Clostridium difficile. Gut 1986; 27(10): 1169-72. [http://dx.doi.org/10.1136/gut.27.10.1169] [PMID: 3781329]

[153] Vuotto C, Moura I, Barbanti F, Donelli G, Spigaglia P. Subinhibitory concentrations of metronidazole increase biofilm formation in Clostridium difficile strains. Pathog Dis 2016; 74(2): ftv114. [http://dx.doi.org/10.1093/femspd/ftv114] [PMID: 26656887]

[154] Perkins HR, Nieto M. The chemical basis for the action of the vancomycin group of antibiotics. Ann N Y Acad Sci 1974; 235(0): 348-63. [http://dx.doi.org/10.1111/j.1749-6632.1974.tb43276.x] [PMID: 4369274]

[155] Yu X, Sun D. Macrocyclic drugs and synthetic methodologies toward macrocycles. Molecules 2013; 18(6): 6230-68. [http://dx.doi.org/10.3390/molecules18066230] [PMID: 23708234]

[156] Nelson RL, Suda KJ, Evans CT. Antibiotic treatment for Clostridium difficile-associated diarrhoea in adults. Cochrane Database Syst Rev 2017; 3: CD004610. [PMID: 28257555]

[157] McDonald LC, Coignard B, Dubberke E, Song X, Horan T, Kutty PK. Ad Hoc Clostridium difficile Surveillance Working Group. Recommendations for surveillance of Clostridium difficile-associated disease. Infect Control Hosp Epidemiol 2007; 28(2): 140-5. [http://dx.doi.org/10.1086/511798] [PMID: 17265394]

[158] Al-Nassir WN, Sethi AK, Li Y, Pultz MJ, Riggs MM, Donskey CJ. Both oral metronidazole and oral vancomycin promote persistent overgrowth of vancomycin-resistant enterococci during treatment of Clostridium difficile-associated disease. Antimicrob Agents Chemother 2008; 52(7): 2403-6. [http://dx.doi.org/10.1128/AAC.00090-08] [PMID: 18443120]

[159] Adler A, Miller-Roll T, Bradenstein R, *et al.* A national survey of the molecular epidemiology of Clostridium difficile in Israel: the dissemination of the ribotype 027 strain with reduced susceptibility to vancomycin and metronidazole. Diagn Microbiol Infect Dis 2015; 83(1): 21-4. [http://dx.doi.org/10.1016/j.diagmicrobio.2015.05.015] [PMID: 26116225]

[160] Dong D, Zhang L, Chen X, *et al.* Antimicrobial susceptibility and resistance mechanisms of clinical Clostridium difficile from a Chinese tertiary hospital. Int J Antimicrob Agents 2013; 41(1): 80-4. [http://dx.doi.org/10.1016/j.ijantimicag.2012.08.011] [PMID: 23148985]

[161] Liao CH, Ko WC, Lu JJ, Hsueh PR. Characterizations of clinical isolates of clostridium difficile by toxin genotypes and by susceptibility to 12 antimicrobial agents, including fidaxomicin (OPT-80) and rifaximin: a multicenter study in Taiwan. Antimicrob Agents Chemother 2012; 56(7): 3943-9. [http://dx.doi.org/10.1128/AAC.00191-12] [PMID: 22508299]

[162] Brouwer MS, Roberts AP, Mullany P, Allan E. In silico analysis of sequenced strains of Clostridium difficile reveals a related set of conjugative transposons carrying a variety of accessory genes. Mob Genet Elements 2012; 2(1): 8-12. [http://dx.doi.org/10.4161/mge.19297] [PMID: 22754747]

[163] Brouwer MS, Warburton PJ, Roberts AP, Mullany P, Allan E. Genetic organisation, mobility and predicted functions of genes on integrated, mobile genetic elements in sequenced strains of Clostridium difficile. PLoS One 2011; 6(8): e23014. [http://dx.doi.org/10.1371/journal.pone.0023014] [PMID: 21876735]

[164] Đapa T, Leuzzi R, Ng YK, *et al.* Multiple factors modulate biofilm formation by the anaerobic pathogen Clostridium difficile. J Bacteriol 2013; 195(3): 545-55. [http://dx.doi.org/10.1128/JB.01980-12] [PMID: 23175653]

[165] Fehér C, Múñez Rubio E, Merino Amador P, *et al.* The efficacy of fidaxomicin in the treatment of

Clostridium difficile infection in a real-world clinical setting: a Spanish multi-centre retrospective cohort. Eur J Clin Microbiol Infect Dis 2017; 36(2): 295-303.
[http://dx.doi.org/10.1007/s10096-016-2802-x] [PMID: 27718071]

[166] Babakhani F, Bouillaut L, Gomez A, Sears P, Nguyen L, Sonenshein AL. Fidaxomicin inhibits spore production in Clostridium difficile. Clin Infect Dis 2012; 55 (Suppl. 2): S162-9.
[http://dx.doi.org/10.1093/cid/cis453] [PMID: 22752866]

[167] Louie TJ, Cannon K, Byrne B, *et al.* Fidaxomicin preserves the intestinal microbiome during and after treatment of Clostridium difficile infection (CDI) and reduces both toxin reexpression and recurrence of CDI. Clin Infect Dis 2012; 55 (Suppl. 2): S132-42.
[http://dx.doi.org/10.1093/cid/cis338] [PMID: 22752862]

[168] Paperno ME, Bakulina TB, Voinova RI. [Clinical and epidemiological peculiarities of intestinal infections caused by Escherichia coli HO-124 in children]. Vopr Okhr Materin Det 1969; 14(7): 86-7. [Clinical and epidemiological peculiarities of intestinal infections caused by Escherichia coli HO-124 in children].
[PMID: 4904592]

[169] Cornely OA, Crook DW, Esposito R, *et al.* OPT-80-004 Clinical Study Group. Fidaxomicin *versus* vancomycin for infection with Clostridium difficile in Europe, Canada, and the USA: a double-blind, non-inferiority, randomised controlled trial. Lancet Infect Dis 2012; 12(4): 281-9.
[http://dx.doi.org/10.1016/S1473-3099(11)70374-7] [PMID: 22321770]

[170] Zhanel GG, Walkty AJ, Karlowsky JA. Fidaxomicin: A novel agent for the treatment of Clostridium difficile infection. Can J Infect Dis Med Microbiol 2015; 26(6): 305-12.
[http://dx.doi.org/10.1155/2015/934594] [PMID: 26744587]

[171] Sears P, Crook DW, Louie TJ, Miller MA, Weiss K. Fidaxomicin attains high fecal concentrations with minimal plasma concentrations following oral administration in patients with Clostridium difficile infection. Clin Infect Dis 2012; 55 (Suppl. 2): S116-20.
[http://dx.doi.org/10.1093/cid/cis337] [PMID: 22752859]

[172] Johnson S, Homann SR, Bettin KM, *et al.* Treatment of asymptomatic Clostridium difficile carriers (fecal excretors) with vancomycin or metronidazole. A randomized, placebo-controlled trial. Ann Intern Med 1992; 117(4): 297-302.
[http://dx.doi.org/10.7326/0003-4819-117-4-297] [PMID: 1322075]

[173] Zhang F, Luo W, Shi Y, Fan Z, Ji G. Should we standardize the 1,700-year-old fecal microbiota transplantation? Am J Gastroenterol 2012; 107(11): 1755-6.
[http://dx.doi.org/10.1038/ajg.2012.251] [PMID: 23160295]

[174] Prohaska JV, Long ET, Nelsen TS. Pseudomembranous enterocolitis; its etiology and the mechanism of the disease process. AMA Arch Surg 1956; 72(6): 977-83.
[http://dx.doi.org/10.1001/archsurg.1956.01270240089013] [PMID: 13312870]

[175] Schwan A, Sjölin S, Trottestam U, Aronsson B. Relapsing clostridium difficile enterocolitis cured by rectal infusion of homologous faeces. Lancet 1983; 2(8354): 845.
[http://dx.doi.org/10.1016/S0140-6736(83)90753-5] [PMID: 6137662]

[176] Bennet JD, Brinkman M. Treatment of ulcerative colitis by implantation of normal colonic flora. Lancet 1989; 1(8630): 164.
[http://dx.doi.org/10.1016/S0140-6736(89)91183-5] [PMID: 2563083]

[177] Abrams RS. Open-label, uncontrolled trial of bower sterilization and repopulation with normal bowel flora for treatment of inflammatory bowel disease. Curr Ther Res Clin Exp 2012; 58(12): 11.

[178] Aas J, Gessert CE, Bakken JS. Recurrent Clostridium difficile colitis: case series involving 18 patients treated with donor stool administered *via* a nasogastric tube. Clin Infect Dis 2003; 36(5): 580-5.
[http://dx.doi.org/10.1086/367657] [PMID: 12594638]

[179] Silverman MS, Davis I, Pillai DR. Success of self-administered home fecal transplantation for chronic

Clostridium difficile infection. Clin Gastroenterol Hepatol 2010; 8(5): 471-3. [http://dx.doi.org/10.1016/j.cgh.2010.01.007] [PMID: 20117243]

[180] Ganc AJ, Ganc RL, Reimão SM, Frisoli Junior A, Pasternak J. Fecal microbiota transplant by push enteroscopy to treat diarrhea caused by Clostridium difficile. Einstein (Sao Paulo) 2015; 13(2): 338-9. [http://dx.doi.org/10.1590/S1679-45082015MD3106] [PMID: 26154556]

[181] Cammarota G, Ianiro G, Tilg H, *et al.* European FMT Working Group. European consensus conference on faecal microbiota transplantation in clinical practice. Gut 2017; 66(4): 569-80. [http://dx.doi.org/10.1136/gutjnl-2016-313017] [PMID: 28087657]

[182] Open-Biome.org. Somerville: The Microbiome Health Research Institute, dba OpenBiome 2012. Fecal transplant pills: Large-scale production begins following successful dosing study; [about 2 screens]. Available from: http://www.openbiome.org/press-releases/2015/10/28/fecal-transplant- pills-larg--scale-production-begins-following-successful-dosing-study

[183] Youngster I, Russell GH, Pindar C, Ziv-Baran T, Sauk J, Hohmann EL. Oral, capsulized, frozen fecal microbiota transplantation for relapsing Clostridium difficile infection. JAMA 2014; 312(17): 1772-8. [http://dx.doi.org/10.1001/jama.2014.13875] [PMID: 25322359]

[184] Bakken JS, Borody T, Brandt LJ, *et al.* Fecal microbiota transplantation workgroup. Treating clostridium difficile infection with fecal microbiota transplantation. Clin Gastroenterol Hepatol 2011; 9(12): 1044-9. [http://dx.doi.org/10.1016/j.cgh.2011.08.014] [PMID: 21871249]

[185] Landy J, Al-Hassi HO, McLaughlin SD, *et al.* Review article: faecal transplantation therapy for gastrointestinal disease. Aliment Pharmacol Ther 2011; 34(4): 409-15. [http://dx.doi.org/10.1111/j.1365-2036.2011.04737.x] [PMID: 21682755]

[186] Khoruts A, Sadowsky MJ, Hamilton MJ. Development of fecal microbiota transplantation suitable for mainstream medicine. Clin Gastroenterol Hepatol 2015; 13(2): 246-50. [http://dx.doi.org/10.1016/j.cgh.2014.11.014] [PMID: 25460566]

[187] Kelly CR, Kahn S, Kashyap P, *et al.* Update on fecal microbiota transplantation 2015: Indications, methodologies, mechanisms and outlook. Gastroenterology 2015; 149(1): 223-37. [http://dx.doi.org/10.1053/j.gastro.2015.05.008] [PMID: 25982290]

[188] UpToDate [homepage on the Internet] Borody TJ, Ramrakha S Fecal microbiota transplantation in the treatment of recurrent Clostridium difficile infection 2016. Available from: http://www.uptodate.com/contents/fecal-microbiota-transplantation-in-the-treatment-of-recurrent-clostridium-difficile-infection?source=search_result&search=Fecal Microbiota Transplantation&selectedTitle=1~27 - references

[189] Lee CH, Steiner T, Petrof EO, *et al.* Frozen *vs* fresh fecal microbiota transplantation and clinical resolution of diarrhea in patients with recurrent clostridium difficile infection: A randomized clinical trial. JAMA 2016; 315(2): 142-9. [http://dx.doi.org/10.1001/jama.2015.18098] [PMID: 26757463]

[190] Petrof EO, Gloor GB, Vanner SJ, *et al.* Stool substitute transplant therapy for the eradication of Clostridium difficile infection: 'RePOOPulating' the gut. Microbiome 2013; 1(1): 3. [http://dx.doi.org/10.1186/2049-2618-1-3] [PMID: 24467987]

[191] O'Donnell MM, Rea MC, O'Sullivan Ó, *et al.* Preparation of a standardised faecal slurry for ex-vivo microbiota studies which reduces inter-individual donor bias. J Microbiol Methods 2016; 129: 109-16. [http://dx.doi.org/10.1016/j.mimet.2016.08.002] [PMID: 27498348]

[192] Persky SE, Brandt LJ. Treatment of recurrent *Clostridium* difficile-associated diarrhea by administration of donated stool directly through a colonoscope. Am J Gastroenterol 2000; 95(11): 3283-5. [PMID: 11095355]

[193] Choi HH, Cho YS. Fecal microbiota transplantation: Current applications, effectiveness and future

perspectives. Clin Endosc 2016; 49(3): 257-65.
[http://dx.doi.org/10.5946/ce.2015.117] [PMID: 26956193]

[194] Bakken JS. Fecal bacteriotherapy for recurrent Clostridium difficile infection. Anaerobe 2009; 15(6): 285-9.
[http://dx.doi.org/10.1016/j.anaerobe.2009.09.007] [PMID: 19778623]

[195] Dutta SK, Girotra M, Garg S, *et al.* Efficacy of combined jejunal and colonic fecal microbiota transplantation for recurrent Clostridium difficile Infection. Clin Gastroenterol Hepatol 2014; 12(9): 1572-6.
[http://dx.doi.org/10.1016/j.cgh.2013.12.032] [PMID: 24440222]

[196] Hirsch BE, Saraiya N, Poeth K, Schwartz RM, Epstein ME, Honig G. Effectiveness of fecal-derived microbiota transfer using orally administered capsules for recurrent Clostridium difficile infection. BMC Infect Dis 2015; 15: 191.
[http://dx.doi.org/10.1186/s12879-015-0930-z] [PMID: 25885020]

[197] Youngster I, Mahabamunuge J, Systrom HK, *et al.* Oral, frozen fecal microbiota transplant (FMT) capsules for recurrent Clostridium difficile infection. BMC Med 2016; 14(1): 134.
[http://dx.doi.org/10.1186/s12916-016-0680-9] [PMID: 27609178]

[198] Vindigni SM, Surawicz CM. Fecal microbiota transplantation. Gastroenterol Clin North Am 2017; 46(1): 171-85.
[http://dx.doi.org/10.1016/j.gtc.2016.09.012] [PMID: 28164849]

[199] Liubakka A, Vaughn BP. Clostridium difficile infection and fecal microbiota transplant. AACN Adv Crit Care 2016; 27(3): 324-37.
[http://dx.doi.org/10.4037/aacnacc2016703] [PMID: 27959316]

[200] Kelly CR, Ihunnah C, Fischer M, *et al.* Fecal microbiota transplant for treatment of Clostridium difficile infection in immunocompromised patients. Am J Gastroenterol 2014; 109(7): 1065-71.
[http://dx.doi.org/10.1038/ajg.2014.133] [PMID: 24890442]

[201] Smith MB, Kelly C, Alm EJ. Policy: How to regulate faecal transplants. Nature 2014; 506(7488): 290-1.
[http://dx.doi.org/10.1038/506290a] [PMID: 24558658]

[202] Lagier JC. Faecal microbiota transplantation: from practice to legislation before considering industrialization. Clin Microbiol Infect 2014; 20(11): 1112-8.
[http://dx.doi.org/10.1111/1469-0691.12795] [PMID: 25273480]

[203] Vyas D, Aekka A, Vyas A. Fecal transplant policy and legislation. World J Gastroenterol 2015; 21(1): 6-11.
[http://dx.doi.org/10.3748/wjg.v21.i1.6] [PMID: 25574076]

[204] U.S. Food and Drug Administration. Silver spring: Draft guidance for industry: Enforcement policy regarding investigational new drug requirements for use of fecal microbiota for transplantation to treat Clostridium difficile infection not responsive to standard therapies 2014. U.S. Department of health and human services; [about 2 screens]. Available from: http://www.fda.gov/BiologicsBloodVaccines/GuidanceComplianceRegulatoryInformation/Guidances/Vaccines/ucm387023.htm

[205] Daloiso V, Minacori R, Refolo P, *et al.* Ethical aspects of Fecal Microbiota Transplantation (FMT). Eur Rev Med Pharmacol Sci 2015; 19(17): 3173-80.
[PMID: 26400519]

[206] Minacori R, Refolo P, Sacchini D, Spagnolo AG. Research ethics committees and clinical research in Italy: where are we going? Eur Rev Med Pharmacol Sci 2015; 19(3): 481-5.
[PMID: 25720722]

[207] Hamilton MJ, Weingarden AR, Sadowsky MJ, Khoruts A. Standardized frozen preparation for transplantation of fecal microbiota for recurrent Clostridium difficile infection. Am J Gastroenterol 2012; 107(5): 761-7.

[http://dx.doi.org/10.1038/ajg.2011.482] [PMID: 22290405]

[208] van Nood E, Vrieze A, Nieuwdorp M, *et al.* Duodenal infusion of donor feces for recurrent Clostridium difficile. N Engl J Med 2013; 368(5): 407-15. [http://dx.doi.org/10.1056/NEJMoa1205037] [PMID: 23323867]

[209] Youngster I, Sauk J, Pindar C, *et al.* Fecal microbiota transplant for relapsing Clostridium difficile infection using a frozen inoculum from unrelated donors: a randomized, open-label, controlled pilot study. Clin Infect Dis 2014; 58(11): 1515-22. [http://dx.doi.org/10.1093/cid/ciu135] [PMID: 24762631]

[210] Kelly CR, Khoruts A, Staley C, *et al.* Effect of fecal microbiota transplantation on recurrence in multiply recurrent clostridium difficile infection: A randomized trial. Ann Intern Med 2016; 165(9): 609-16. [http://dx.doi.org/10.7326/M16-0271] [PMID: 27547925]

[211] Tremaroli V, Bäckhed F. Functional interactions between the gut microbiota and host metabolism. Nature 2012; 489(7415): 242-9. [http://dx.doi.org/10.1038/nature11552] [PMID: 22972297]

[212] Chang JY, Antonopoulos DA, Kalra A, *et al.* Decreased diversity of the fecal Microbiome in recurrent Clostridium difficile-associated diarrhea. J Infect Dis 2008; 197(3): 435-8. [http://dx.doi.org/10.1086/525047] [PMID: 18199029]

[213] Khoruts A, Dicksved J, Jansson JK, Sadowsky MJ. Changes in the composition of the human fecal microbiome after bacteriotherapy for recurrent Clostridium difficile-associated diarrhea. J Clin Gastroenterol 2010; 44(5): 354-60. [PMID: 20048681]

[214] Hamilton MJ, Weingarden AR, Unno T, Khoruts A, Sadowsky MJ. High-throughput DNA sequence analysis reveals stable engraftment of gut microbiota following transplantation of previously frozen fecal bacteria. Gut Microbes 2013; 4(2): 125-35. [http://dx.doi.org/10.4161/gmic.23571] [PMID: 23333862]

[215] Buffie CG, Bucci V, Stein RR, *et al.* Precision microbiome reconstitution restores bile acid mediated resistance to Clostridium difficile. Nature 2015; 517(7533): 205-8. [http://dx.doi.org/10.1038/nature13828] [PMID: 25337874]

[216] Wilson KH, Silva J, Fekety FR. Suppression of Clostridium difficile by normal hamster cecal flora and prevention of antibiotic-associated cecitis. Infect Immun 1981; 34(2): 626-8. [PMID: 7309245]

[217] Corthier G, Dubos F, Raibaud P. Modulation of cytotoxin production by Clostridium difficile in the intestinal tracts of gnotobiotic mice inoculated with various human intestinal bacteria. Appl Environ Microbiol 1985; 49(1): 250-2. [PMID: 3977313]

[218] Itoh K, Lee WK, Kawamura H, Mitsuoka T, Magaribuchi T. Intestinal bacteria antagonistic to Clostridium difficile in mice. Lab Anim 1987; 21(1): 20-5. [http://dx.doi.org/10.1258/002367787780740662] [PMID: 3560860]

[219] Lawley TD, Clare S, Walker AW, *et al.* Targeted restoration of the intestinal microbiota with a simple, defined bacteriotherapy resolves relapsing Clostridium difficile disease in mice. PLoS Pathog 2012; 8(10): e1002995. [http://dx.doi.org/10.1371/journal.ppat.1002995] [PMID: 23133377]

[220] Reeves AE, Theriot CM, Bergin IL, Huffnagle GB, Schloss PD, Young VB. The interplay between microbiome dynamics and pathogen dynamics in a murine model of Clostridium difficile Infection. Gut Microbes 2011; 2(3): 145-58. [http://dx.doi.org/10.4161/gmic.2.3.16333] [PMID: 21804357]

[221] Khoruts A, Sadowsky MJ. Understanding the mechanisms of faecal microbiota transplantation. Nat

Rev Gastroenterol Hepatol 2016; 13(9): 508-16.
[http://dx.doi.org/10.1038/nrgastro.2016.98] [PMID: 27329806]

[222] Merrigan MM, Sambol SP, Johnson S, Gerding DN. Prevention of fatal Clostridium difficile-associated disease during continuous administration of clindamycin in hamsters. J Infect Dis 2003; 188(12): 1922-7.
[http://dx.doi.org/10.1086/379836] [PMID: 14673773]

[223] Sambol SP, Merrigan MM, Tang JK, Johnson S, Gerding DN. Colonization for the prevention of Clostridium difficile disease in hamsters. J Infect Dis 2002; 186(12): 1781-9.
[http://dx.doi.org/10.1086/345676] [PMID: 12447764]

[224] Britton RA, Young VB. Interaction between the intestinal microbiota and host in Clostridium difficile colonization resistance. Trends Microbiol 2012; 20(7): 313-9.
[http://dx.doi.org/10.1016/j.tim.2012.04.001] [PMID: 22595318]

[225] Hibbing ME, Fuqua C, Parsek MR, Peterson SB. Bacterial competition: surviving and thriving in the microbial jungle. Nat Rev Microbiol 2010; 8(1): 15-25.
[http://dx.doi.org/10.1038/nrmicro2259] [PMID: 19946288]

[226] Wilson KH, Perini F. Role of competition for nutrients in suppression of Clostridium difficile by the colonic microflora. Infect Immun 1988; 56(10): 2610-4.
[PMID: 3417352]

[227] Rea MC, Sit CS, Clayton E, *et al.* Thuricin CD, a posttranslationally modified bacteriocin with a narrow spectrum of activity against Clostridium difficile. Proc Natl Acad Sci USA 2010; 107(20): 9352-7.
[http://dx.doi.org/10.1073/pnas.0913554107] [PMID: 20435915]

[228] Le Lay C, Dridi L, Bergeron MG, Ouellette M, Fliss IL. Nisin is an effective inhibitor of Clostridium difficile vegetative cells and spore germination. J Med Microbiol 2016; 65(2): 169-75.
[http://dx.doi.org/10.1099/jmm.0.000202] [PMID: 26555543]

[229] May T, Mackie RI, Fahey GC Jr, Cremin JC, Garleb KA. Effect of fiber source on short-chain fatty acid production and on the growth and toxin production by Clostridium difficile. Scand J Gastroenterol 1994; 29(10): 916-22.
[http://dx.doi.org/10.3109/00365529409094863] [PMID: 7839098]

[230] Rolfe RD. Role of volatile fatty acids in colonization resistance to Clostridium difficile. Infect Immun 1984; 45(1): 185-91.
[PMID: 6735467]

[231] Su WJ, Waechter MJ, Bourlioux P, Dolegeal M, Fourniat J, Mahuzier G. Role of volatile fatty acids in colonization resistance to Clostridium difficile in gnotobiotic mice. Infect Immun 1987; 55(7): 1686-91.
[PMID: 3596806]

[232] Allegretti JR, Kearney S, Li N, *et al.* Recurrent Clostridium difficile infection associates with distinct bile acid and microbiome profiles. Aliment Pharmacol Ther 2016; 43(11): 1142-53.
[http://dx.doi.org/10.1111/apt.13616] [PMID: 27086647]

[233] Lawley TD, Clare S, Walker AW, *et al.* Antibiotic treatment of clostridium difficile carrier mice triggers a supershedder state, spore-mediated transmission, and severe disease in immunocompromised hosts. Infect Immun 2009; 77(9): 3661-9.
[http://dx.doi.org/10.1128/IAI.00558-09] [PMID: 19564382]

[234] Jarchum I, Liu M, Lipuma L, Pamer EG. Toll-like receptor 5 stimulation protects mice from acute Clostridium difficile colitis. Infect Immun 2011; 79(4): 1498-503.
[http://dx.doi.org/10.1128/IAI.01196-10] [PMID: 21245274]

[235] Ekmekciu I, von Klitzing E, Fiebiger U, *et al.* Immune responses to broad-spectrum antibiotic treatment and fecal microbiota transplantation in mice. Front Immunol 2017; 8: 397.

[http://dx.doi.org/10.3389/fimmu.2017.00397] [PMID: 28469619]

[236] Ianiro G, Bibbò S, Gasbarrini A, Cammarota G. Therapeutic modulation of gut microbiota: current clinical applications and future perspectives. Curr Drug Targets 2014; 15(8): 762-70. [http://dx.doi.org/10.2174/1389450115666140606111402] [PMID: 24909808]

[237] Aroniadis OC, Brandt LJ. Fecal microbiota transplantation: past, present and future. Curr Opin Gastroenterol 2013; 29(1): 79-84. [http://dx.doi.org/10.1097/MOG.0b013e32835a4b3e] [PMID: 23041678]

[238] Biondo-Simões MdLP. Opções terapêuticas para as doenças inflamatórias intestinais: Revisão. Rev bras Coloproct 2003; 23(3): 10.

[239] Gong D, Gong X, Wang L, Yu X, Dong Q. Involvement of reduced microbial diversity in inflammatory bowel disease. Gastroenterol Res Pract 2016; 2016: 6951091. https://www.hindawi.com/journals/grp/2016/6951091/ [cited 8th July 2017]. [http://dx.doi.org/10.1155/2016/6951091] [PMID: 28074093]

[240] Sheehan D, Shanahan F. The gut microbiota in inflammatory bowel disease. Gastroenterol Clin North Am 2017; 46(1): 143-54. [http://dx.doi.org/10.1016/j.gtc.2016.09.011] [PMID: 28164847]

[241] Lopez J, Grinspan A. Fecal microbiota transplantation for inflammatory bowel disease. Gastroenterol Hepatol (N Y) 2016; 12(6): 374-9. [PMID: 27493597]

[242] Colman RJ, Rubin DT. Fecal microbiota transplantation as therapy for inflammatory bowel disease: a systematic review and meta-analysis. J Crohn's Colitis 2014; 8(12): 1569-81. [http://dx.doi.org/10.1016/j.crohns.2014.08.006] [PMID: 25223604]

[243] Moayyedi P, Surette MG, Kim PT, *et al.* Fecal microbiota transplantation induces remission in patients With active ulcerative colitis in a randomized controlled trial. Gastroenterology 2015; 149(1): 102-109.e6. [http://dx.doi.org/10.1053/j.gastro.2015.04.001] [PMID: 25857665]

[244] Rossen NG, Fuentes S, van der Spek MJ, *et al.* Findings from a randomized controlled trial of fecal transplantation for patients with ulcerative colitis. Gastroenterology 2015; 149(1): 110-118.e4. [http://dx.doi.org/10.1053/j.gastro.2015.03.045] [PMID: 25836986]

[245] Scaldaferri F, Pecere S, Petito V, *et al.* Efficacy and mechanisms of action of fecal microbiota transplantation in ulcerative colitis: pitfalls and promises from a first meta-analysis. Transplant Proc 2016; 48(2): 402-7. [http://dx.doi.org/10.1016/j.transproceed.2015.12.040] [PMID: 27109966]

[246] Pardi DS, D'Haens G, Shen B, Campbell S, Gionchetti P. Clinical guidelines for the management of pouchitis. Inflamm Bowel Dis 2009; 15(9): 1424-31. [http://dx.doi.org/10.1002/ibd.21039] [PMID: 19685489]

[247] Cohen NA, Maharshak N. Novel indications for fecal microbial transplantation: Update and review of the literature. Dig Dis Sci 2017; 62(5): 1131-45. [http://dx.doi.org/10.1007/s10620-017-4535-9] [PMID: 28315032]

[248] Fang S, Kraft CS, Dhere T, *et al.* Successful treatment of chronic Pouchitis utilizing fecal microbiota transplantation (FMT): a case report. Int J Colorectal Dis 2016; 31(5): 1093-4. [http://dx.doi.org/10.1007/s00384-015-2428-y] [PMID: 26525055]

[249] Landy J, Walker AW, Li JV, *et al.* Variable alterations of the microbiota, without metabolic or immunological change, following faecal microbiota transplantation in patients with chronic pouchitis. Sci Rep 2015; 5: 12955. [http://dx.doi.org/10.1038/srep12955] [PMID: 26264409]

[250] Drossman DA. The functional gastrointestinal disorders and the Rome III process. Gastroenterology 2006; 130(5): 1377-90.

[http://dx.doi.org/10.1053/j.gastro.2006.03.008] [PMID: 16678553]

[251] Thompson WG, Longstreth GF, Drossman DA, Heaton KW, Irvine EJ, Müller-Lissner SA. Functional bowel disorders and functional abdominal pain. Gut 1999; 45 (Suppl. 2): II43-7. [http://dx.doi.org/10.1136/gut.45.2008.ii43] [PMID: 10457044]

[252] Lovell RM, Ford AC. Global prevalence of and risk factors for irritable bowel syndrome: a meta-analysis. Clin Gastroenterol Hepatol 2012; 10(7): 712-721.e4. [http://dx.doi.org/10.1016/j.cgh.2012.02.029] [PMID: 22426087]

[253] Camilleri M, Lasch K, Zhou W. Irritable bowel syndrome: methods, mechanisms, and pathophysiology. The confluence of increased permeability, inflammation, and pain in irritable bowel syndrome. Am J Physiol Gastrointest Liver Physiol 2012; 303(7): G775-85. [http://dx.doi.org/10.1152/ajpgi.00155.2012] [PMID: 22837345]

[254] Distrutti E, Monaldi L, Ricci P, Fiorucci S. Gut microbiota role in irritable bowel syndrome: New therapeutic strategies. World J Gastroenterol 2016; 22(7): 2219-41. [http://dx.doi.org/10.3748/wjg.v22.i7.2219] [PMID: 26900286]

[255] Konturek PC, Haziri D, Brzozowski T, *et al.* Emerging role of fecal microbiota therapy in the treatment of gastrointestinal and extra-gastrointestinal diseases. J Physiol Pharmacol 2015; 66(4): 483-91. [PMID: 26348073]

[256] Pinn DM, Aroniadis OC, Brandt LJ. Is fecal microbiota transplantation the answer for irritable bowel syndrome? A single-center experience. Am J Gastroenterol 2014; 109(11): 1831-2. [http://dx.doi.org/10.1038/ajg.2014.295] [PMID: 25373585]

[257] ClinicalTrials.gov [homepage from the internet]. U.S. National Library of Medicine [updated 18th Dec 2017; cited 20th Dec 2017]. http://www.clinicaltrials.gov/

[258] Manges AR, Steiner TS, Wright AJ. Fecal microbiota transplantation for the intestinal decolonization of extensively antimicrobial-resistant opportunistic pathogens: a review. Infect Dis (Lond) 2016; 48(8): 587-92. [http://dx.doi.org/10.1080/23744235.2016.1177199] [PMID: 27194400]

[259] Kronman MP, Zerr DM, Qin X, *et al.* Intestinal decontamination of multidrug-resistant Klebsiella pneumoniae after recurrent infections in an immunocompromised host. Diagn Microbiol Infect Dis 2014; 80(1): 87-9. [http://dx.doi.org/10.1016/j.diagmicrobio.2014.06.006] [PMID: 25041704]

[260] Saidel-Odes L, Polachek H, Peled N, *et al.* A randomized, double-blind, placebo-controlled trial of selective digestive decontamination using oral gentamicin and oral polymyxin E for eradication of carbapenem-resistant Klebsiella pneumoniae carriage. Infect Control Hosp Epidemiol 2012; 33(1): 14-9. [http://dx.doi.org/10.1086/663206] [PMID: 22173517]

[261] Kamada N, Chen GY, Inohara N, Núñez G. Control of pathogens and pathobionts by the gut microbiota. Nat Immunol 2013; 14(7): 685-90. [http://dx.doi.org/10.1038/ni.2608] [PMID: 23778796]

[262] Oostdijk EA, de Smet AM, Blok HE, *et al.* Ecological effects of selective decontamination on resistant gram-negative bacterial colonization. Am J Respir Crit Care Med 2010; 181(5): 452-7. [http://dx.doi.org/10.1164/rccm.200908-1210OC] [PMID: 19965807]

[263] Millan B, Park H, Hotte N, *et al.* Fecal microbial transplants reduce antibiotic-resistant genes in patients with recurrent clostridium difficile infection. Clin Infect Dis 2016; 62(12): 1479-86. [http://dx.doi.org/10.1093/cid/ciw185] [PMID: 27025836]

[264] Ponte A, Pinho R, Mota M. Fecal microbiota transplantation: is there a role in the eradication of carbapenem-resistant Klebsiella pneumoniae intestinal carriage? Rev Esp Enferm Dig 2017; 109(5): 392.

[PMID: 28196423]

[265] Crum-Cianflone NF, Sullivan E, Ballon-Landa G. Fecal microbiota transplantation and successful resolution of multidrug-resistant-organism colonization. J Clin Microbiol 2015; 53(6): 1986-9. [http://dx.doi.org/10.1128/JCM.00820-15] [PMID: 25878340]

[266] Biliński J, Grzesiowski P, Muszyński J, *et al.* Fecal microbiota transplantation inhibits multidrug-resistant gut pathogens: Preliminary report performed in an immunocompromised host. Arch Immunol Ther Exp (Warsz) 2016; 64(3): 255-8. [http://dx.doi.org/10.1007/s00005-016-0387-9] [PMID: 26960790]

[267] Ubeda C, Bucci V, Caballero S, *et al.* Intestinal microbiota containing Barnesiella species cures vancomycin-resistant Enterococcus faecium colonization. Infect Immun 2013; 81(3): 965-73. [http://dx.doi.org/10.1128/IAI.01197-12] [PMID: 23319552]

[268] Mogna L, Deidda F, Nicola S, Amoruso A, Del Piano M, Mogna G. *In vitro* inhibition of klebsiella pneumoniae by lactobacillus delbrueckii subsp. delbrueckii LDD01 (DSM 22106): An innovative strategy to possibly counteract such infections in humans? J Clin Gastroenterol 2016; 50 (Suppl 2, Proceedings from the 8th probiotics, Prebiotics & New foods for microbiota and Human health meeting held in rome, Italy on September 13-15, 2015): S136-9. [http://dx.doi.org/10.1097/MCG.0000000000000680] [PMID: 27741158]

[269] Jensen MD, Ryan DH, Apovian CM, *et al.* American college of cardiology/american heart association task force on practice guidelines; Obesity society. 2013 AHA/ACC/TOS guideline for the management of overweight and obesity in adults: a report of the American college of cardiology/american heart association task force on practice guidelines and The obesity society. J Am Coll Cardiol 2014; 63 (25 Pt B): 2985-3023. [http://dx.doi.org/10.1016/j.jacc.2013.11.004] [PMID: 24239920]

[270] Ng M, Fleming T, Robinson M, *et al.* Global, regional, and national prevalence of overweight and obesity in children and adults during 1980-2013: a systematic analysis for the Global Burden of Disease Study 2013. Lancet 2014; 384(9945): 766-81. [http://dx.doi.org/10.1016/S0140-6736(14)60460-8] [PMID: 24880830]

[271] Li J, Zhao F, Wang Y, *et al.* Gut microbiota dysbiosis contributes to the development of hypertension. Microbiome 2017; 5(1): 14. [http://dx.doi.org/10.1186/s40168-016-0222-x] [PMID: 28143587]

[272] Hur KY, Lee MS. Gut microbiota and metabolic disorders. Diabetes Metab J 2015; 39(3): 198-203. [http://dx.doi.org/10.4093/dmj.2015.39.3.198] [PMID: 26124989]

[273] Woting A, Blaut M. The intestinal microbiota in metabolic disease. Nutrients 2016; 8(4): 202. [http://dx.doi.org/10.3390/nu8040202] [PMID: 27058556]

[274] Petra AI, Panagiotidou S, Hatziagelaki E, Stewart JM, Conti P, Theoharides TC. Gut-microbiota-brain axis and Its effect on neuropsychiatric disorders with suspected immune dysregulation. Clin Ther 2015; 37(5): 984-95. [http://dx.doi.org/10.1016/j.clinthera.2015.04.002] [PMID: 26046241]

[275] Evrensel A, Ceylan ME. Fecal microbiota transplantation and Its usage in neuropsychiatric disorders. Clin Psychopharmacol Neurosci 2016; 14(3): 231-7. [http://dx.doi.org/10.9758/cpn.2016.14.3.231] [PMID: 27489376]

[276] Mangiola F, Ianiro G, Franceschi F, Fagiuoli S, Gasbarrini G, Gasbarrini A. Gut microbiota in autism and mood disorders. World J Gastroenterol 2016; 22(1): 361-8. [http://dx.doi.org/10.3748/wjg.v22.i1.361] [PMID: 26755882]

[277] Rosenfeld CS. Microbiome disturbances and autism spectrum disorders. Drug Metab Dispos 2015; 43(10): 1557-71. [http://dx.doi.org/10.1124/dmd.115.063826] [PMID: 25852213]

[278] Dinan TG, Cryan JF. Melancholic microbes: a link between gut microbiota and depression?

Neurogastroenterol Motil 2013; 25(9): 713-9. [http://dx.doi.org/10.1111/nmo.12198] [PMID: 23910373]

[279] Borody TJ, Nowak A, Finlayson S. The GI microbiome and its role in Chronic Fatigue Syndrome: A summary of bacteriotherapy. Journal of the Australasian College of Nutritional and Environmental Medicine 2012; 31(3)

[280] Betrapally NS, Gillevet PM, Bajaj JS. Gut microbiome and liver disease. Transl Res 2017; 179: 49-59. [http://dx.doi.org/10.1016/j.trsl.2016.07.005] [PMID: 27477080]

[281] Cusi K. Role of obesity and lipotoxicity in the development of nonalcoholic steatohepatitis: pathophysiology and clinical implications. Gastroenterology 2012; 142(4): 711-725.e6. [http://dx.doi.org/10.1053/j.gastro.2012.02.003] [PMID: 22326434]

[282] Boursier J, Diehl AM. Implication of gut microbiota in nonalcoholic fatty liver disease. PLoS Pathog 2015; 11(1): e1004559. [http://dx.doi.org/10.1371/journal.ppat.1004559] [PMID: 25625278]

[283] Mouzaki M, Comelli EM, Arendt BM, *et al.* Intestinal microbiota in patients with nonalcoholic fatty liver disease. Hepatology 2013; 58(1): 120-7. [http://dx.doi.org/10.1002/hep.26319] [PMID: 23401313]

[284] Mutlu EA, Gillevet PM, Rangwala H, *et al.* Colonic microbiome is altered in alcoholism. Am J Physiol Gastrointest Liver Physiol 2012; 302(9): G966-78. [http://dx.doi.org/10.1152/ajpgi.00380.2011] [PMID: 22241860]

[285] Wiest R, Garcia-Tsao G. Bacterial translocation (BT) in cirrhosis. Hepatology 2005; 41(3): 422-33. [http://dx.doi.org/10.1002/hep.20632] [PMID: 15723320]

[286] Kao D, Roach B, Park H, *et al.* Fecal microbiota transplantation in the management of hepatic encephalopathy. Hepatology 2016; 63(1): 339-40. [http://dx.doi.org/10.1002/hep.28121] [PMID: 26264779]

[287] Ren YD, Ye ZS, Yang LZ, *et al.* Fecal microbiota transplantation induces hepatitis B virus e-antigen (HBeAg) clearance in patients with positive HBeAg after long-term antiviral therapy. Hepatology 2017; 65(5): 1765-8. [http://dx.doi.org/10.1002/hep.29008] [PMID: 28027582]

[288] Wang W, Xu S, Ren Z, Jiang J, Zheng S. Gut microbiota and allogeneic transplantation. J Transl Med 2015; 13: 275. [http://dx.doi.org/10.1186/s12967-015-0640-8] [PMID: 26298517]

[289] Kakihana K, Fujioka Y, Suda W, *et al.* Fecal microbiota transplantation for patients with steroid-resistant acute graft-*versus*-host disease of the gut. Blood 2016; 128(16): 2083-8. [http://dx.doi.org/10.1182/blood-2016-05-717652] [PMID: 27461930]

[290] Li Q, Wang C, Tang C, *et al.* Therapeutic modulation and reestablishment of the intestinal microbiota with fecal microbiota transplantation resolves sepsis and diarrhea in a patient. Am J Gastroenterol 2014; 109(11): 1832-4. [http://dx.doi.org/10.1038/ajg.2014.299] [PMID: 25373588]

[291] Guo Y, Qi Y, Yang X, *et al.* Association between polycystic ovary syndrome and gut microbiota. PLoS One 2016; 11(4): e0153196. [http://dx.doi.org/10.1371/journal.pone.0153196] [PMID: 27093642]

[292] Kabeerdoss J, Sandhya P, Danda D. Gut inflammation and microbiome in spondyloarthritis. Rheumatol Int 2016; 36(4): 457-68. [http://dx.doi.org/10.1007/s00296-015-3414-y] [PMID: 26719306]

[293] Zheng H, Liang H, Wang Y, *et al.* Altered gut microbiota composition associated with eczema in infants. PLoS One 2016; 11(11): e0166026. [http://dx.doi.org/10.1371/journal.pone.0166026] [PMID: 27812181]

[294] Abrahamsson TR, Jakobsson HE, Andersson AF, Björkstén B, Engstrand L, Jenmalm MC. Low gut microbiota diversity in early infancy precedes asthma at school age. Clin Exp Allergy 2014; 44(6): 842-50.
[http://dx.doi.org/10.1111/cea.12253] [PMID: 24330256]

[295] Bisgaard H, Li N, Bonnelykke K, *et al.* Reduced diversity of the intestinal microbiota during infancy is associated with increased risk of allergic disease at school age. J Allergy Clin Immunol 2011; 128(3): 646-52.
[http://dx.doi.org/10.1016/j.jaci.2011.04.060] [PMID: 21782228]

[296] Garrett WS. Cancer and the microbiota. Science 2015; 348(6230): 80-6.
[http://dx.doi.org/10.1126/science.aaa4972] [PMID: 25838377]

CHAPTER 3

Alternative Approaches to Antimicrobials

Ayhan Filazi[*] and **Begum Yurdakok-Dikmen**

Department of Pharmacology and Toxicology, Faculty of Veterinary Medicine, Ankara University, Ankara/Turkey

Abstract: Historically, infectious diseases has been a major threat to human and other animals health and an important cause of morbidity and mortality. The introduction of antimicrobials in the first half of the twentieth century revolutionized medicine by substantially reducing morbidity and mortality rates from infectious diseases. Nevertheless, it was soon observed that bacteria could become resistant to antimicrobials, and resistant strains emerged shortly after the introduction of every new antimicrobial drug. Unfortunately, resistance is a natural and unavoidable consequence of antimicrobial use. For this reason, new antimicrobials are urgently needed, but so are additional approaches to protect the value of available antimicrobials.

The discovery of a new antimicrobial is not an easy task where the scene is further complicated by a variety of interacting factors. In order to eliminate the spread of antimicrobial resistance, firstly, the inappropriate use should be reduced both in human and veterinary applications and alternative approaches should be considered.

Until now, many alternatives, including plant-derived compounds, bacteriophages and phage lysins, probiotics, and into antimicrobial peptides from a variety of sources have been tested especially against resistant strains. These approaches are currently lacking in antimicrobial management, meanwhile, they have demonstrated considerable potential for application in other diseases.

The review presents an insight into antimicrobials, particularly direct-food microbial, as well as other alternative products such as plant-derived compounds, bacteriophages and phage lysins, and antimicrobial peptides along with other alternative products, including novel approaches applicable to the field.

Keywords: Antimicrobials, Probiotics, Prebiotics, Synbiotics, Organic acids, Bacteriophages, Antimicrobial peptides, Bacteriocins, Fecal therapy, Nanoparticles.

[*] **Corresponding author Ayhan Filazi:** Department of Pharmacology and Toxicology, Faculty of Veterinary Medicine, Ankara University, Ankara/Turkey; Tel: +90-312-3170315/4435, Fax: +90-312-3164472; E-mail: filazi@veterinary.ankara.edu.tr

Atta-ur-Rahman (Ed.)

INTRODUCTION

The origin of antimicrobial therapy in the west dates as early as the 1600's to the use of cinchona tree bark to treat malaria by South American natives. This discovery made by native South Americans has been used to develop the drug quinine, which is still used to treat malaria after it was brought to Europe in the seventeenth century. On the other hand, it is known that herbal remedies have been used for nearly 2,000 years in Chinese medicine and described in ancient Sumerian and Egyptian records [1].

The first antimicrobial agents used commercially were developed by German scientists that demonstrated the antimicrobial effects of dye by injecting them into mice infected with streptococci. This *in vivo* experiment led to the discovery of the compound called Prontosil which is used to treat not only streptococci but also staphylococci along with a variety of gram-negative and -positive organisms. In the 1930's, Prontosil, as an antibacterial agent, was widely used in many countries to treat various diseases. The first antibiotics were isolated from bacteria. The first of these drugs was streptomycin; which was isolated from bacteria in the soil in the laboratory of Selman Waksman who is the founder of the term antibiotic. Waksman defined antibiotics as "*any substances produced by microorganisms that inhibit another microorganism at high dilutions*." The inhibition of bacterial growth by an accidental fungal contamination led to the discovery of the miracle drug penicillin, which was isolated from the *Penicillium* fungus. Penicillins, cephalosporins, aminoglycosides, and carbapenems are all natural products isolated from microorganisms and, therefore, they are called antibiotics. Since sulfonamides, quinolones, and oxazolidinones are synthetic chemicals, they are called antibacterial agents, not antibiotics [2].

Today antimicrobials can be prepared synthetically or semi-synthetically, to be used as antibacterials, antifungals, antiparasitics, or antivirals. The future of humanity depends on the evolving capability of the antimicrobial to prevent and treat illnesses. Antibacterials have been the most preferred group of medicines for many years and are used especially in food-producing animals to promote growth, increase food conversion and to produce healthier, more affordable food with better quality by preventing subclinical diseases and treating bacterial diseases [3]. However, the widespread use of antimicrobials has led to the gradual development of resistance [4], the formation of residues in food offered for consumption [5 - 12], changes in the beneficial/pathogenic microorganism ratio in the gastrointestinal system [13], suppression of the immune system [14], host organ toxicity [15] and the emergence of environmental problems [16]. The prevalence of resistance genes that emerge in pathogenic organisms has caused the future effectiveness of current antimicrobials to be questioned [17]. According to the Centers for Disease Control (CDC) in the United States of America (USA) and the European Centre for Disease Preventions and Control (ECDC) in Europe, it has been reported that nearly 25,000 people died due to antibiotic-resistant infections [18, 19]. It has also been reported that the ratio is not lower in other countries and that many of the procedures implemented to manage antibiotic resistance have failed [20]. The most recent research towards the development of new antibiotics, through the discovery and isolation of DNA gyrase inhibitors, provide new opportunities to manage bacterial resistance without repeating previous mistakes [21, 22]. However, microorganisms might also become resistant to these newly developed antimicrobials if they are not used rationally [23].

The use of antibiotics to promote growth has been prohibited, national antimicrobial resistance and residue monitoring programs have been established, preventive medicine has been promoted, antimicrobial medicine use policies have been developed, and rational use has started to be encouraged in many countries to prevent and minimize problems caused by antimicrobials [24]. Within this context, especially in European countries, the use of some antibiotics in the feed of food-producing animals as growth promoters was first limited in June 1999 and, since 2006, it has become completely prohibited [25]. However, these antibiotics can still be used freely for therapeutic purposes. The ban on using antibiotics as growth promoters in farm animals has increased the need for substances that will replace antibiotics, especially in animal husbandry. Therefore, multifocal approaches have started to be implemented to prevent diseases in food animals, including vaccinations, supporting immunity, the development of nutritional strategies that contribute to the regulation of hygiene and the intestinal flora and farm management procedures [26].

Due to the reasons mentioned above, there is an urgent need for new antimicrobials. Meanwhile, conventional methods for the discovery of these compounds are insufficient to complement the demand; and yet strategies for the prudent use along with complementary and supplementary alternatives should be introduced in the field in advance. Since current antimicrobials are used to treat or prevent illnesses and to promote growth in food animals, the alternative or supplementary compounds to be developed should also carry these properties. Within this context, it can be seen that there is an increasing number of studies that mainly investigate the use of vaccines, immunomodulators, probiotics, prebiotics, enzymes, and organic acids in the prevention of many diseases that were considered as an impasse in the past [27]. It has been stated that probiotics, as direct-fed microbes that contain the spores of living or non-living organisms, unconventional chemicals, bacteriocins, bacteriophages and antimicrobial peptides receive importance within the context of studies on alternatives to antimicrobials, especially in food animal production [28, 29]. Moreover, in recent years, it has been reported that different treatment strategies such as nanoparticles, fecal bacteriotherapy and fecal transplantation practices have a potential to become alternatives. It is impossible to predict and control the emergence of new and complex diseases, with cross or multidrug-resistant microorganisms causing clinical failures. Therefore, the increase in the demand for drugs against these organisms would lead to the search and production of new groups of antimicrobials to be always on the agenda [30].

In order to describe the alternative approaches to antimicrobials, first the modulation of the intestinal microbiota, then the alternative compounds for the treatment and last, the proper strategies to better target the organism are discussed.

PRACTICES BASED ON THE MODULATION OF INTESTINAL MICROBIOTA

It has been reported that there are more than 1,000 species of microorganisms in the intestines that are very important to human and food-producing animal health which comprise approximately 10^{10}-10^{12} microorganisms in 1 g of intestinal content [31, 32]. Microorganisms found in the intestines include *Lactobacillus, Bifidobacterium*, coliforms, *Enterococcus, Clostridium, Staphylococcus, Pseudomonas*, yeast, and fungi and are considered as the residential flora [33]. Microorganisms (except pathogens) contribute to the health of mammals by modulating the immune system and by helping food to be digested better [34]. Since the host and intestinal microorganisms evolve together, the interactions between them that occur during health and disease conditions need to be understood [35, 36].

To treat some of the infections, antimicrobials are required to be delivered systematically (through oral or parenteral routes). In such cases, the population of the intestinal microbiota usually decreases no matter where the infection is [37, 38]. Since this decrease changes the diversity of intestinal bacteria, irritable bowel syndrome or *Clostridium difficile* infections may develop [37]. The formation of reservoirs of antibiotic resistance genes in the intestinal microbiota, the increase in horizontal gene transfer between bacterial strains, the population rise of some bacteria due to the change in the carbohydrate composition, changes in metabolic activity that contributes to digestion, and changes in the immune responses of distal organs can be counted among the undesirable results of systemic antimicrobial use [39 - 41]. The only way to reduce or eliminate such potential undesirable adverse effects in the intestinal microbiota is to use products that are more specific against certain infectious agents. Probiotics, prebiotics, synbiotics, enzymes and organic acids are widely used in this context.

Probiotics

Despite complicated results, exogenous bacteria (probiotics) have been used for years to maintain health and prevent diseases [42]. Probiotics have been defined by the Food and Drug Administration (FAO) and the World Health Organization (WHO) [43] as "*living microorganisms that have beneficial effects on health when administered in sufficient doses.*" The most commonly used probiotics are lactobacilli and bifidobacteria, and it has also been reported that bacteria such as *Escherichia coli,* and *Bacillus coagulans* and the yeast *Saccharomyces cerevisiae* var. *boulardii* are commercially available. Besides these microorganisms, new beneficial microbes such as *Akkermansia muciniphila*, *Eubacterium halii,* and *Faecalibacterium prausnitzii* have been discovered and their potential to be used as probiotics is currently being investigated. Nonetheless, it is an important feature for any probiotic strain to receive a safety approval at the species level by competent authorities [44].

Probiotics may exert their beneficial health effects on the digestive and immune systems, the metabolism and the mental health of the host. The beneficial health effects of probiotics are usually strain-specific, and even if they belong to the same species, one strain might not exhibit the same effect as another strain [45]. Besides these traits, the organisms may have common characteristics due to similarities in their metabolism [46].

Studies show that an ideal probiotic must have the following characteristics [47, 48];

- They must not be affected by the activities of the normal microflora in the digestive system, and they should be able to adapt,
- They must have the ability to exhibit beneficial effects on the host,
- They must easily and rapidly reproduce in the environment they are in,
- They must be tolerant to different pH values, bile, and organic acids,
- They must be able to influence local metabolic activity,
- They must be able to send signals to or interact with the immune system cells,
- They must be able to attach to intestinal cells,
- They must not be pathogenic and must not secrete toxic substances,
- They must not disrupt the tissues and the permeability of the intestines,
- Their effectiveness must not decrease when they are added to water, feed or food,
- They must not lose their effectiveness during production and storage.

Therefore, probiotics which have a direct inhibitory effect on pathogens or that have high resistance to intestinal colonization by pathogens are considered as important factors. In such cases, probiotics inhibit the activities of pathogenic bacteria by means of different mechanisms, and they may reduce the frequency and duration of diseases.

Other effects of probiotics are described as follows [30, 47 - 49]:

- Their competitive and antagonistic effects in order to protect the normal intestinal microflora from pathogenic bacteria,
- They alter metabolism by increasing the activity of the digestive system and by decreasing bacterial enzyme activity and the production of ammonia,
- They reorganize feed consumption and regulate digestion,
- They contribute to the absorption of organic substances that are decomposed by bacteria,
- They stimulate the immune system and facilitate immunomodulation by increasing antibody levels and macrophage activity against pathogenic microorganisms or by promoting cytokine release,
- They inhibit the production of bacterial toxins.

Apart from these effects, it has also been stated that probiotics positively affect the water quality in the intestines [49] and that they can also be effective in the presence of microorganisms such as protozoa (*Giardia* spp., *Cryptosporidium* spp.,), coccidia (*Eimeria* spp.) [48], and viruses (*Hematopoetic necrosis* virus, *Oncorhyncus masou* virus) [49].

In recent years, many studies on probiotics have been conducted in different animal species. Currently, probiotics in food-producing animals (such as poultry, cattle, sheep, goats, fish and pigs) are intensely used against pathogenic microorganisms to prevent infections or provide protection, to increase weight gain, to regulate feed/food consumption and digestibility, to improve immune responses, and to decrease the mortality rates due to their beneficial effects [27, 29, 50]. In humans, it has been observed that probiotics have a potential effectiveness in autoimmune diseases, urogenital, dental, and respiratory system infections, rheumatoid arthritis, allergies and vaginal candidiasis. Despite the negative preconceptions regarding the effectiveness of probiotics in the human digestive system, it has been stated that probiotics can be used for many conditions such as constipation, *Helicobacter pylori* infections, and inflammatory and chronic intestinal diseases [51].

Although it has been expressed in recent years that commercial products which contain various combinations of probiotics that act by means of different mechanisms can be used to enhance the effectiveness of probiotic products and offer more than one effect together, it has also been stated that the fact that bacterial strains used as probiotics may exhibit host or strain-specific differences in combination. These strain differences should not be neglected [48].

The main focus points in the use of such products are that probiotic microorganisms are not effective in all diseases, and the dose (in humans at least $1x10^8$ cfu/day) along with the duration of the use required to obtain the desired benefit from disease-specific strains must also be known. It is also stated that probiotic microorganisms might also be antibiotic resistant, which may cause superinfections, and that the adjuvant used in their preparation might have side effects or harmful metabolic activity [51].

Although there are many studies on beneficial and culturable intestinal bacteria such as *E. coli* and *Bacteroides fragilis*, the current knowledge about many of the intestinal bacteria is limited. In addition, sequencing technologies with high efficiency may facilitate the identification of both culturable bacteria and bacteria that have not been cultured yet. By this means, prevalent models of the microbiota in the intestines of mammals may be defined [52]. Most of the models that have been defined until today are at relatively higher taxonomic classes and they cannot reach the levels of subspecies diversity that is believed to be important for the functions of the ecosystem. In fact, the importance of the diversity of subspecies has been known for a very long time in some bacteria such as *E. coli.* Future studies may discover significant subspecies that have an ecological importance to mammals. For example, it has been reported that some strains of *Propionibacterium acnes* are more associated with acne than other strains [53]. Therefore, it is believed that efforts toward defining various members and functions of intestinal bacteria will increase the ability of scientists and clinicians to modulate them.

Our knowledge about why bacterial populations sometimes end with instability even when only a single probiotic is added is still limited. Therefore, in order to enhance the natural bacterial population and to competitively eliminate pathogens, mostly complex microbial mixtures are used. It has been reported that this approach is particularly successful in poultry. Furthermore, a significant decrease in the population of intestinal *Salmonella* treated with mixtures of anaerobic commensal bacteria is also described [27]. In addition, the fact that it is difficult to administer anaerobes to animals in large shelters and that their beneficial effects cannot be sustained during the proceeding production processes such as transport are the most important disadvantages of this approach.

The most important concern about probiotics is the fact that they might also become antibiotic-resistant and that resistance can be transmitted to the intestinal bacteria [54]. Competent authorities have started to publish reports that assess the risk of transferring antibiotic resistance from probiotics and expressed their concerns [55].

Probiotics become resistant by means of various mechanisms, some of these are well known, while others are unknown. In probiotics, acquired genes that can be transferred through conjugative plasmids, transposases, and prophage/ bacteriophage elements which are currently at the center of safety concerns [54, 56, 57]. It has also been reported that vancomycin and tetracycline resistance can form intrinsically in lactobacilli [58]. On the other hand, beta-lactam (ampicillin) resistance in lactobacilli and bifidobacteria is atypical [57]. Intrinsic resistance to aminoglycosides (kanamycin, gentamicin, streptomycin) has been observed in some strains of *Lactobacillus* [59]. The advantages and disadvantages of some probiotics are shown in Table **1**.

Table 1. Advantages and disadvantages of some probiotics.

Some Probiotics	Advantages	Disadvantages	References
Lactobacillus spp. *Bacillus* spp. *Bifidobacterium* spp. *Pedioccus* spp. *Streptococcus* spp. *Vibrio* spp. *Bacteriodes* spp. *Enterococcus* spp. *Clostridium butyricum* *Aspergillus* spp. *Saccharomyces* spp. *Rhodotorula rubra* *Torulopsis candida*	-Improve commensal gut microflora health -Prevent colonization of pathogen in gut colonization -Treat *Clostridium difficile* or antibiotic-associated diarrhea -Combined with other antibiotics or other alternatives to antibiotics	Antibiotic resistant genes in probiotic organism and transmissible to gut organisms	[27 - 29, 42 - 59]

Prebiotics

The term prebiotic was first used by Gibson and Robrefroid, and, in 2010, prebiotics were defined by the International Scientific Association for Probiotics and Prebiotics (ISAPP) as "*selectively fermented compounds that produce beneficial effects on the health of the host by making changes specific to the activity or composition of the gastrointestinal microbiota*" [61]. Although the debate on the final definition continues, it has been defined that these products, also known as the diet fibers, cause specific changes in the composition or activity of the gastrointestinal microbiota, which induce beneficial physiological effects and reduce the risk of various intestinal and systemic pathogens not only in the colon but also in the entire body [62]. Although all prebiotics can be classified as fibers, not all fibers are prebiotics. Fiber has to specifically stimulate microbiota to be prebiotic [63]. Most of the research has focused on bifidobacteria and lactobacilli that can selectively change the composition of the intestinal microbiota; however, current research topics have expanded towards other species such as *Eubacterium* spp., *Faecalibacterium* spp. and *Roseburia* spp [61].

Most prebiotics are derived from plants. Many vascular plants accumulate glucose polymers as energy reserves and structural components in some of their tissues. Additionally, in approximately 10% of all vascular plants (and in some bacteria), the polysaccharide reserve polymers accumulated are bound to fructose chains with glycosidic bonds. These uncommon polymers, called fructans, contain two compounds, namely fructooligosaccharides (FOS) or inulin [64]. Non-starch polysaccharides are regarded as the main fiber components of the diet and contrary to starch they can be digested completely in the intestines, and they are absorbed as free glucose. On the other hand, fructans and other polysaccharides cannot be digested by the digestive enzymes of mammals, and they reach the colon unchanged. Their function in the colon is for the production of the main short-chain fatty acids such as acetate, butyrate, valerate, isovalerate and caproate to provide energy for various groups of beneficial bacteria and to promote their growth by doing so. Therefore, indigestible oligosaccharides and polysaccharides are called prebiotics [60, 65]. For a compound to be called prebiotic, it must be fermentable by bacteria, it must selectively stimulate the activity or development of intestinal bacteria, it must not be influenced by gastric activity, it should not be hydrolyzed by digestive system enzymes, it should not be digested in the intestines at all, and it should be effective at very low levels [66, 67]. Contrary to probiotics, prebiotics are non-living compounds and cannot be used by the host cells. Probiotics are more effective in the posterior intestines sections and prebiotics are mainly more effective in the colonic anterior portion of the intestines [47]. The most commonly used prebiotics that carries these features today are inulin, FOS, lactulose, galactooligosaccharides (GOS), mannan oligosaccharides (MOS) and human milk oligosaccharides (HMO) [68].

Among prebiotics, chicory (*Cichorum intybus*) was found to accumulate the most inulin in its tissues (especially in roots). Some other examples of inulin-accumulating plants are Jerusalem artichoke (*Helianthus tuberosus,* roots), blue agave (*Agave tequilana*, leaves), asparagus (*Asparagus officinalis,* stem), garlic (*Allium sativum*, bulbs), artichoke (*Cynara scolimus,* flowers), and leek (*Allium ampeloprasum*, leaves) [69]. Industrially, inulin is produced from these plants by water extraction at high temperatures [70]. Lactulose is another synthetic prebiotic disaccharide that is usually obtained by the industrial isomerization of lactose present in whey [71]. GOS are found in mammalian milk, including human milk and contain 2 to 10 galactose rings attached to a terminal glucose. However, at an industrial level (as in the production of baby formulas), GOS is obtained from lactose in whey through transglycosylation catalyzed by a lactase hydrolysis of lactose [72]. Human milk also contains other carbohydrates called HMO (10 g/L) besides lactose (70 g/L) and GOS (5 g/L) [73]. The HMO ring structure is formed by the combination of glucose, galactose, fructose, N-acetylglucosamine and N-acetylneuraminic acid. In humans, these rings make it possible for HMO to remain in the intestines without being digested. The main components of HMO are 2-fucosyl-lactose, lacto-N-tetraose and lacto-N-neotetraose. HMO is responsible for the growth of the high number of *Bifidobacterium* spp. in the feces of babies [74]

The recommended daily intake level of prebiotics is 25 g/day for women and 38 g/day for men younger than 50 [75]. It has also been reported that prebiotics can be effective even when they are taken at doses of 2.5-5 g/day [66]. Due to the total low prebiotic intake in western diets, in order to increase the consumption, prebiotics are recommended to be taken with functional prebiotic foods. A variety of foods enriched with prebiotic fibers such as inulin, FOS, or GOS are available for this purpose. All these prebiotic fibers are considered safe for human consumption and, therefore, they are easily used in a variety of foods such as yogurt, drinks, biscuits, grains and bakery products without any limitations [75].

Probiotics and prebiotics share common mechanisms in the modulation of the endogenous flora. In contrast, the composition and metabolism of prebiotics are rather unique, and their effects on the gastrointestinal system have been explained better than the effects of probiotics. It has been reported that prebiotics have osmotic effects when they are non-fermented, and that they selectively stimulate the activity and development of one or a limited number of bacteria, increase the formation of short-chain fatty acids, reduce the ratio of nitrogenous products and the activities of reductive enzymes, and regulate the immune system when they are fermented. It has been reported that probiotics increase gas production (carbon dioxide, methane and hydrogen) and fecal weight, prevent constipation, regulate the intestinal habitat, slightly reduce the luminal pH levels in the colon, increase the synthesis of proteins that facilitate the absorption of minerals (such as Ca^{2+}, Mg^{2+}) or the ratio of active transport proteins, and reduce the ratio of toxic, mutagenic, genotoxic metabolites, bile acids and cancer-inducing enzymes by means of their fundamental effects [61]. By reducing the intestinal pH, probiotics regulate the intestinal flora and lead to the stimulation of the activity or development of carbohydrate-fermenting bacteria. By means of the subsequent increase in bacterial effectiveness (bifidobacteria and lactobacilli), they reduce beta-glucuronidase and nitroreductase activity. It has been reported that increased ratios of short-chain fatty acids increase the production of secretions, heal the structure of the mucosa, increase the use of energy and regulate its balance that is dependent on the intestinal microbiota by binding to the G-protein-dependent

Grp-41 receptor. It has been identified that butyrate, which is the energy source of colonic cells, is particularly responsible for this effect [62].

The osmotic effect of prebiotics is expected to lead to the softening of feces, and, due to this effect, it is favored by patients experiencing irritable bowel syndrome. Meanwhile, prebiotics are also used to prevent or alleviate various clinical conditions of diarrhea and to prevent constipation by slightly softening the feces [76]. Prebiotics also regulate the barrier functions of the intestinal epithelia, provide energy for colonic cells, prevent antibiotic-related diarrhea, reduce the load on kidneys by increasing the absorption of nitrogen by the intestines, reduce the concentrations of the very low density lipoproteins (VLDL) and the serum triglycerides in the liver, prevent atopic dermatitis, increase immunoglobulin-A (IgA) production, and even reduces the risk of cancer in the colon [61]. Due to their array of effects, probiotics are used in food animal production to increase the feed conversion ratio and milk production, to produce immunomodulatory effects (humoral and cellular), to increase production quality (such as to lower the cholesterol level in eggs, increase the ratio of components in milk fat), regulate the intestinal microbiota (such as *Bifidobacterium* spp. and *Lactobacillus* spp.) and to reduce the effectiveness of pathogenic microorganisms (such as *Salmonella* spp. and *E. coli*) especially in poultry and also pigs, ruminants and fish [77].

Depending on the dose and the duration of the treatment, prebiotics might exert undesirable effects such as intestinal swelling, pain, excessive gas formation or diarrhea. Also, they are not as effective as probiotics for removing specific pathogens or neutralizing their effects. These undesirable side effects might increase even more due to the absorption of simple sugars during active diarrhea conditions (Table **2**) [47].

Table 2. Advantages and disadvantages of some prebiotics.

Some Prebiotics	Advantages	Disadvantages	References
Fructooligosaccharides Galactooligosaccharides Mannan oligosaccarides Inulin Lactulose	-Improve commensal gut microflora health -Prevent constipation -Treatment of hepatic encephalopathy -Treatment of irritable bowel syndrome -Prevent the cholesterol gallstone disease. -Prevent the gut-associated infections	-Dyspepsia, spasm, tympanitis or diarrhea in overdoses. -Low activity against specific pathogens -Improves the absorption of sugars in the active diarrhea	[47, 60 - 77]

Synbiotics

The term synbiotic was first used by Gibson and Roberfroid [60] together with the term prebiotic and is defined as a "*mixture of probiotics and prebiotics that beneficially affect the host by facilitating the survival and implantation of live microbial dietary supplements in the gastrointestinal system, that selectively stimulate the growth or activate the metabolism of one or more than one beneficial bacteria that improve the conditions of the host.*"

Synbiotics have two types of effects, complementary and synergistic [67]. Although prebiotics are selected independently for their complementary effects, probiotics need to be selected in a specific way that will produce the desired beneficial effects in the host. For synergistic effects, the selected probiotic for its beneficial effects are complemented by the prebiotic for an enhanced activation of the probiotic through its growth or other mechanisms. In this synergy, much lower concentrations of probiotics would suffice for the desired effect. Therefore, the selected prebiotic must have a higher affinity and selectivity and it should also promote the development or growth of the chosen probiotic microorganism [47, 50].

Some authors have argued that probiotics are not activated by prebiotics and they are effective in the small intestines, while prebiotics are only effective in the large intestines, and, therefore, a symbiotic effect between these two components cannot be mentioned, but that the symbiotic relationship that emerges by means of individual mechanisms could only be called a synergism [78]. However, no matter what the relationship between them, many studies have shown that various combinations of prebiotics and probiotics have beneficial effects on the regulation of intestinal flora, prevention of sepsis, inflammatory and irritable intestinal diseases, colonic cancer and bone diseases (such as osteoporosis) and some surgical diseases (such as intestinal and liver). Today, the use of synbiotics is rather limited in food animal production and it is mostly preferred in fish, poultry, and pigs [67]. Such examples are the combination of *Bifidobacterium* spp. + FOS (or GOS), *Lactobacillus* spp. + inulin, or *Bifidobacterium* spp. + *Lactobacillus* spp. + FOS (or inulin) [79]. The use of synbiotics is rather limited by the density and diversity of the rumen microbial flora in ruminants [27]. However, two studies conducted in calves have shown that the combination of *Enterococcus faecium* + lactulose [80] and in dairy cattle the combination of *Lactobacillus casei* + dextran [81] had limited but positive effects on production (Table **3**).

Prior to the selection of the synbiotics, the desired effect in the host should be well defined and the potency of the pro/prebiotics should be clarified. In order to increase these desired effects, different clinical studies based on age, the breeding period and disease groups are often required to clearly establish the safety of these combinations [82]. It has also been proven that such products used as alternatives to antimicrobials today increase the efficiency of animal production in the short-term through the prevention of bacterial resistance, reduction of some undesirable effects (*e.g.* residues, allergy risk) in the consumed products (meat, milk, egg, yogurt, and cheese), and potentially increase the product quality and safety.

Table 3. Advantages and disadvantages of some synbiotics (combination of probiotic and prebiotics).

Some Synbiotics	Advantages	Disadvantages	References
Bifidobacterium animalis + FOS[1] *Bifidobacterium breve*+ *Lactobacillus casei* + GOS[2] *Bifidobacterium clausii* + FOS[1] *Bifidobacterium longum* + inulin *Bifidobacterium subtilis* + FOS[1] *Enterococcus faecium* + FOS[1]	-Regulation of intestinal flora -Prevent the sepsis -Prevent the inflammatory and irritable bowel syndrome	Reliability is different depending on host age and species	[27, 47, 50, 60, 67, 78 - 82]

[1]Fructooligosaccharides, [2]Galactooligosaccharides

Enzymes

Enzymes can influence all the chemical events that occur in the body, and are defined as "*biological catalysts that regulate the speed and direction of chemical reactions in the organism by being present without entering reactions or being decomposed*" [83]. The practice of administering enzymes is particularly used in the feed of food-producing animals with the aim to gain financial profit (by supplementing rations that contain corn, wheat, soybean meals and other food products with exogenous enzymes) through the increase of the feed conversion ratio, the nutritional value or digestibility of the feed. The most prominent of these enzymes are of fungal or bacterial origin and are obtained from *Bacillus subtilis, Lactobacillus acidophilus*, *L. plantarum* and *Streptococcus faecium* bacteria, *Trichoderma longibrachiatum, T. reeseii* and *Aspergillus oryzae* fungi and *Saccharomyces cerevisiae* yeast [84].

By adding enzymes to food alone or in combination with probiotics it is possible to decompose indigestible polysaccharides in cell membranes (particularly in wheat and barley), increase the intestinal viscosity through lowered polysaccharides, and increase the digestibility and metabolic energy values of food [47]. It has also been found that exogenous enzymes prevent colonization by reducing the growth and fermentation processes of harmful microorganisms in the intestines. In this sense, enzymes have particularly been considered as feed supplements in food animal production that are alternative to antibiotics used as growth promoters and regulators [85].

Enzyme activity may be influenced by the pH of the intestinal canal, temperature, moisture, the composition of the feed, and the age and species of the animal. Although feed is mostly digested by microbial activity in the rumen in ruminants and because rumen fermentation is underdeveloped in newborn and young ruminants, it has been reported that the use of exogenous enzymes is more useful to increase feed conversion. Feed supplements are preferred more in poultry due to the faster passage of feed through the digestive systems compared to other species, and the lack of enzymes to decompose the plant cell membrane. Similarly, it has been observed that it is more beneficial to use glucanase in poultry feed predominantly containing barley and wheat, cellulase in roughage that contains high levels of raw cellulose, and amylase in starter feeds for sheep and calves that contain high ratios of starch. On the other hand, it has been reported that non-starch polysaccharide enzymes added to corn/soybean-based feed were not able to produce the effects anticipated (Table **4**) [83]

Table 4. Some practices based on the modulation of intestinal microbiota.

Application	Advantages	Disadvantages	References
Enzymes (α-Amylase, α-Galactosidase, Phytase, Lipase, Cellulase, Pectinase, Xylanase, Protease)	-Regulate digestion, -Reduce the viscosity of the intestine caused by polysaccharides	Activity is different depending on species and food consumed	[47, 83 - 85]
Organic acids (Sorbic acid, Lactic acid, Propionic acid, Acetic acid, Butyric acid, Formic acid, Tartaric acid, Fumaric acid, Malic acid, Citric acid)	-Easily pass through the cell membrane of gram-negative bacteria -Bacteriostatic or bactericide	-No effect on acid-resisting bacteria -Antagonistic when used together with some antibiotics.	[86 - 90]

Organic Acids

Organic acids are defined as "*products of microbial metabolism that contain carbon elements, which frequently occur in plant and animal tissue and form during the microbial fermentation of carbohydrates particularly in the large intestines.*" Organic acids may be formed by normal biochemical processes (*e.g.*, acidity, hydrolysis, bacterial growth) and they are also added to food products directly or indirectly [86, 87].

It has been found that not every organic acid has an effect on microflora and that each acid has a unique spectrum of antimicrobial activity. It has been identified that organic acids with specific antimicrobial effects are short-chain simple monocarboxylic acids (formic, acetic, propionic and butyric acid). It has been reported that the pKa value of most organic acids with antimicrobial activity ranges between 3 and 5. Among these acids, lactic acid is more effective on bacteria, and acetic, formic and propionic acids have a wider range of antimicrobial activity that includes yeast and fungi. Substances such as sorbic and fumaric acids are most effective as antifungals [87, 88]. Organic acids were found to have an important effect against bacterial skin infections in broilers and that they positively affect the performance of different animal groups [88]. These organic acids are also defined as "good food" or "feed preservatives" and are used as a protective measure for the storage of food and feed [89]. Some of the fundamental effects of organic acids added to the diet are namely: 1) Ease of stomach acidification; 2) Prevention of the formation of biological amines; 3) Increase metabolism in the liver and kidneys; 4) Increase the secretion of active components from the body, facilitate the optimal use of acids and their salts in the digestive system; 5) Alter the balance of the digestive tract microflora in favor of beneficial microorganisms, particularly helping in the growth of *Lactobacillus* spp. in the digestive system and preventing the colonization of pathogenic microorganisms by preventing irregularities in this system; 6) Enhances the flavor and indirectly increase the intake of food; 7) Facilitate the better digestion of foods by increasing salivation and the production of enzymes in the digestive system; 8) Increase live weight gain and the food conversion ratio by increasing feed consumption; 9) Produce complexes with Ca, P, Mg, and Zn and increase their absorption; and 10) Reduce the risk for decomposition while food is being stored [87, 89]. Organic acids were also found to be effective in reducing the number of bacterial species that are not particularly acid-tolerant, such as *E. coli, Salmonella* spp. and *Campylobacter* spp., for the prevention of their colonization in the cecum [90]. Therefore, organic acids are widely used in animal production (Table **4**).

Depending on the length of the organic acid, carbon chain, their pH values and under the condition that the environment contains adequate levels of dissociated acid molecules that have been with the bacteria for a long time, *in vitro* studies have shown that organic acids easily penetrate through the cellular membrane of bacteria (particularly gram-negatives). Following their uptake, these compounds were found to lower the intracellular pH, disrupt intracellular amino acid metabolism, cytoplasmic membrane structure, the protein and electron transport systems, reduce intracellular ATP production and exert bacteriostatic and bactericidal effects [86, 88]. Since these effects of organic acids develop due to anion accumulation that occurs during their dissolution, they were found to be ineffective in acid-resistant bacteria. Additionally, clinical studies have reported that organic acids have antagonistic effects with some antibiotics that prevent the transportation of nutrients into the cell, increase the permeability of the cytoplasmic cell and disrupt the synthesis of some macromolecules [86]. It has been reported that organic acids widely used in food-producing animals may exhibit varying effectiveness against intestinal bacteria depending on their doses. Most of the organic acids are effective in the anterior portions of the digestive system (the crop and gizzard of poultry, the stomach and small intestines of pigs) and they were found to predominantly inhibit the growth of coliform bacteria in these portions [87].

ALTERNATIVE APPROACHES DEVELOPED TOWARDS TARGET MICROORGANISMS

Various approaches have been introduced for the decrease of the use of antimicrobials causing resistance or even entirely replacing them. These approaches include, but are not limited to bacteriophages (phages), antimicrobial peptides, bacteriocins, predator bacteria, fecal bacteriotherapy and nanoparticles. Their only difference from classic antimicrobials is that these approaches only target specific microorganisms and do not affect organisms that are not targeted, and, therefore, they do not lead to any sort of resistance.

Bacteriophages

The use of bacteriophages (phage in short) is not a new concept. First observations were made in 1896 then, between 1915 and 1917, they were introduced to treat dysentery in humans and poultry typhus caused by *Salmonella gallinarum* [91]. Although it was not widely accepted at first, phage preparations were produced by the Pasteur Institute in France until 1974 and until the 1990's in the USA [92]. In fact, although they were used continuously in Western European countries and in Georgia, the interest in phages has decreased outside the post-Soviet Union countries after the discovery of antibiotics [93].

The decrease in antibiotics being discovered today and the increase in multi-drug resistant bacteria led phages to be considered as alternative therapies for certain types of bacterial pathogens. Bacteriophages are defined as "*viruses that can be found in any environment that do not infect mammalian or plant cells and only infect bacterial cells all living beings are exposed to at high ratios through water or food,*" or "*viruses that infect bacterial cells by transferring their own genetic material (DNA or RNA)*" or shortly as "*bacterial viruses*" [47].

Phages have particularly drawn attention as an alternative in food safety, agriculture and farming practices, especially those that lead to the spread of antibiotic-resistant superbug bacteria [94]. In the USA, the FDA has approved the use of commercial phage preparations against widespread bacterial pathogens such as *Listeria monocytogenes* and *Salmonella* spp. in ready-to-use food [95].

In Georgia (Eliava Institute and The Center for Phage Therapy in Tbilisi) and Poland (Institute of Immunology and Experimental Therapy in Warsaw), specific studies on phages and phage therapies yielded important clinical results about their use as alternatives to antibiotics. Among these studies, a limited number of studies on humans led to their approved use by health authorities such as the FDA and the European Medicines Agency (EMA) [96]. It has been reported that more than 5,100 phages have been studied since the 1960's and that nearly 96% of these phages belonged to the *Myoviridae, Siphoviridae* and *Podoviridae* families [97].

Depending on the type of the phage being used, they demonstrate lysogenic or lytic effects. Lysogenic responses are obtained from temperate phages. In such cases, phage reproduction occurs in a later period because the genetic material of the virus is integrated into the genome of the bacterium. In this case, cell lysis does not occur and the bacteria become immune to attacks by other phages of the same strain, and it becomes lysogenic (generally more virulent) bacterium [96, 97]. Such bacteria are characterized by the presence of a prophage, in other words, an inactive phage integrated into the genome and it remains in a latent state over a few bacterial cell divisions. When the host is exposed to stress or cellular damage the prophage becomes activated by exiting the bacterial genome [96 - 98]. Therefore, lysogenic phages are not preferred much for phage therapies. In addition, because of their ability to transmit antibiotic-resistant genes to non-resistant bacteria, it has been possible to obtain genetically modified lysogenic phages by this means. By doing so, the goal is to increase the susceptibility of bacteria to a class of antibiotics by inserting specific genes into the bacterial genomes [97, 98].

Lytic responses are obtained from lytic phages that are also called virulent phages. In such cases, the target is the metabolism of the host bacteria, and the goal is to produce new phage particles. The genetic material of the virus is replicated in the cytoplasm. This process stimulates the production of holins and lysines, and lytic phage particles that are produced in cycles of 30 minutes. Eventually, the bacteria undergo lysis. These phages are preferred more often for phage therapy because they reproduce their own DNA, and particularly because they lyse the bacterial cell within the short replication period and transfer their own genetic material into the host genome easily without integrating, while it can also be transmitted to other cells [96, 97]. Although many phages have been named within this context, the number of phages that have been studied clinically in humans and food animal production that have been regarded as usable is rather limited. It has been expressed that phage therapies are used more predominantly in the prevention of infections or reduction of infectious effectiveness of *Streptococcus pneumoniae, Mycobacterium tuberculosis* (lung infections), *Haemophilus influenza type-B* (pneumoniae, bacteremia and meningitis) and *Shigella* spp. (dysentery), diarrhea caused by some types of *Vibrio cholerae* and *E. coli (*gastrointestinal), skin infections in humans, neonatal *E. coli (*septicemia*)* in cattle, sheep and poultry production, *S. aureus (*mastitis*), Salmonella pullorum* (pullorum in poultry*), S. enterica* var. Typhimurium and *Campylobacter jejuni (*gastroenteritis in animals) [91, 99, 100]. The φNK5 lytic phage has also been found to be very effective against *Klebsiella pneumoniae* in rats [47]. Some phages (such as LMP 102, Listex P100) are also used to process and store ready-to-eat meat and poultry-sourced foods [101].

The lytic phages of some pathogens are cultured and used in the treatment of infections, both in humans and food-producing animals [91, 102]. Although it is suggested that this treatment method is effective against systemic infections [103, 104], it has been reported that it is still in the process of development and it is more often used in the treatment of accessible infections such as infections of the paranasal sinuses or the skin [105]. While lytic phages are used in the USA for the treatment of food-derived infections in food animals [106] and for the biocontrol of plant pathogens [107], their use in humans is rather limited in European countries [104]. Today, there are claims that competent authorities in western countries are prohibiting the use of phage therapies in humans [105].

At present, some experimental models with the objective to deliver phages intravenously or intraperitoneally have also been tried and were found to exhibit therapeutic effectiveness. The intravenous delivery route was not found to be ideal since it is impossible to ensure that the bacteriophage solution is completely free of pyrogens [96]. It has also been reported that the delivery of the PAK-P1 phage and the SS-phage intranasally in that order in rats to treat pulmonary infections caused by *Pseudomonas aeruginosa* and *Klebsiella pneumoniae* is relatively less effective than intraperitoneal delivery [108]. Therefore, choosing the delivery route is also an important aspect of phage therapy [109].

It is believed that the use of phages is limited because their effectiveness and specificity generally vary when they encounter antibiotics and this result is hard to explain [93, 110]. To perform phage therapy, the bacterial pathogen must be identified correctly after which its safety must be determined by assessing its susceptibility to current phages *in vitro* [92, 110]. Additionally, perhaps, the most important obstacles are the lack of a specific regulatory framework to evaluate individual therapies and the difficulty in registering patents. It is believed that these uncertainties might cause the drug industry to hesitate from investing in phage development research efforts [110, 111]. Although phages have a promising future as alternatives to antimicrobials, there are a limited number of controlled studies towards their safety and effectiveness [112].

In contrast to antimicrobial treatment, phage therapy targets specific bacteria, which minimizes their effects on untargeted bacteria. Empirical studies claim that phages could also be of the narrow or broad spectrum and that this range effect is mostly related to the phage titer, and makes it important to measure the side effects of phage therapy on commensal bacteria [113]. Another important factor is the potential of the microorganism to become resistant to these specific phages. However, this condition causes a temporary problem for the treatment. Cellular modifications that might occur during phage resistance may also lead to a long-lasting weakness in the host bacteria (change in the microbiota) [114]. The development of resistance can be reduced by using a cocktail containing more than one phage [106]. Although specificity is advantageous in terms of reducing the development of resistance, it is rather difficult to treat different pathogenic subtypes in humans and other animals using one phage [115].

To adapt phage therapy, the effectiveness of specific subunits of the phage needs to be regulated [116]. For example, endolysins produced by phages cause lysis by passing through the cellular wall and breaking the bonds in the peptidoglycan layer. As a group, endoglycans exhibit their effects by targeting at least five different areas in the peptidoglycan matrix [117]. This diversity allows bioengineering to expand the scope of substrates and specificity to species in a single phage therapy. It is being suggested that resistance can be overcome by using genetically modified phages [116]. However, endolysins are ineffective against gram-positive bacteria because their peptidoglycan layer is immune to endolysins and they are only used in the treatment of gram-negative bacteria [118]. It has also been reported that one other type of peptidoglycan hydrolases that includes exolysins also has an additional therapeutic potential [119].

To overcome the obstacles mentioned above, some researchers argue that treatment will be more successful and resistance will be reduced if phages are used in combination with antibiotics instead of being used separately [111, 120]. The main reason for the combination of phage and antibiotics is the perception that the two selective effects will be more effective together than they would be alone. It has been reported that the interaction between the two treatments could be additive, antagonistic or synergistic [121]. Although synergistic effects are preferred for treatment success, it has been expressed that additive effects may also be useful if the immune system of the host is effective [122]. Today, there are examples of *in vivo* and *in vitro* positive interactions between antibiotics and lytic phages in the control of bacterial pathogens. For example, phages combined with antibiotics may eradicate biofilm-forming *Klebsiella pneumoniae* strains and prevent resistant variants from emerging *in vitro*. Such combinations may also stop methicillin-resistant *Staphylococcus aureus* (MRSA) infections in the hind paws of diabetic rats and completely protect broiler chickens from *E. coli* infections [123]. In such cases, when compared to treatment with single antimicrobials, administering antibiotics in combination with a lytic phage not only results in the eradication and better control of the bacteria but also completely prevents the emergence of resistant variants [124, 125]. One study reported that subinhibitory concentrations of the beta-lactam antimicrobial penicillin are effective on phage-resistant *Streptococcus* spp. strains that are rendered sensitive to exogenous phages [126]. A similar observation has been reported for *E. coli*, and this activity is known as phage-antibiotic synergy (PAS). It has also been shown that different beta-lactam antibiotics (cephalosporins) increase the phage lysis of bacteria at subinhibitory concentrations [127]. It is known that subinhibitory concentrations of antibiotics cause DNA damage [128] and that DNA damage induces prophages. This event has been demonstrated *in vitro* for a few different combinations of bacterial pathogens and their phages with various antibiotics such as the combination of T4 phage +cefotaxime against *E. coli* biofilms [129], phage MR-5 and tetracycline, linezolid or ketolide antibiotics against MRSA [130], ceftriaxone and three phages against *P. aeruginosa* [131]. The bactericidal effects of penicillin and gentamicin against *Streptococcus pneumoniae* increase in the presence of phage lysin Clp-1 [132]. In addition to the fact that phage-antibiotic combinations increase the effectiveness against the bacteria targeted, it has also been shown that they could reduce phage and antibiotic resistance [133]. Thus, phage-antibiotic therapy is being presented as a promising approach to control bacterial pathogens and to limit the development of resistance. Even infections caused by multiple-antibiotic-resistant bacteria such as *S. aureus, Burkholderia cenocepacia* and *P. aeruginosa* have been treated successfully using some combinations of antibiotics and phages. However, the mechanisms behind the synergy obtained are still unknown [130, 131, 134, 135].

In conclusion, when compared to conventional antibiotic treatments phage treatments have the following advantages (Table **5**) [47, 97, 109, 135]:

Table 5. Primary advantages and disadvantages of bacteriophages.

Some Bacteriophages	Advantages	Disadvantages	References
B44/1 + B44/2 for ETEC[1] S13 for ETEC[1] AB2 for *Salmonella typhimurium*	-Species-specific -Easily isolated - High tissue permeability - Do not affect eukaryotic cells - Penetrate bacterial biofilms -Non-allergenic - Easily cross the blood brain barrier	-Requisite to identify the microbial agent responsible for the infection -Potential for resistance development -Insufficient clinical trials	[47, 91 - 135]

[1]Enterotoxigenic *Escherichia coli*

- Phage particles are specific to certain types or strains of bacteria depending on the receptor type that recognizes them,
- Phages are natural predators,
- They are widespread in the environment,
- They penetrate into the skin well, hence they can reach chronically infected wounds that are located deeper in the skin,
- They can be used to eradicate pathogenic organisms without affecting commensal bacteria, and also to prevent secondary infections.
- They have no known adverse effect on eukaryotic cells,
- They exponentially grow inside the bacteria and therefore, accumulate in high concentrations as long as the bacteria are in the infected area,
- There is no need for continuous phage administration because the phage particle in the target bacteria will continue to replicate until the bacterial load drops to levels where they are no longer dangerous for the organism in question,
- They can penetrate through bacterial biofilms well,
- They can overcome bacterial resistance,
- Procedures to isolate new bacteriophages are simpler and more economical than developing new antibiotics,
- They offer a safe therapeutic option for patients with allergies to antibiotics,
- Phage particles can easily pass the blood-brain barrier, and they can be used to treat bacterial infections of the central nervous system,

Phage therapy also has some disadvantages [47, 97, 109, 135]:

- It is necessary to identify the infectious agent causing the infection and its specific phage needs to be isolated from the environment,
- When phages are administered systematically, they induce the immune response that leads to antibody production, and this response reduces the effectiveness of antimicrobial treatment,
- Its ideal route of administration, optimal dose, frequency of administration and average treatment duration have to be determined beforehand,
- The new genes of some bacteriophages need to be identified in an organized way,
- Bacteria can become resistant to phages by means of different mechanisms,
- Inadequate purification during phage preparation may manifest as inadequate stability and low viability rates,
- The heterogeneity and species-specific mechanisms of phages have not been explained entirely by scientific authorities yet.

Antimicrobial Peptides

Antimicrobial peptides (AMPs) were first recognized when Alexander Fleming discovered the enzyme lysozyme that has antibacterial effects and is found in the tears and urine of humans. It has been reported that more than 1,700 AMPs are known to date [136]. AMPs are low molecular weight proteins that are secreted continuously or in response to induction from epithelial cells, liver, lymphoid tissue, phagocytes, gills and skin of mammals, plants and insects. They usually contain less than 100 amino acids, and their molecular weight varies between 1 to 5 kDa. AMPs are mostly cationic (due to the presence of amino acids such as cysteine and lysine that are required for them to attach to the membrane), and they may separate into their hydrophobic and hydrophilic domains. AMPs cause cell death by damaging the cell membranes of microorganisms due to their amphipathic nature. Although most AMPs have a cationic structure, some of them, such as maximin-H5 (secreted from amphibians) and dermcidin (secreted from human eccrine sweat glands), may also have an anionic structure rich in aspartic and glutamic acids [109]. AMPs kill microorganisms directly and are also regarded as protective components of the host defense system and the natural immune response [137]. In addition, these peptides stimulate the immune system to act against invading pathogens. It has been shown that AMPs demonstrate their immunomodulatory activity by changing the gene expression of the host, by producing chemokine-like effects and inducing chemokine production, by inhibiting the production of pro-inflammatory cytokines induced by lipopolysaccharides or by supporting wound healing [138]. AMPs carry out the following functions [109]:

- They stimulate the accumulation of immune cells (neutrophils, macrophages, and lymphocytes) in the infected area,
- They neutralize lipopolysaccharide endotoxins produced by gram-negative bacteria,
- They stimulate angiogenesis,
- They act as immunomodulators and control the immune system response to certain microorganisms,
- They have anti-inflammatory characteristics.

It has been reported that AMPs have inhibitory effects on pathogenic organisms, and a broad spectrum of activity including some yeasts and filamentous fungi, and that they might have antifungal, antibacterial (*E. coli, Staphylococcus aureus, Pseudomonas aeruginosa, Mannheimia (Pasteurella) haemolytica, Rhodococcus equi, Listeria* spp., *Salmonella* spp. and *Bacillus* spp.), in fish antiparasitic (against the protozoan ectoparasite *Amyloodinium ocellatum*) effects even at micromolar doses and that they prevent food decomposition [136]. It has also been stated that some AMPs (SMAP29 and OV-1, -2, -3) are also active against some multidrug-resistant bacterial strains (*P. aeruginosa, Burkholderia cepacia, Achromobacter xylosoxidans* and *Stenotrophomonas maltophilia*) [139].

AMPs produce their effects by damaging the cellular membrane, by accumulating inside the cell they can lead to the death of the bacterial cell before reaching minimal effective concentrations. Such AMPs can influence the following processes that are important for cell survival [109, 140, 141]:

- When AMPs reach minimal inhibitor concentration they can inhibit DNA replication or RNA synthesis (*e.g.*, buforin II, pleurocidin and dermaceptin),
- They can inhibit protein synthesis (*e.g.*, indolicidin and PR-39),
- By changing the permeability of ion channels (*e.g.*, cecropin P1, defensins) they attach to chaperone heat shock protein-70 and prevent adequate protein folding,
- They reorganize the synthesis of proteins which lead to the accumulation of inactive proteins that cause cell death (*e.g.*, the highly reactive antibacterial peptide pyrrhocoricin that is isolated from insects),
- They can disrupt or inhibit the transport and energy metabolism of the cytoplasmic membrane [*e.g.*, bactenecins (Bac5 and Bac7)]. Highly positive charged AMPs can inhibit the activity of some anionic cytoplasmic enzymes,
- They can facilitate the aggregation or concentration of intracellular contents (*e.g.*, anionic peptides),
- They can cause apoptosis by inducing the production of oxygen free radicals that are responsible for damage to the DNA, cellular membrane and mitochondria (*e.g.*, papiliocin),
- They can inhibit lipid II transglycosylation which is an important process for the synthesis of peptidoglycan.

In addition, AMPs can exhibit more than one of these mechanisms mentioned above in bacterial infections.

Today it is seen that AMPs are widely used to prevent infections and to strengthen the immune system mostly in livestock and fish [137, 142]. Besides the potential therapeutic effectiveness of AMPs, it has been reported that some of them (especially SMAP29 and sheep cathelicidin) can also be used as biopreservatives in frozen sheep/lamb meat [139, 141]. Despite the many advantages of AMPs until today only a few AMPs have been approved by the FDA and the EMEA and only for topical use in humans. One reason for this lack of approval is high concentrations of AMPs required in the infected area to obtain therapeutic effects in *in vivo* clinical studies. Such high concentrations are close to toxic doses. A point of support is that AMPs can be filtered through the kidneys easily because they are small molecules and, therefore, their half-lives are rather short [143, 144].

In conclusion, although AMPS have stronger effects than some antimicrobials, the high cost of their production and purification and their requirements for preservation and storage after they are produced limit their use (Table **6**).

Although AMPs are drawing increasing attention as alternatives to traditional antimicrobials, some studies have reported that they are not good candidates for treatment due to their toxic effects on mammalian cells and similar bacteriocins have already been developed as a replacement [109].

Table 6. Primary advantages and disadvantages of antimicrobial peptides.

Some Antimicrobial peptids	Advantages	Disadvantages	References
Buforin II Pleurocidin Dermaseptin Sekropin P1 Defensin Pyrrocorocin Bactenecins Papiliocin	-Immunomodulatory -Anti-inflammatory -Bactericide	-Production and purification are expensive -Storage conditions are difficult. -Toxic on eukaryotic cells	[109, 136 - 144]

Bacteriocins

These compounds are defined as peptides generally secreted by bacteria and synthesized by small ribosomes. Bacteriocins act by inserting themselves into the plasma membranes of target bacteria, forming pores and causing lysis. They exert this effect on susceptible microorganisms by disrupting the permeability of the cytoplasmic membrane, inhibiting the synthesis of the cellular membrane and by leading to energy consumption [109]. Bacteria can produce bacteriocins in nearly every generation, and some estimates suggest that at least 99% of all bacteria produce at least one bacteriocin. There are a variety of such compounds that potentially could of benefit with therapeutic purposes [145].

Most commensal bacteria produce endogenous bacteriocins, which offer a great potential for alternatives to antimicrobials [146]. It has been reported that the bacteriocins synthesized by lactic acid bacteria (LAB), called LAB bacteriocins or "lanbiotics", are the bacteriocins most frequently used as biopreservatives in the food industry. These bacteria occur in cheese, yogurt, and other fermented milk products and are generally accepted to be safe. It has been reported that nisin A has bactericidal effects and it is used as a food preservative in more than fifty countries [147]. Some LAB bacteriocins can be used to treat infections caused by *Helicobacter pylori*, *E. coli* and *Salmonella* in the gastrointestinal canal [47]. Another biopreservative widely used in the food industry is the bacteriocin RM6 that is isolated from *Enterococcus faecalis* found in raw milk. This bacteriocin was found to prevent the proliferation of *Listeria monocytogenes*, *Bacillus cereus* and MRSA strains [148].

Besides the antimicrobial activity of bacteriocins, since they are colorless, tasteless, odorless and heat resistant, they can be used in meat and milk products which are processed at high temperatures. For this use, they can also be added to foodstuffs alone or in combination (such as nisin and pediocin) as biopreservatives. They were also found to cause no adverse effects in the host when they were used with other compounds (such as AMPs, prebiotics, proteins, salt, nitrite, organic acid, chelating agents, essential fats) to produce possible synergistic effects. Therefore, bacteriocins are more suitable as a replacement of chemical preservatives in the preservation of normal and fermented foods rather than for the control of pathogenic microorganism [149]. However, since bacteriocins have a peptide structure and they are found in the gastrointestinal canal, they can easily be metabolized into amino acids by proteases [147].

Among bacteriosins, it is suggested that some lantibiotics, such as nisin A and F, mersacidin, mutacin 1140, lactacin 3147 and pediocin AcH/PA-1, that are commonly used in the food industry as biopreservatives can also be effective against MRSA and vancomycin-resistant *Enterococcus* strains and they could potentially be used in the treatment of multiple-drug-resistant and other bacterial infections [147]. For example, nisin F was found to inhibit the bacteria within at least 15 minutes following injection into rats infected with *Staphylococcus aureus* [150]. Also, nisin A could be used as an alternative to antibiotics in women with mastitis caused by *S. aureus* [151]. It has been reported that mersacidin synthesized by *Bacillus* spp. is effective against some strains of MRSA and that it eliminates these bacteria from the nasal mucosa [152].

Similar to phage therapies, an advantage of bacteriocins is their lower potential to elicit the development of resistance than classic antimicrobial agents. For example, in routine use, it has been observed that nearly no resistance develops even with the widespread use of nisin A [153]. This lack of resistance is mainly related to their narrow target spectrum. Another example is the bacteriocin thuricin CD that selectively kills *C. difficile*, and it does not affect the non-pathogenic microbiota around it [154]. It has been shown that resistance to bacteriocins could only develop under *in vitro* conditions. In one study, resistance to bacteriocins was found to develop in *E. coli* and *Listeria monocytogenes* strains when they were exposed to bacteriocins at increasing concentrations over long periods [155]. Therefore, in order to limit the development of resistance, careful and controlled practices should be implemented and, when needed, bacteriocin cocktails should be used just as they are in phage therapies.

Although bacteriocins have the potential to be used as alternatives, the drug industry is unwilling to allocate resources to clinical research and the production of bacteriocin-containing preparations. It has been reported that the main reason for this lack of commitment is the low bacteriocin production efficiency during the fermentation procedure, the production of unstable products, costly and time-consuming purification procedure, and unclarified legislations concerning these products [109].

Similar to phage therapy, it has been shown that bacteriocins also increase the antimicrobial activity of antibiotics. In particular, the synergy between bacteriocins and polymyxin antibiotics is effective against many challenging gram-negative pathogens. Colistin, also known as polymyxin E, has been used widely, but its use has been limited by its neurotoxicity and nephrotoxicity. However, currently, colistin is still known as the last resort treatment for the infections such as *P. aeruginosa* that are multi-drug-resistant [103]. It has also been shown that colistin produces a strong synergy when it is combined with the bacteriocins nisin A and pediocin PA against *E. coli* O157:H7. Additionally, there are many surprising findings that show bacteriocins could reduce the cytotoxicity of colistin under *in vitro* conditions [156]. In combination with other bacteriocins and other antimicrobial peptides, bacteriocins act synergistically. It has been reported that lactic acid bacteria bacteriocins pediocin-PA, sackacin P and curvacin A cannot inhibit the growth of *E. coli* when used alone, but they increase the inhibitory potency of the antimicrobial peptide pleurocidin by four times [157]. Similarly, when the bacteriocin nisin A and the antimicrobial peptide polymyxin B are used in combination they produce a synergy against both gram-positive and gram-negative bacteria (*L. innocua* HPB1 or *E. coli* RR1, respectively) [158]. The combination of various bacteriocins, antimicrobial peptides, and conventional antibiotics have the potential to increase the variety of treatment options and the efficiency against potential pathogens (Table **7**).

Table 7. Primary advantages and disadvantages of bacteriocins.

Some Bacteriocins	Advantages	Disadvantages	References
Nisin A Mersacidin Pediocin PA-1 Plantaricin S Thuricin CD	-Synergistic when used together with antibiotics -Low potential for resistance development	-Production and purification are expensive -Legislation is uncertain	[47, 103, 109, 145 - 160]

Currently, bacteriocins are being genetically modified to enhance their potency, stability and broaden their spectrum of effect, to make their use possible in many different pathological conditions, and to increase their *in vivo* usability. For example, by using the genetic engineering approach, a lysine ring has been added to the N-terminal domain of sakacin 44K and the neutral ring of sakacin T20K has been replaced with a different cationic ring. This approach has enhanced the potency and therapeutic effectiveness of both bacteriocins [159]. In recent years, it has also been reported that some bacteriocins have been isolated from the marine bacteria such as *Vibrio, Pseudoalteromonas, Aeromonas, Alteromonas.* Therefore, the high biodiversity of this ecosystem still to be discovered would, in the near future, be expected to lead to the discovery of many bacteriocins with strong antimicrobial activity and broader spectrum of effect [160].

Predator Bacteria

Predator bacteria offer an option different from the alternative practices mentioned above. Many different types of predator bacteria have been identified. Among those, *Bdellovibrio* and similar organisms (BALOs) were found to be rather promising as alternatives. BALOs are motile Deltaproteobacteria that strictly consume gram-negative bacteria to fulfill their need for energy and nutrients [161]. The genomes of many BALOs code hydrolase enzymes that help in digestion and also affect bacterial biofilms [162]. Since the biofilm-coated bacteria are nearly 1,000 times less susceptible to antimicrobials than those found in planktonic cells, biofilms cause challenges during the treatment of infections in both humans and other animals. Also, because BALOs affect bacterial biofilms, they have a therapeutic advantage as an alternative to antimicrobials [163]. BALOs are particularly useful in controlling diseases caused by complex microbial structures that are difficult to be reached by antimicrobials, such as polymicrobial infections in patients with cystic fibrosis, *Proteus* spp. biofilms in catheters [164] and microorganisms that cause periodontal diseases [165]. Clinical research indicates that BALOs can be effective against multidrug-resistant pathogens such as *Acinetobacter baumannii, E. coli, K. pneumoniae, P. aeruginosa* and *P. putida*, and other persistent infections [166]. Although the clinical use of BALOs to treat persistent infections has not yet become prevalent, when their characteristics and potential targets are taken into consideration further studies are necessary.

Some researchers have proposed that BALOs and other predator organisms can colonize the intestinal canal of mammals and they have amphipathic (both antibiotic and probiotic) effects [161]. For example, in the chicken, it was shown that the oral administration of BALOs reduced the cecal *S. enterica* population and inflammation in an experimental infection induced by *S. enterica* [167]. However, some researchers have stated that BALOs cannot form *in vivo* colonies. It can be seen that the collateral effects of BALOs have still not been investigated in detail in non-targeted bacteria.

BALOs have a unique lipopolysaccharide (LPS) structure that is less endotoxigenic than the lipopolysaccharides of *E. coli.* This structure has a low affinity for the human LPS receptor and produces a limited immune response. In addition, macrophages exposed to BALOs induce the inflammatory cytokines TNF-α and IL-6 at lower rates [161]. Despite this promising preliminary data, more comprehensive studies are required to show the effectiveness of BALOs and their collateral effects on microbial populations and to understand how and when resistance develops.

Fecal Bacteriotherapy

Fecal bacteriotherapy, or with its current name fecal transplantation therapy (FTT), is the procedure of applying feces obtained from healthy donors to patients. It has been reported that FTT can particularly be used in humans to protect the microflora during colonic diseases, in irritable bowel diseases and in the treatment of diarrhea caused by *Clostridium difficile* [168]. FTT is actually an old procedure. It has been used in China since the fourth century to treat intestinal diseases. In the western world, it had been used at the beginning of the seventeenth century to treat rumen acidosis in cattle [169]. Although its mechanics are not understood completely, it is accepted that FTT usually helps patients regain the bacterial diversity that is changed by antibacterial treatment, infections or other factors that cause dysbiosis. For instance, antibacterials that are used to treat recurring *C. difficile* infections may actually increase the recurrence rate. Therefore, FTT is a valuable option of treatment for protection against recurring *C. difficile* infections [169, 170]. However, this treatment method requires donor standardization and measures such as protection against unknown pathogens to succeed [171].

Nanoparticles

The use of nanotechnology in modern medicine has been defined as the greatest engineering innovation of recent times. The fact that nanoparticles increase durability, performance, strengthening and flexibility and their unique physicochemical characteristics have particularly increased the demand for their derivative products by modern medicine ever since they were discovered by the health industry [172]. Nanoparticles (NPs) are widely used in treatment methods such as targeted medicine transport systems, prognostic observation of treatment and the identification of tumors [173, 174]. Meanwhile, it is known that humans continuously exposed to NPs in their work environments may develop unpredictable health problems. The air pollution may also disrupt the ecosystem and negatively impact other biological species in the environment [175].

The use of NPs is an important alternative strategy to antimicrobials, and they have emerged as new antimicrobial agents. NPs usually have a size of 0.2-100 nanometers, and they have a high surface-to-volume ratio [176]. These features increase their interactions with microorganisms and, subsequently, their antimicrobial effects. Transmission electron microscopy (TEM), low-resolution TEM (LRTEM) and high-resolution TEM (HRTEM) technologies help characterize nanoparticles and facilitate their use in various fields. The chemical, electrical, mechanic, optic, magnetic and electrochemical characteristics of NPs differ in bulk material that has high surface-to-volume rations [177]. The physicochemical and biological characteristics of NPs can be manipulated depending on the procedure selected [178]. They may be of the organic or inorganic structure. It has also been shown that they are more resistant to hostile reaction conditions of inorganic NPs [176].

It has been reported that NPs were of use in many areas and especially in the development of new diagnostic methods (imaging), targeted vaccines and therapeutic agents. Currently, there are many commercial NP products suitable for species-specific use in humans and other animals and it has been reported that the effects of these products have been established scientifically [177]. However, it has also been reported that the use of these products has been rather limited due to their undesirable side effects and low bioavailability [172]. To eliminate such negative effects and to improve their therapeutic indexes and safety profiles, different NP medicine transportation systems (such as polymeric, solid fat, inorganic, ceramic, carbon, metallic, nanoparticles, mycelles, nanoemulsions and liposomes) have been used. Currently, it has been reported that among more than 200 products designed for targeted treatments in humans, only about 30 nanoparticles with anti-neoplastic, antibiotic, analgesic, and anti-inflammatory effects have been developed that have been approved for use in clinical trials [176]. Nanoparticles were also found to be effective in some tumoral diseases in cats and dogs (breast adenocarcinoma, oral melanoma, hemangiosarcoma, osteosarcoma or soft tissue sarcoma), in painful conditions in dogs, in leishmaniasis and anemia, in *E. coli* infections and foot-and-mouth diseases in pigs, in brucellosis and leukemia in cattle, in babesiosis and imaging methods in horses, in *Staphylococcus* mastitis and *Fasciola hepatica* infections in sheep, in colibacillosis and *Salmonella* infections in poultry and bursal infections caused by viruses [178]. It can be noted that concerns about the expensive cost, the variability of the sensitivity of NP formulations in different cells, tissues or liquids, and the fact that their species-specific pharmacokinetic and toxicokinetic differences have not yet been entirely established and have significantly limited their use in veterinary practice [172].

OTHER PRACTICES

Although the effects of some metabolites/compounds with biological activity that are used against microbial pathogens as alternatives to current antimicrobials have not yet been established entirely, it is seen that some researchers continue to develop new experimental treatment strategies and to research their usability.

It has been known for some time that "natural products" have been an important starting point for antimicrobial treatments and drug discoveries. Once it was realized that natural products have antibacterial effects, compounds such as tetramate (tetramic acid derivatives), equisetin [179], reutericyclin [180], streptolydigin [181] and new agents such as kibdelomycin [182] have been developed and introduced to the markets. These compounds exhibit wide biological effects by means of their antibacterial characteristics. Among those compounds, reutericyclin, vermisporin, equisetin, BU4514N, delaminomycin C, PF1052, oxacetin and vancoresmycin were found to exhibit antibacterial effects [183]. Products containing tetramate enter the target cell much easier since they create pores allowing the compound to penetrate through the external membrane [184].

Similarly, it has been reported that lactobacilli from natural compounds that are used as probiotics can produce antimicrobial products effective against the pathogen *Candida.* Some of these products are organic acids (lactic acid and acetic acid), hydrogen peroxide (H_2O_2), bacteriocins and antifungal compounds with low molecular weight. Lactobacilli generally produce lactic acid which weakly inhibits the metabolic activity of *Candida* spp [185], while the concentration of H_2O_2 produced is not at levels high enough to be effective against fungi [186].

Lactic acid bacteria produce bacteriocins, which are antimicrobial proteins with a molecular weight of a few thousand daltons or more. Bacteriocin L23 produced from *Lactobacillus fermentum L23* [187], plantaricin produced from *L. plantarum* [188] and *pentocin* TV35b produced from *L. pentosus* [189] were found to be rather effective against the yeast *Candida.* It has also been shown that the low molecular weight substances of lactobacilli such as reuterin [190], reutericyclin [180] and dyacetyl [191] are effective against *Candida* [192]. These antimicrobial substances produced by lactic acid bacteria are also active at low pH levels and are resistant to heat [193]. These antimicrobials also have a broad spectrum of effect. Although they have not yet been presented in clinical trials, currently, the most prominent ones among these substances with defined antimicrobial characteristics are reuterin, reutericyclin, and pyroglutamic acid.

Reuterin, which is obtained from *Lactobacillus reuteri,* is produced by the combination of glucose and glycerol or glyceraldehyde during anaerobic growth of the bacterium. This substance inhibits the enzyme by acting as the substrate-binding subunit of the ribonucleoreductase enzyme due to its strong affinity for sulfhydryl enzymes. Therefore, reuterin exhibits broad spectrum (antibacterial, antifungal, antiprotozoal and antiviral) activity by disrupting DNA synthesis [149]. It has been observed that the clinical use of reuterin is rather limited and that the *L. reuteri* strain is preferred mostly as a probiotic bacterium in poultry, pigs, and turkeys, which reduces the mortality of infections with *Salmonella* spp [194]. The organic compound reutericyclin, which is also a tetramic acid derivative, is produced from a strain of *Lactobacillus reuteri* (LTH2584) and its structure significantly differs from reuterin. Reutericyclin antimicrobial activity is also different than bacteriocins, reuterin and organic acids. In cells, reutericyclins selectively disrupt transmembrane proton potentials. Therefore, by acting as proton ionophores, reutericyclins exhibit bacteriostatic or bactericidal effects against gram-positive bacteria and some food-related pathogens (*Lactobacillus* spp., *Bacillus subtilis*, *B. cereus*, *Enterococcus faecalis*, *Staphylococcus aureus* and *Listeria innocua*) [195 - 197]. It has been shown that the minimal inhibitory concentration of reutericyclin for gram-positive bacteria is 0.05-1 mg/L, where its inhibitory activity was found to increase with high salt concentrations (2%) and low pH values (4.5). Reutericyclin does not affect the growth of gram-negative bacteria and yeast (>100 mg/L) [46], while it exhibits inhibitory effects on the mutant lipopolysaccharide strains of *E. coli*. It has also been shown that reutericyclins dose-dependently induce the lysis of *L. sanfranciscensis* cells and inhibit the germination of *B. subtilis* spores (the transformation of spore into a bacterium), but that the spore is not affected when germination is not allowed [180]. Reutericyclin does not have a wide field of use with its known characteristics. Today, reutericyclin is mostly used to treat staphylococcal skin infections [198] and *Clostridium difficile* in humans [199] and as a biopreservative in foods [195].

Another compound, pyroglutamic acid, has a low molecular weight and it naturally occurs in some vegetables, fruits, and plants. Pyroglutamic acid is also formed in soybean and grain products as a result of fermentation. This organic compound has a 2-pyrrolidone-5-carboxylic acid (PCA) structure, which exhibits more potent antimicrobial activity than the products from lactic acid bacteria. The mechanism of action is similar to that of organic acids. PCA exhibits inhibitory effects, especially against *B. subtilis, Enterobacter cloacae, Pseudomonas putida* and *P. fluorescens*. This compound currently does not have an extensive range of use, but it is suggested to be used in milk production where PCA-producing starter cultures have benefits for the preservation of various foods and increase the sensory quality [200].

The strategy to use the alternative compounds in combination with antimicrobials to strengthen their effects would potentially lead to the preservation of antimicrobials for use by future generations [201]. The benefit of this approach is to prolong the longevity of approved antimicrobials. As time has passed since the discovery of the first antimicrobial substance, and the discovery of resistance, the use of adjuvants has received much importance [202]. New generations of the most commonly used class of antibiotics, beta-lactams, are no longer being developed. Instead, studies have put more focus on the development and discovery of beta-lactamase inhibitors responsible for beta-lactam resistance. Penicillin-type beta-lactams are generally delivered together with beta-lactamase inhibitors, and similar to the use of amoxicillin and clavulanic acid being delivered together [203].

The bacterial efflux pump is generally known to be a great obstacle to treatment with antimicrobial agents. The efflux pump in the bacterial membrane renders antimicrobials and proteins ineffective by pumping them outside the cell. There are a few efflux pump inhibitors that have been approved for clinical use in eukaryotic cells [204]. It has been demonstrated that the efflux inhibitor thioridazine has effectively treated multidrug-resistant *Mycobacterium tuberculosis* in mice [205] and in humans [206], and it has been approved as an antipsychotic. In addition, the inhibition of the efflux pump CmeABC has also been targeted to control the food-derived pathogen *Campylobacter jejuni* in poultry because it contributes to antibiotic resistance and colonization [207].

When discussing the rational use of antimicrobials, vaccines are the most important instruments that are rarely remembered. Preventing many viral and bacterial diseases that occur in humans and other animals by means of vaccination is regarded as a rather successful approach with current technological advancements. For example, in a comprehensive review of weekly reports of some diseases in the USA, it has been estimated that more than 100 million disease cases have been prevented by means of vaccination for smallpox, measles, mumps, rubella, diphtheria, hepatitis A, whooping cough and paralysis [208]. Although there are no effective vaccinations for many of the infectious diseases of humans and other animals, additional research and new vaccination approaches may lead to the discovery of important alternatives to antibiotic therapy. Vaccinations targeting secondary opportunistic pathogens and vaccinations targeting primary infections could reduce the need for antibiotics used to prevent secondary infections [209].

CONCLUSION

Factors such as globalization, climate change, the increasing population density, and the use of land and the management of natural resources play effective roles in the spread and increase in incidences of current pathogens and diseases. On the basis of these outcomes, today's human and other animal health problems can be grouped under three headings: infectious diseases (the risk for a pandemic), more virulent microorganisms (superbugs) and the transportation of current diseases to new areas. The FAO emphasizes that such problems can be overcome by global and innovative health management, natural resource management, increasing the importance given to public health in the food chain and agriculture, performing complete analyses of old and new disease-producing agents, increasing the capacity to identify, diagnose and monitor pathogens particularly those that affect our lives, strongly supporting joint global studies and using technology the best way possible. When all of these factors are taken into consideration, increasing the safety and activity of the current drugs against microbial pathogens with new technological advancements should be one of today's goals. Furthermore, the definition and application of the rational use of antimicrobials are important challenges partially due to empirical use. The empirical use of antimicrobial substances generally limits the identification of the etiological cause of the disease. Diagnostic instruments are rather important for doctors and veterinary surgeons who make decisions about the causative agent of the disease and deciding about the best way to stop it. Although unique diagnostic tools are available for certain types of pathogens, they are insufficient, particularly for less threatening viruses and certain types of microbes. For example, viruses are the cause of many of the acute upper respiratory system infections and can be resolved without any interventions. Despite this restriction and the lack of epidemiological data regarding their benefits in the long term, antibiotics are often used to treat symptoms or secondary bacterial infections. Therefore, advanced diagnostic instruments are expected to increase the treatment rates of diseases and provide information about the most suitable antibiotic selection options.

The antimicrobials that are used today have been exceptionally successful in the management of infections. However, horizontal gene transfer that results in an increase in the number of multiple-drug resistant organisms and acquired mutations, along with a decrease in the discovery of new antimicrobial molecules, have led to undesirable resistance issues. Therefore, alternative approaches are required to preserve the effectiveness of the molecules present today and to maintain the high potency of future discoveries. Within this context, there are many sources such as probiotics, prebiotics, bacteriophages, and bacteriocins that could be alternatives to antimicrobials. However, it must not be forgotten that resistance may also be developed to alternative antimicrobial substances that target a specific pathogen just as it can to antibiotics. Adaptation to new environments over eons may always result in horizontal transfer as a bacterial response to such selective pressure. If used inappropriately, resistance to current therapeutic phages, bacteriocins, predators and other inhibitor compounds will become inevitable. In summing things up, all methods that specifically kill or inhibit the growth of microbes should be used rationally.

CONSENT FOR PUBLICATION

Not applicable.

CONFLICT OF INTEREST

The author (editor) declares no conflict of interest, financial or otherwise.

ACKNOWLEDGMENT

Declared none.

REFERENCES

[1] Filazi A. Antibiotic residues in food of animal origin and evaluating of their risks. Turkiye Klinikleri J Vet Sci 2012; 3(3): 1-7.

[2] Kaya S. Kemoterapotikler.Veteriner Farmakoloji. 5th ed. Ankara: Medisan Press 2013; pp. 323-665.

[3] Sekkin S, Kum C. Antibacterial drugs in fish farms: Application and its effects.Recent Advances in Fish Farms. Croatia: InTech 2011; pp. 217-50.
[http://dx.doi.org/10.5772/26919]

[4] Filazi A, Yurdakok-Dikmen B, Kuzukiran O. Antibiotic resistance in poultry. Turkiye Klinikleri J Vet Sci-Pharmacology and Toxicology-Special Topics 2015; 1(2): 42-51.

[5] Ergin-Kaya S, Filazi A. Determination of antibiotic residues in milk samples. Kafkas Univ Vet Fak Derg 2010; 16(Suppl-A): S31-5.
[http://dx.doi.org/http://dx.doi.org/10.9775/kvfd.2009.1174]

[6] Ince S, Filazi A. Bound residues. Turk vet hekim bir derg 2007; 7: 79-85.

[7] Filazi A, Sireli UT, Cadirci O. Residues of gentamicin in eggs following medication of laying hens. Br Poult Sci 2005; 46(5): 580-3.
[http://dx.doi.org/10.1080/00071660500273243] [PMID: 16359111]

[8] Filazi A, Sireli UT, Yurdakok B, Aydin FG, Kucukosmanoglu AG. Depletion of florfenicol and florfenicol amine residues in chicken eggs. Br Poult Sci 2014; 55(4): 460-5.
[http://dx.doi.org/10.1080/00071668.2014.935701] [PMID: 24945307]

[9] Filazi A, Yurdakok B. Residue problems in milk after antibiotic treatment and tests used for detection of this problem. Turkiye Klinikleri J Vet Sci 2010; 1(1): 34-43.

[10] Kaya S, Yavuz H, Akar F, Liman BC, Filazi A. Antibiotic residues in meat, liver and kidney samples from slaughter cattle. Ankara Univ Vet Fak Derg 1992; 39(1-2): 13-29.

[11] Sen F, Filazi A. Detection of penicillin residues in cow milk using high performance liquid chromatography with UV-diode array detection. J Anim Vet Adv 2014; 13(7): 477-83.

[12] Yikilmaz Y, Filazi A. Detection of florfenicol residues in salmon trout *via* GC–MS. Food Anal Methods 2015; 8(4): 1027-33.
[http://dx.doi.org/10.1007/s12161-014-9982-8]

[13] Sireli UT, Filazi A, Onaran B, Artik N, Ulker H. Residual concerns in meat. Turkiye klinikleri J Food Hyg Technol-Special Topics 2015; 1(2): 7-16.

[14] Sanli Y, Asti R, Yardımcı H, *et al.* The effects of some antibacterial drugs on the immune system of poultry. Turk J Vet Anim Sci 1999; 23(6): 547-55.

[15] Yavuz H, Filazi A, Bilgili A, *et al.* Study of the immuno-toxicological effects of cyfluthrin. Ankara

Univ Vet Fak Derg 1996; 43: 209-13.

[16] Filazi A, Yurdakok-Dikmen B, Kuzukıran O. Poisoning cases originated from environmental pollutants. Turkiye Klinikleri J Vet Sci Pharmacol Toxicol-Special Topics 2015; 1(3): 45-52.

[17] Projan SJ, Shlaes DM. Antibacterial drug discovery: is it all downhill from here? Clin Microbiol Infect 2004; 10 (Suppl. 4): 18-22.
[http://dx.doi.org/10.1111/j.1465-0691.2004.1006.x] [PMID: 15522036]

[18] Laxminarayan R, Duse A, Wattal C, *et al.* Antibiotic resistance-the need for global solutions. Lancet Infect Dis 2013; 13(12): 1057-98.
[http://dx.doi.org/10.1016/S1473-3099(13)70318-9] [PMID: 24252483]

[19] Reardon S. 2015.Dramatic rise seen in antibiotic use http://www.nature.com/news/dramatic-rise-seen-
[http://dx.doi.org/10.1038/nature.2015.18383]

[20] Reardon S. Antibiotic resistance sweeping developing world. Nature 2014; 509(7499): 141-2.
[http://dx.doi.org/10.1038/509141a] [PMID: 24805322]

[21] Ling LL, Schneider T, Peoples AJ, *et al.* A new antibiotic kills pathogens without detectable resistance. Nature 2015; 517(7535): 455-9.
[http://dx.doi.org/10.1038/nature14098] [PMID: 25561178]

[22] Cociancich S, Pesic A, Petras D, *et al.* The gyrase inhibitor albicidin consists of p-aminobenzoic acids and cyanoalanine. Nat Chem Biol 2015; 11(3): 195-7.
[http://dx.doi.org/10.1038/nchembio.1734] [PMID: 25599532]

[23] Baumann S, Herrmann J, Raju R, *et al.* Cystobactamids: myxobacterial topoisomerase inhibitors exhibiting potent antibacterial activity. Angew Chem Int Ed Engl 2014; 53(52): 14605-9.
[http://dx.doi.org/10.1002/anie.201409964] [PMID: 25510965]

[24] World Organization for Animal Health (OIE). Chapter 6. Responsible and prudent use of antimicrobial agents in veterinary medicine. Terrestrial Animal Health Code 2016 .http://www.oie.int/international-standard-setting/terrestrial-code/access-online/ [cited: 24th October 2016].

[25] Nollet L. AGP alternatives-part I. EU close to a future without antibiotic growth promoters. World Poult 2005; 21(6): 14-5.

[26] Conraths FJ, Schwabenbauer K, Vallat B, *et al.* Animal health in the 21st century-a global challenge. Prev Vet Med 2011; 102(2): 93-7.
[http://dx.doi.org/10.1016/j.prevetmed.2011.04.003] [PMID: 21570729]

[27] Callaway TR, Edrington TS, Anderson RC, *et al.* Probiotics, prebiotics and competitive exclusion for prophylaxis against bacterial disease. Anim Health Res Rev 2008; 9(2): 217-25.
[http://dx.doi.org/10.1017/S1466252308001540] [PMID: 19102792]

[28] Ricke SC, Kundinger MM, Miller DR, Keeton JT. Alternatives to antibiotics: chemical and physical antimicrobial interventions and foodborne pathogen response. Poult Sci 2005; 84(4): 667-75.
[http://dx.doi.org/10.1093/ps/84.4.667] [PMID: 15844827]

[29] Tellez G, Pixley C, Wolfenden RE, Layton SL, Hargis BM. Probiotics/direct fed microbials for Salmonella control in poultry. Food Res Int 2012; 45(2): 628-33.
[http://dx.doi.org/10.1016/j.foodres.2011.03.047]

[30] Edens FW. An alternative for antibiotic use in poultry: probiotics. Rev Bras Cienc Avic 2003; 5(2): 1-17.
[http://dx.doi.org/10.1590/S1516-635X2003000200001]

[31] Ley RE, Peterson DA, Gordon JI. Ecological and evolutionary forces shaping microbial diversity in the human intestine. Cell 2006; 124(4): 837-48.
[http://dx.doi.org/10.1016/j.cell.2006.02.017] [PMID: 16497592]

[32] Whitman WB, Coleman DC, Wiebe WJ. Prokaryotes: the unseen majority. Proc Natl Acad Sci USA 1998; 95(12): 6578-83.

[http://dx.doi.org/10.1073/pnas.95.12.6578] [PMID: 9618454]

[33] Guarner F, Malagelada JR. Gut flora in health and disease. Lancet 2003; 361(9356): 512-9. [http://dx.doi.org/10.1016/S0140-6736(03)12489-0] [PMID: 12583961]

[34] Zoetendal EG, Cheng B, Koike S, Mackie RI. Molecular microbial ecology of the gastrointestinal tract: from phylogeny to function. Curr Issues Intest Microbiol 2004; 5(2): 31-47. [PMID: 15460065]

[35] Kau AL, Ahern PP, Griffin NW, Goodman AL, Gordon JI. Human nutrition, the gut microbiome and the immune system. Nature 2011; 474(7351): 327-36. [http://dx.doi.org/10.1038/nature10213] [PMID: 21677749]

[36] McFall-Ngai M, Hadfield MG, Bosch TC, *et al.* Animals in a bacterial world, a new imperative for the life sciences. Proc Natl Acad Sci USA 2013; 110(9): 3229-36. [http://dx.doi.org/10.1073/pnas.1218525110] [PMID: 23391737]

[37] Rashid MU, Weintraub A, Nord CE. Effect of new antimicrobial agents on the ecological balance of human microflora. Anaerobe 2012; 18(2): 249-53. [http://dx.doi.org/10.1016/j.anaerobe.2011.11.005] [PMID: 22155131]

[38] Sullivan A, Edlund C, Nord CE. Effect of antimicrobial agents on the ecological balance of human microflora. Lancet Infect Dis 2001; 1(2): 101-14. [http://dx.doi.org/10.1016/S1473-3099(01)00066-4] [PMID: 11871461]

[39] Allen HK, Trachsel J, Looft T, Casey TA. Finding alternatives to antibiotics. Ann N Y Acad Sci 2014; 1323: 91-100. [http://dx.doi.org/10.1111/nyas.12468] [PMID: 24953233]

[40] Frei R, Akdis M, O'Mahony L. Prebiotics, probiotics, synbiotics, and the immune system: experimental data and clinical evidence. Curr Opin Gastroenterol 2015; 31(2): 153-8. [http://dx.doi.org/10.1097/MOG.0000000000000151] [PMID: 25594887]

[41] Ng KM, Ferreyra JA, Higginbottom SK, *et al.* Microbiota-liberated host sugars facilitate post-antibiotic expansion of enteric pathogens. Nature 2013; 502(7469): 96-9. [http://dx.doi.org/10.1038/nature12503] [PMID: 23995682]

[42] Huyghebaert G, Ducatelle R, Van Immerseel F. An update on alternatives to antimicrobial growth promoters for broilers. Vet J 2011; 187(2): 182-8. [http://dx.doi.org/10.1016/j.tvjl.2010.03.003] [PMID: 20382054]

[43] Joint FAO/WHO Working Group. Report on Drafting Guidelines for the Evaluation of Probiotics in Food London, Ontario, Canada 2002.http://www.who.int/foodsafety/fs_management/en/probiotic_guidelines.pdf?ua=1 [cited: 24th October 2016].

[44] Bourdichon F, Casaregola S, Farrokh C, *et al.* Food fermentations: microorganisms with technological beneficial use. Int J Food Microbiol 2012; 154(3): 87-97. [http://dx.doi.org/10.1016/j.ijfoodmicro.2011.12.030] [PMID: 22257932]

[45] Fijan S. Microorganisms with claimed probiotic properties: an overview of recent literature. Int J Environ Res Public Health 2014; 11(5): 4745-67. [http://dx.doi.org/10.3390/ijerph110504745] [PMID: 24859749]

[46] Ouwehand AC, Vesterlund S. Antimicrobial components from lactic acid bacteria.Lactic acid bacteria microbiological and functional aspects. 3rd ed. New York: Marcel Dekker Inc 2004; pp. 375-98. [http://dx.doi.org/10.1201/9780824752033.ch11]

[47] Kum C, Sekkin S. Alternative approaches to antibiotics. Turkiye Klinikleri J Vet Sci 2012; 3(3): 84-116.

[48] Musa HH, Wu SL, Zhu CH, Seri HI, Zhu GQ. The potential benefits of probiotics in animal production and health. J Anim Vet Adv 2009; 8(2): 313-21.

[49] Balcázar JL, de Blas I, Ruiz-Zarzuela I, Cunningham D, Vendrell D, Múzquiz JL. The role of

probiotics in aquaculture. Vet Microbiol 2006; 114(3-4): 173-86. [http://dx.doi.org/10.1016/j.vetmic.2006.01.009] [PMID: 16490324]

[50] Armut M, Filazi A. Evaluation of the effects produced by the addition of growth- promoting products to broiler feed. Turk J Vet Anim Sci 2012; 36(4): 330-7.

[51] Mikelsaar M. Human microbial ecology: lactobacilli, probiotics, selective decontamination. Anaerobe 2011; 17(6): 463-7. [http://dx.doi.org/10.1016/j.anaerobe.2011.07.005] [PMID: 21787875]

[52] Arumugam M, Raes J, Pelletier E, *et al.* MetaHIT Consortium. Enterotypes of the human gut microbiome. Nature 2011; 473(7346): 174-80. [http://dx.doi.org/10.1038/nature09944] [PMID: 21508958]

[53] Fitz-Gibbon S, Tomida S, Chiu BH, *et al.* Propionibacterium acnes strain populations in the human skin microbiome associated with acne. J Invest Dermatol 2013; 133(9): 2152-60. [http://dx.doi.org/10.1038/jid.2013.21] [PMID: 23337890]

[54] Nawaz M, Wang J, Zhou A, *et al.* Characterization and transfer of antibiotic resistance in lactic acid bacteria from fermented food products. Curr Microbiol 2011; 62(3): 1081-9. [http://dx.doi.org/10.1007/s00284-010-9856-2] [PMID: 21212956]

[55] EFSA Panel on Additives and Products or Substances used in Animal Feed (FEEDAP). Guidance on the assessment of bacterial susceptibility to antimicrobials of human and veterinary importance. EFSA J 2012; 10(6): 2740.

[56] Mathur S, Singh R. Antibiotic resistance in food lactic acid bacteria--a review. Int J Food Microbiol 2005; 105(3): 281-95. [http://dx.doi.org/10.1016/j.ijfoodmicro.2005.03.008] [PMID: 16289406]

[57] Ouoba LI, Lei V, Jensen LB. Resistance of potential probiotic lactic acid bacteria and bifidobacteria of African and European origin to antimicrobials: determination and transferability of the resistance genes to other bacteria. Int J Food Microbiol 2008; 121(2): 217-24. [http://dx.doi.org/10.1016/j.ijfoodmicro.2007.11.018] [PMID: 18063151]

[58] Klare I, Konstabel C, Werner G, *et al.* Antimicrobial susceptibilities of Lactobacillus, Pediococcus and Lactococcus human isolates and cultures intended for probiotic or nutritional use. J Antimicrob Chemother 2007; 59(5): 900-12. [http://dx.doi.org/10.1093/jac/dkm035] [PMID: 17369278]

[59] Marshall BM, Ochieng DJ, Levy SB. Commensals: under-appreciated reservoir of antibiotic resistance. Microbes 2009; 4: 231-8.

[60] Gibson GR, Roberfroid MB. Dietary modulation of the human colonic microbiota: introducing the concept of prebiotics. J Nutr 1995; 125(6): 1401-12. [PMID: 7782892]

[61] Roberfroid M, Gibson GR, Hoyles L, *et al.* Prebiotic effects: metabolic and health benefits. Br J Nutr 2010; 104 (Suppl. 2): S1-S63. [http://dx.doi.org/10.1017/S0007114510003363] [PMID: 20920376]

[62] Fernández J, Redondo-Blanco S, Gutiérrez-del-Río I, Miguélez EM, Villar CJ, Lombó F. Colon microbiota fermentation of dietary prebiotics towards short-chain fatty acids and their roles as anti-inflammatory and antitumour agents: A review. J Funct Foods 2016; 25: 511-22. [http://dx.doi.org/10.1016/j.jff.2016.06.032]

[63] Meyer D. Health benefits of prebiotic fibers. Adv Food Nutr Res 2015; 74: 47-91. [http://dx.doi.org/10.1016/bs.afnr.2014.11.002] [PMID: 25624035]

[64] Vijn I, Smeekens S. Fructan: more than a reserve carbohydrate? Plant Physiol 1999; 120(2): 351-60. [http://dx.doi.org/10.1104/pp.120.2.351] [PMID: 10364386]

[65] Quirós-Sauceda AE, Palafox-Carlos H, Sáyago-Ayerdi SG, *et al.* Dietary fiber and phenolic

compounds as functional ingredients: interaction and possible effect after ingestion. Food Funct 2014; 5(6): 1063-72.
[http://dx.doi.org/10.1039/C4FO00073K] [PMID: 24740575]

[66] Anadón A, Martínez-Larrañaga MR, Caballero V, Castellano V. Chapter 2. Assessment of prebiotics and probiotics: An overview.Bioactive foods in promoting health: probiotics and prebiotics. 1st ed. London: Academic Press 2010; pp. 19-41.
[http://dx.doi.org/10.1016/B978-0-12-374938-3.00002-5]

[67] Kolida S, Gibson GR. Synbiotics in health and disease. Annu Rev Food Sci Technol 2011; 2: 373-93.
[http://dx.doi.org/10.1146/annurev-food-022510-133739] [PMID: 22129388]

[68] Gibson PR, Barrett JS. The concept of small intestinal bacterial overgrowth in relation to functional gastrointestinal disorders. Nutrition 2010; 26(11-12): 1038-43.
[http://dx.doi.org/10.1016/j.nut.2010.01.005] [PMID: 20418060]

[69] Chi ZM, Zhang T, Cao TS, Liu XY, Cui W, Zhao CH. Biotechnological potential of inulin for bioprocesses. Bioresour Technol 2011; 102(6): 4295-303.
[http://dx.doi.org/10.1016/j.biortech.2010.12.086] [PMID: 21247760]

[70] Borromei C, Careri M, Cavazza A, *et al.* Evaluation of fructooligosaccharides and inulins as potentially health benefiting food ingredients by HPAEC-PED and MALDI-TOF MS. Int J Anal Chem 2009; 530639.
[http://dx.doi.org/10.1155/2009/530639]

[71] Padilla B, Frau F, Ruiz-Matute AI, *et al.* Production of lactulose oligosaccharides by isomerisation of transgalactosylated cheese whey permeate obtained by β-galactosidases from dairy Kluyveromyces. J Dairy Res 2015; 82(3): 356-64.
[http://dx.doi.org/10.1017/S0022029915000217] [PMID: 26004434]

[72] Sangwan V, Tomar SK, Ali B, Singh RR, Singh AK. Production of β-galactosidase from streptococcus thermophilus for galactooligosaccharides synthesis. J Food Sci Technol 2015; 52(7): 4206-15.
[http://dx.doi.org/10.1007/s13197-014-1486-4] [PMID: 26139885]

[73] Ballard O, Morrow AL. Human milk composition: nutrients and bioactive factors. Pediatr Clin North Am 2013; 60(1): 49-74.
[http://dx.doi.org/10.1016/j.pcl.2012.10.002] [PMID: 23178060]

[74] Ruiz-Moyano S, Totten SM, Garrido DA, *et al.* Variation in consumption of human milk oligosaccharides by infant gut-associated strains of Bifidobacterium breve. Appl Environ Microbiol 2013; 79(19): 6040-9.
[http://dx.doi.org/10.1128/AEM.01843-13] [PMID: 23892749]

[75] Brownawell AM, Caers W, Gibson GR, *et al.* Prebiotics and the health benefits of fiber: current regulatory status, future research, and goals. J Nutr 2012; 142(5): 962-74.
[http://dx.doi.org/10.3945/jn.112.158147] [PMID: 22457389]

[76] Marteau P, Boutron-Ruault MC. Nutritional advantages of probiotics and prebiotics. Br J Nutr 2002; 87 (Suppl. 2): S153-7.
[http://dx.doi.org/10.1079/BJN2002531] [PMID: 12088512]

[77] Guclu BK. Effects of probiotic and prebiotic (mannanoligosaccharide) supplementation on performance, egg quality and hatchability in quail breeders. Ankara Univ Vet Fak Derg 2011; 58(1): 27-32.
[http://dx.doi.org/10.1501/Vetfak_0000002445]

[78] Cerezuela R, Meseguer J, Esteban MA. Current knowledge in synbiotic use for fish aquaculture: A review. J Aquac Res Development 2011; S1: 008.
[http://dx.doi.org/http://dx.doi.org/10.4172/2155-9546.S1-008]

[79] Iannitti T, Palmieri B. Therapeutical use of probiotic formulations in clinical practice. Clin Nutr 2010; 29(6): 701-25.

[http://dx.doi.org/10.1016/j.clnu.2010.05.004] [PMID: 20576332]

[80] Fleige S, Preißinger W, Meyer HH, Pfaffl MW. Effect of lactulose on growth performance and intestinal morphology of pre-ruminant calves using a milk replacer containing Enterococcus faecium. Animal 2007; 1(3): 367-73.
[http://dx.doi.org/10.1017/S1751731107661850] [PMID: 22444334]

[81] Yasuda K, Hashikawa S, Sakamoto H, Tomita Y, Shibata S, Fukata T. A new synbiotic consisting of Lactobacillus casei subsp. casei and dextran improves milk production in Holstein dairy cows. J Vet Med Sci 2007; 69(2): 205-8.
[http://dx.doi.org/10.1292/jvms.69.205] [PMID: 17339767]

[82] Roberfroid MB. Prebiotics and probiotics: are they functional foods? Am J Clin Nutr 2000; 71(6) (Suppl.): 1682S-7S.
[PMID: 10837317]

[83] Slominski BA. Recent advances in research on enzymes for poultry diets. Poult Sci 2011; 90(9): 2013-23.
[http://dx.doi.org/10.3382/ps.2011-01372] [PMID: 21844268]

[84] EU (European Union). Regulation (EC) No: 1831/2003 of the European Parliament and of the Council of 22 September 2003 on additives for use in animal nutrition OJ L 2003 ; 268: 29-43. http://eur-lex.europa.eu/legal-content/EN/TXT/?uri=CELEX%3A32

[85] Report of the Joint Expert Advisory Committee on Antibiotic Resistance (JETACAR).. The use of antibiotics in food producing animals: antibiotic resistant bacteria in animals and humans. Commonwealth Department of Agriculture, Fisheries and Forestry-Australia 1999. http://www.health.gov.au/internet/main/publishing.nsf/Content/health-pubs-

[86] Dibner JJ, Buttin P. Use of organic acids as a model to study the impact of gut microflora on nutrition and metabolism. J Appl Poult Res 2002; 11(4): 453-63.
[http://dx.doi.org/10.1093/japr/11.4.453]

[87] Kum S, Eren U, Onol AG, Sandikci M. Effects of dietary organic acid supplementation on the intestinal mucosa in broilers. Rev Med Vet 2010; 161(10): 463-8.

[88] Ricke SC. Perspectives on the use of organic acids and short chain fatty acids as antimicrobials. Poult Sci 2003; 82(4): 632-9.
[http://dx.doi.org/10.1093/ps/82.4.632] [PMID: 12710485]

[89] Theron MM, Lues JF. Organic acids and food preservation. Boca Raton: CRC Press 2011; pp. 1-342.

[90] Hsiao CP, Siebert KJ. Modeling the inhibitory effects of organic acids on bacteria. Int J Food Microbiol 1999; 47(3): 189-201.
[http://dx.doi.org/10.1016/S0168-1605(99)00012-4] [PMID: 10359489]

[91] Johnson RP, Gyles CL, Huff WE, *et al.* Bacteriophages for prophylaxis and therapy in cattle, poultry and pigs. Anim Health Res Rev 2008; 9(2): 201-15.
[http://dx.doi.org/10.1017/S1466252308001576] [PMID: 19102791]

[92] Kutateladze M, Adamia R. Bacteriophages as potential new therapeutics to replace or supplement antibiotics. Trends Biotechnol 2010; 28(12): 591-5.
[http://dx.doi.org/10.1016/j.tibtech.2010.08.001] [PMID: 20810181]

[93] Chanishvili N. Phage therapy--history from Twort and d'Herelle through Soviet experience to current approaches. Adv Virus Res 2012; 83: 3-40.
[http://dx.doi.org/10.1016/B978-0-12-394438-2.00001-3] [PMID: 22748807]

[94] Endersen L, O'Mahony J, Hill C, Ross RP, McAuliffe O, Coffey A. Phage therapy in the food industry. Annu Rev Food Sci Technol 2014; 5: 327-49.
[http://dx.doi.org/10.1146/annurev-food-030713-092415] [PMID: 24422588]

[95] Oliveira M, Abadias M, Colás-Medà P, Usall J, Viñas I. Biopreservative methods to control the

growth of foodborne pathogens on fresh-cut lettuce. Int J Food Microbiol 2015; 214: 4-11. [http://dx.doi.org/10.1016/j.ijfoodmicro.2015.07.015] [PMID: 26210531]

[96] Maura D, Debarbieux L. Bacteriophages as twenty-first century antibacterial tools for food and medicine. Appl Microbiol Biotechnol 2011; 90(3): 851-9. [http://dx.doi.org/10.1007/s00253-011-3227-1] [PMID: 21491205]

[97] Wittebole X, De Roock S, Opal SM. A historical overview of bacteriophage therapy as an alternative to antibiotics for the treatment of bacterial pathogens. Virulence 2014; 5(1): 226-35. [http://dx.doi.org/10.4161/viru.25991] [PMID: 23973944]

[98] Lu TK, Collins JJ. Engineered bacteriophage targeting gene networks as adjuvants for antibiotic therapy. Proc Natl Acad Sci USA 2009; 106(12): 4629-34. [http://dx.doi.org/10.1073/pnas.0800442106] [PMID: 19255432]

[99] Rodríguez-Rubio L, Martínez B, Rodríguez A, Donovan DM, Götz F, García P. The phage lytic proteins from the Staphylococcus aureus bacteriophage vB_SauS-phiIPLA88 display multiple active catalytic domains and do not trigger staphylococcal resistance. PLoS One 2013; 8(5): e64671. [http://dx.doi.org/10.1371/journal.pone.0064671] [PMID: 23724076]

[100] Sørensen MC, Gencay YE, Birk T, *et al.* Primary isolation strain determines both phage type and receptors recognised by Campylobacter jejuni bacteriophages. PLoS One 2015; 10(1): e0116287. [http://dx.doi.org/10.1371/journal.pone.0116287] [PMID: 25585385]

[101] Lang LH. FDA approves use of bacteriophages to be added to meat and poultry products. Gastroenterology 2006; 131(5): 1370. [PMID: 17067600]

[102] Abedon ST, Kuhl SJ, Blasdel BG, Kutter EM. Phage treatment of human infections. Bacteriophage 2011; 1(2): 66-85. [http://dx.doi.org/10.4161/bact.1.2.15845] [PMID: 22334863]

[103] Biswas S, Brunel JM, Dubus JC, Reynaud-Gaubert M, Rolain JM. Colistin: an update on the antibiotic of the 21st century. Expert rev Anti infect Ther 2012; 10(8): 917-34. [http://dx.doi.org/10.1586/eri.12.78] [PMID: 23030331]

[104] Międzybrodzki R, Borysowski J, Weber-Dąbrowska B, *et al.* Clinical aspects of phage therapy. Adv Virus Res 2012; 83: 73-121. [http://dx.doi.org/10.1016/B978-0-12-394438-2.00003-7] [PMID: 22748809]

[105] Chan BK, Abedon ST, Loc-Carrillo C. Phage cocktails and the future of phage therapy. Future Microbiol 2013; 8(6): 769-83. [http://dx.doi.org/10.2217/fmb.13.47] [PMID: 23701332]

[106] Goodridge LD, Bisha B. Phage-based biocontrol strategies to reduce foodborne pathogens in foods. Bacteriophage 2011; 1(3): 130-7. [http://dx.doi.org/10.4161/bact.1.3.17629] [PMID: 22164346]

[107] Balogh B, Jones JB, Iriarte FB, Momol MT. Phage therapy for plant disease control. Curr Pharm Biotechnol 2010; 11(1): 48-57. [http://dx.doi.org/10.2174/138920110790725302] [PMID: 20214607]

[108] Carmody LA, Gill JJ, Summer EJ, *et al.* Efficacy of bacteriophage therapy in a model of Burkholderia cenocepacia pulmonary infection. J Infect Dis 2010; 201(2): 264-71. [http://dx.doi.org/10.1086/649227] [PMID: 20001604]

[109] Rios AC, Moutinho CG, Pinto FC, *et al.* Alternatives to overcoming bacterial resistances: State-of-t-e-art. Microbiol Res 2016; 191: 51-80. [http://dx.doi.org/10.1016/j.micres.2016.04.008] [PMID: 27524653]

[110] Young R, Gill JJ. MICROBIOLOGY. Phage therapy redux--What is to be done? Science 2015; 350(6265): 1163-4. [http://dx.doi.org/10.1126/science.aad6791] [PMID: 26785457]

[111] Nobrega FL, Costa AR, Kluskens LD, Azeredo J. Revisiting phage therapy: new applications for old resources. Trends Microbiol 2015; 23(4): 185-91.
[http://dx.doi.org/10.1016/j.tim.2015.01.006] [PMID: 25708933]

[112] Kingwell K. Bacteriophage therapies re-enter clinical trials. Nat Rev Drug Discov 2015; 14(8): 515-6.
[http://dx.doi.org/10.1038/nrd4695] [PMID: 26228748]

[113] Koskella B, Meaden S. Understanding bacteriophage specificity in natural microbial communities. Viruses 2013; 5(3): 806-23.
[http://dx.doi.org/10.3390/v5030806] [PMID: 23478639]

[114] Levin BR, Bull JJ. Population and evolutionary dynamics of phage therapy. Nat Rev Microbiol 2004; 2(2): 166-73.
[http://dx.doi.org/10.1038/nrmicro822] [PMID: 15040264]

[115] Pirnay JP, De Vos D, Verbeken G, *et al.* The phage therapy paradigm: prêt-à-porter or sur-mesure? Pharm Res 2011; 28(4): 934-7.
[http://dx.doi.org/10.1007/s11095-010-0313-5] [PMID: 21063753]

[116] Fischetti VA. Bacteriophage lysins as effective antibacterials. Curr Opin Microbiol 2008; 11(5): 393-400.
[http://dx.doi.org/10.1016/j.mib.2008.09.012] [PMID: 18824123]

[117] Schmelcher M, Donovan DM, Loessner MJ. Bacteriophage endolysins as novel antimicrobials. Future Microbiol 2012; 7(10): 1147-71.
[http://dx.doi.org/10.2217/fmb.12.97] [PMID: 23030422]

[118] Walmagh M, Boczkowska B, Grymonprez B, Briers Y, Drulis-Kawa Z, Lavigne R. Characterization of five novel endolysins from Gram-negative infecting bacteriophages. Appl Microbiol Biotechnol 2013; 97(10): 4369-75.
[http://dx.doi.org/10.1007/s00253-012-4294-7] [PMID: 22832988]

[119] Vollmer W, Joris B, Charlier P, Foster S. Bacterial peptidoglycan (murein) hydrolases. FEMS Microbiol Rev 2008; 32(2): 259-86.
[http://dx.doi.org/10.1111/j.1574-6976.2007.00099.x] [PMID: 18266855]

[120] Pelfrene E, Willebrand E, Cavaleiro Sanches A, Sebris Z, Cavaleri M. Bacteriophage therapy: a regulatory perspective. J Antimicrob Chemother 2016; 71(8): 2071-4.
[http://dx.doi.org/10.1093/jac/dkw083] [PMID: 27068400]

[121] Crain CM, Kroeker K, Halpern BS. Interactive and cumulative effects of multiple human stressors in marine systems. Ecol Lett 2008; 11(12): 1304-15.
[http://dx.doi.org/10.1111/j.1461-0248.2008.01253.x] [PMID: 19046359]

[122] Lebeis SL, Kalman D. Aligning antimicrobial drug discovery with complex and redundant host-pathogen interactions. Cell Host Microbe 2009; 5(2): 114-22.
[http://dx.doi.org/10.1016/j.chom.2009.01.008] [PMID: 19218083]

[123] Verma V, Harjai K, Chhibber S. Restricting ciprofloxacin-induced resistant variant formation in biofilm of Klebsiella pneumoniae B5055 by complementary bacteriophage treatment. J Antimicrob Chemother 2009; 64(6): 1212-8.
[http://dx.doi.org/10.1093/jac/dkp360] [PMID: 19808232]

[124] Lehman SM, Donlan RM. Bacteriophage-mediated control of a two-species biofilm formed by microorganisms causing catheter-associated urinary tract infections in an *in vitro* urinary catheter model. Antimicrob Agents Chemother 2015; 59(2): 1127-37.
[http://dx.doi.org/10.1128/AAC.03786-14] [PMID: 25487795]

[125] McVicker G, Prajsnar TK, Williams A, *et al.* Clonal expansion during Staphylococcus aureus infection dynamics reveals the effect of antibiotic intervention. PLoS Pathog 2014; 10(2): e1003959.
[http://dx.doi.org/10.1371/journal.ppat.1003959] [PMID: 24586163]

[126] Verhue WM. Interaction of bacteriophage infection and low penicillin concentrations on the performance of yogurt cultures. Appl Environ Microbiol 1978; 35(6): 1145-9. [PMID: 16345294]

[127] Comeau AM, Tétart F, Trojet SN, Prère MF, Krisch HM. Phage-Antibiotic Synergy (PAS): beta-lactam and quinolone antibiotics stimulate virulent phage growth. PLoS One 2007; 2(8): e799. [http://dx.doi.org/10.1371/journal.pone.0000799] [PMID: 17726529]

[128] Gutierrez A, Laureti L, Crussard S, *et al.* β-Lactam antibiotics promote bacterial mutagenesis *via* an RpoS-mediated reduction in replication fidelity. Nat Commun 2013; 4: 1610. [http://dx.doi.org/10.1038/ncomms2607] [PMID: 23511474]

[129] Ryan EM, Alkawareek MY, Donnelly RF, Gilmore BF. Synergistic phage-antibiotic combinations for the control of Escherichia coli biofilms *in vitro*. FEMS Immunol Med Microbiol 2012; 65(2): 395-8. [http://dx.doi.org/10.1111/j.1574-695X.2012.00977.x] [PMID: 22524448]

[130] Kaur S, Harjai K, Chhibber S. Methicillin-resistant Staphylococcus aureus phage plaque size enhancement using sublethal concentrations of antibiotics. Appl Environ Microbiol 2012; 78(23): 8227-33. [http://dx.doi.org/10.1128/AEM.02371-12] [PMID: 23001655]

[131] Knezevic P, Curcin S, Aleksic V, Petrusic M, Vlaski L. Phage-antibiotic synergism: a possible approach to combatting Pseudomonas aeruginosa. Res Microbiol 2013; 164(1): 55-60. [http://dx.doi.org/10.1016/j.resmic.2012.08.008] [PMID: 23000091]

[132] Djurkovic S, Loeffler JM, Fischetti VA. Synergistic killing of Streptococcus pneumoniae with the bacteriophage lytic enzyme Cpl-1 and penicillin or gentamicin depends on the level of penicillin resistance. Antimicrob Agents Chemother 2005; 49(3): 1225-8. [http://dx.doi.org/10.1128/AAC.49.3.1225-1228.2005] [PMID: 15728935]

[133] Zhang QG, Buckling A. Phages limit the evolution of bacterial antibiotic resistance in experimental microcosms. Evol Appl 2012; 5(6): 575-82. [http://dx.doi.org/10.1111/j.1752-4571.2011.00236.x] [PMID: 23028398]

[134] Kamal F, Dennis JJ. Burkholderia cepacia complex Phage-Antibiotic Synergy (PAS): antibiotics stimulate lytic phage activity. Appl Environ Microbiol 2015; 81(3): 1132-8. [http://dx.doi.org/10.1128/AEM.02850-14] [PMID: 25452284]

[135] Torres-Barceló C, Hochberg ME. Evolutionary rationale for phages as complements of antibiotics. Trends Microbiol 2016; 24(4): 249-56. [http://dx.doi.org/10.1016/j.tim.2015.12.011] [PMID: 26786863]

[136] Bruhn O, Grötzinger J, Cascorbi I, Jung S. Antimicrobial peptides and proteins of the horse--insights into a well-armed organism. Vet Res (Faisalabad) 2011; 42: 98. [http://dx.doi.org/10.1186/1297-9716-42-98] [PMID: 21888650]

[137] Cheema UB, Younas M, Sultan JI, Iqbal A, Tariq M, Waheed A. Antimicrobial peptides: An alternative of antibiotics in ruminants. Adv Agric Biotechnol 2011; 2: 15-21.

[138] Uppu DS, Ghosh C, Haldar J. Surviving sepsis in the era of antibiotic resistance: are there any alternative approaches to antibiotic therapy? Microb Pathog 2015; 80: 7-13. [http://dx.doi.org/10.1016/j.micpath.2015.02.001] [PMID: 25677832]

[139] Brogden KA, Ackermann M, McCray PB Jr, Tack BF. Antimicrobial peptides in animals and their role in host defences. Int J Antimicrob Agents 2003; 22(5): 465-78. [http://dx.doi.org/10.1016/S0924-8579(03)00180-8] [PMID: 14602364]

[140] Brogden KA. Antimicrobial peptides: pore formers or metabolic inhibitors in bacteria? Nat Rev Microbiol 2005; 3(3): 238-50. [http://dx.doi.org/10.1038/nrmicro1098] [PMID: 15703760]

[141] Brogden NK, Brogden KA. Will new generations of modified antimicrobial peptides improve their

potential as pharmaceuticals? Int J Antimicrob Agents 2011; 38(3): 217-25. [PMID: 21733662]

[142] Bruhn O, Cauchard J, Schlusselhuber M, *et al.* Antimicrobial properties of the equine alpha-defensin DEFA1 against bacterial horse pathogens. Vet Immunol Immunopathol 2009; 130(1-2): 102-6. [http://dx.doi.org/10.1016/j.vetimm.2009.01.005] [PMID: 19211153]

[143] Park SC, Park Y, Hahm KS. The role of antimicrobial peptides in preventing multidrug-resistant bacterial infections and biofilm formation. Int J Mol Sci 2011; 12(9): 5971-92. [http://dx.doi.org/10.3390/ijms12095971] [PMID: 22016639]

[144] Yeung AT, Gellatly SL, Hancock RE. Multifunctional cationic host defence peptides and their clinical applications. Cell Mol Life Sci 2011; 68(13): 2161-76. [http://dx.doi.org/10.1007/s00018-011-0710-x] [PMID: 21573784]

[145] Snyder AB, Worobo RW. Chemical and genetic characterization of bacteriocins: antimicrobial peptides for food safety. J Sci Food Agric 2014; 94(1): 28-44. [http://dx.doi.org/10.1002/jsfa.6293] [PMID: 23818338]

[146] Cotter PD, Ross RP, Hill C. Bacteriocins - a viable alternative to antibiotics? Nat Rev Microbiol 2013; 11(2): 95-105. [http://dx.doi.org/10.1038/nrmicro2937] [PMID: 23268227]

[147] Nishie M, Nagao J, Sonomoto K. Antibacterial peptides "bacteriocins": an overview of their diverse characteristics and applications. Biocontrol Sci 2012; 17(1): 1-16. [http://dx.doi.org/10.4265/bio.17.1] [PMID: 22451427]

[148] Huang E, Zhang L, Chung YK, Zheng Z, Yousef AE. Characterization and application of enterocin RM6, a bacteriocin from Enterococcus faecalis. Biomed Res Int 2013; 206917. [http://dx.doi.org/http://dx.doi.org/10.1155/2013/206917]

[149] Dicks LM, Botes M. Probiotic lactic acid bacteria in the gastro-intestinal tract: health benefits, safety and mode of action. Benef Microbes 2010; 1(1): 11-29. [http://dx.doi.org/10.3920/BM2009.0012] [PMID: 21831747]

[150] Brand AM, de Kwaadsteniet M, Dicks LM. The ability of nisin F to control Staphylococcus aureus infection in the peritoneal cavity, as studied in mice. Lett Appl Microbiol 2010; 51(6): 645-9. [http://dx.doi.org/10.1111/j.1472-765X.2010.02948.x] [PMID: 21029139]

[151] Fernández L, Delgado S, Herrero H, Maldonado A, Rodríguez JM. The bacteriocin nisin, an effective agent for the treatment of staphylococcal mastitis during lactation. J Hum Lact 2008; 24(3): 311-6. [http://dx.doi.org/10.1177/0890334408317435] [PMID: 18689718]

[152] Kruszewska D, Sahl HG, Bierbaum G, Pag U, Hynes SO, Ljungh A. Mersacidin eradicates methicillin-resistant Staphylococcus aureus (MRSA) in a mouse rhinitis model. J Antimicrob Chemother 2004; 54(3): 648-53. [http://dx.doi.org/10.1093/jac/dkh387] [PMID: 15282239]

[153] Zendo T. Screening and characterization of novel bacteriocins from lactic acid bacteria. Biosci Biotechnol Biochem 2013; 77(5): 893-9. [http://dx.doi.org/10.1271/bbb.130014] [PMID: 23649268]

[154] Rea MC, Alemayehu D, Ross RP, Hill C. Gut solutions to a gut problem: bacteriocins, probiotics and bacteriophage for control of Clostridium difficile infection. J Med Microbiol 2013; 62(Pt 9): 1369-78. [http://dx.doi.org/10.1099/jmm.0.058933-0] [PMID: 23699066]

[155] Naghmouchi K, Belguesmia Y, Baah J, Teather R, Drider D. Antibacterial activity of class I and IIa bacteriocins combined with polymyxin E against resistant variants of Listeria monocytogenes and Escherichia coli. Res Microbiol 2011; 162(2): 99-107. [http://dx.doi.org/10.1016/j.resmic.2010.09.014] [PMID: 20868743]

[156] Naghmouchi K, Baah J, Hober D, *et al.* Synergistic effect between colistin and bacteriocins in controlling Gram-negative pathogens and their potential to reduce antibiotic toxicity in mammalian

epithelial cells. Antimicrob Agents Chemother 2013; 57(6): 2719-25. [http://dx.doi.org/10.1128/AAC.02328-12] [PMID: 23571533]

[157] Lüders T, Birkemo GA, Fimland G, Nissen-Meyer J, Nes IF. Strong synergy between a eukaryotic antimicrobial peptide and bacteriocins from lactic acid bacteria. Appl Environ Microbiol 2003; 69(3): 1797-9. [http://dx.doi.org/10.1128/AEM.69.3.1797-1799.2003] [PMID: 12620872]

[158] Naghmouchi K, Drider D, Baah J, Teather R. Nisin A and Polymyxin B as Synergistic Inhibitors of Gram-positive and Gram-negative Bacteria. Probiotics Antimicrob Proteins 2010; 2(2): 98-103. [http://dx.doi.org/10.1007/s12602-009-9033-8] [PMID: 26781118]

[159] Kazazic M, Nissen-Meyer J, Fimland G. Mutational analysis of the role of charged residues in target-cell binding, potency and specificity of the pediocin-like bacteriocin sakacin P. Microbiology 2002; 148(Pt 7): 2019-27. [http://dx.doi.org/10.1099/00221287-148-7-2019] [PMID: 12101290]

[160] Desriac F, Defer D, Bourgougnon N, Brillet B, Le Chevalier P, Fleury Y. Bacteriocin as weapons in the marine animal-associated bacteria warfare: inventory and potential applications as an aquaculture probiotic. Mar Drugs 2010; 8(4): 1153-77. [http://dx.doi.org/10.3390/md8041153] [PMID: 20479972]

[161] Dwidar M, Monnappa AK, Mitchell RJ. The dual probiotic and antibiotic nature of Bdellovibrio bacteriovorus. BMB Rep 2012; 45(2): 71-8. [http://dx.doi.org/10.5483/BMBRep.2012.45.2.71] [PMID: 22360883]

[162] Pasternak Z, Pietrokovski S, Rotem O, Gophna U, Lurie-Weinberger MN, Jurkevitch E. By their genes ye shall know them: genomic signatures of predatory bacteria. ISME J 2013; 7(4): 756-69. [http://dx.doi.org/10.1038/ismej.2012.149] [PMID: 23190728]

[163] Costerton J, Keller D. Oral periopathogens and systemic effects. Gen Dent 2007; 55(3): 210-5. [PMID: 17511362]

[164] Sockett RE, Lambert C. Bdellovibrio as therapeutic agents: a predatory renaissance? Nat Rev Microbiol 2004; 2(8): 669-75. [http://dx.doi.org/10.1038/nrmicro959] [PMID: 15263901]

[165] Van Essche M, Quirynen M, Sliepen I, *et al.* Killing of anaerobic pathogens by predatory bacteria. Mol Oral Microbiol 2011; 26(1): 52-61. [http://dx.doi.org/10.1111/j.2041-1014.2010.00595.x] [PMID: 21214872]

[166] Kadouri DE, To K, Shanks RM, Doi Y. Predatory bacteria: a potential ally against multidrug-resistant Gram-negative pathogens. PLoS One 2013; 8(5): e63397. [http://dx.doi.org/10.1371/journal.pone.0063397] [PMID: 23650563]

[167] Atterbury RJ, Hobley L, Till R, *et al.* Effects of orally administered Bdellovibrio bacteriovorus on the well-being and Salmonella colonization of young chicks. Appl Environ Microbiol 2011; 77(16): 5794-803. [http://dx.doi.org/10.1128/AEM.00426-11] [PMID: 21705523]

[168] Smits LP, Bouter KE, de Vos WM, Borody TJ, Nieuwdorp M. Therapeutic potential of fecal microbiota transplantation. Gastroenterology 2013; 145(5): 946-53. [http://dx.doi.org/10.1053/j.gastro.2013.08.058] [PMID: 24018052]

[169] Borody TJ, Paramsothy S, Agrawal G. Fecal microbiota transplantation: indications, methods, evidence, and future directions. Curr Gastroenterol Rep 2013; 15(8): 337. [http://dx.doi.org/10.1007/s11894-013-0337-1] [PMID: 23852569]

[170] Vyas D, L'esperance HE, Vyas A. Stool therapy may become a preferred treatment of recurrent Clostridium difficile? World J Gastroenterol 2013; 19(29): 4635-7. [http://dx.doi.org/10.3748/wjg.v19.i29.4635] [PMID: 23922461]

[171] Gupta S, Allen-Vercoe E, Petrof EO. Fecal microbiota transplantation: in perspective. Therap Adv

Gastroenterol 2016; 9(2): 229-39.
[http://dx.doi.org/10.1177/1756283X15607414] [PMID: 26929784]

[172] Rai M, Ingle AP, Gaikwad S, Gupta I, Gade A, Silvério da Silva S. Nanotechnology based anti-infectives to fight microbial intrusions. J Appl Microbiol 2016; 120(3): 527-42.
[http://dx.doi.org/10.1111/jam.13010] [PMID: 26599354]

[173] Jena M, Mishra S, Jena S, Mishra SS. Nanotechnology-future prospect in recent medicine: A review. Int J Basic Clin Pharmacol 2013; 2(4): 353-9.
[http://dx.doi.org/10.5455/2319-2003.ijbcp20130802]

[174] Wong IY, Bhatia SN, Toner M. Nanotechnology: emerging tools for biology and medicine. Genes Dev 2013; 27(22): 2397-408.
[http://dx.doi.org/10.1101/gad.226837.113] [PMID: 24240230]

[175] Fröhlich E, Salar-Behzadi S. Toxicological assessment of inhaled nanoparticles: role of *in vivo*, *ex vivo*, *in vitro*, and *in silico* studies. Int J Mol Sci 2014; 15(3): 4795-822.
[http://dx.doi.org/10.3390/ijms15034795] [PMID: 24646916]

[176] Rudramurthy GR, Swamy MK, Sinniah UR, Ghasemzadeh A. Nanoparticles: alternatives against drug-resistant pathogenic microbes. Molecules 2016; 21(7): E836.
[http://dx.doi.org/10.3390/molecules21070836] [PMID: 27355939]

[177] Hajipour MJ, Fromm KM, Ashkarran AA, *et al.* Antibacterial properties of nanoparticles. Trends Biotechnol 2012; 30(10): 499-511.
[http://dx.doi.org/10.1016/j.tibtech.2012.06.004] [PMID: 22884769]

[178] Underwood C, van Eps AW. Nanomedicine and veterinary science: the reality and the practicality. Vet J 2012; 193(1): 12-23.
[http://dx.doi.org/10.1016/j.tvjl.2012.01.002] [PMID: 22365842]

[179] Singh SB, Zink DL, Goetz MA, Dombrowski AW, Polishook JD, Hazuda DJ. Equisetin and a novel opposite stereochemical homolog phomasetin, two fungal metabolites as inhibitors of HIV-1 Rintegrase. Tetrahedron Lett 1998; 39(16): 2243-6.
[http://dx.doi.org/10.1016/S0040-4039(98)00269-X]

[180] Gänzle MG, Höltzel A, Walter J, Jung G, Hammes WP. Characterization of reutericyclin produced by Lactobacillus reuteri LTH2584. Appl Environ Microbiol 2000; 66(10): 4325-33.
[http://dx.doi.org/10.1128/AEM.66.10.4325-4333.2000] [PMID: 11010877]

[181] Tuske S, Sarafianos SG, Wang X, *et al.* Inhibition of bacterial RNA polymerase by streptolydigin: stabilization of a straight-bridge-helix active-center conformation. Cell 2005; 122(4): 541-52.
[http://dx.doi.org/10.1016/j.cell.2005.07.017] [PMID: 16122422]

[182] Phillips JW, Goetz MA, Smith SK, *et al.* Discovery of kibdelomycin, a potent new class of bacterial type II topoisomerase inhibitor by chemical-genetic profiling in Staphylococcus aureus. Chem Biol 2011; 18(8): 955-65.
[http://dx.doi.org/10.1016/j.chembiol.2011.06.011] [PMID: 21867911]

[183] Yendapally R, Hurdle JG, Carson EI, Lee RB, Lee RE. N-substituted 3-acetyltetramic acid derivatives as antibacterial agents. J Med Chem 2008; 51(5): 1487-91.
[http://dx.doi.org/10.1021/jm701356q] [PMID: 18281930]

[184] Jeong YC, Moloney MG. Equisetin, reutericyclin and streptolodygin as natural product lead structures for novel antibiotic libraries. Future Med Chem 2015; 7(14): 1861-77.
[http://dx.doi.org/10.4155/fmc.15.97] [PMID: 26431450]

[185] Köhler GA, Assefa S, Reid G. Probiotic interference of Lactobacillus rhamnosus GR-1 and Lactobacillus reuteri RC-14 with the opportunistic fungal pathogen Candida albicans. Infect Dis Obstet Gynecol 2012; 636474.
[http://dx.doi.org/http://dx.doi.org/10.1155/2012/636474]

[186] Shokryazdan P, Sieo CC, Kalavathy R, *et al.* Probiotic potential of Lactobacillus strains with

antimicrobial activity against some human pathogenic strains. Biomed Res Int 2014; 927268. [http://dx.doi.org/http://dx.doi.org/10.1155/2014/927268]

[187] Pascual LM, Daniele MB, Giordano W, Pájaro MC, Barberis IL. Purification and partial characterization of novel bacteriocin L23 produced by Lactobacillus fermentum L23. Curr Microbiol 2008; 56(4): 397-402. [http://dx.doi.org/10.1007/s00284-007-9094-4] [PMID: 18172715]

[188] Sharma A, Srivastava S. Anti-Candida activity of two-peptide bacteriocins, plantaricins (Pln E/F and J/K) and their mode of action. Fungal Biol 2014; 118(2): 264-75. [http://dx.doi.org/10.1016/j.funbio.2013.12.006] [PMID: 24528647]

[189] Okkers DJ, Dicks LM, Silvester M, Joubert JJ, Odendaal HJ. Characterization of pentocin TV35b, a bacteriocin-like peptide isolated from Lactobacillus pentosus with a fungistatic effect on Candida albicans. J Appl Microbiol 1999; 87(5): 726-34. [http://dx.doi.org/10.1046/j.1365-2672.1999.00918.x] [PMID: 10594714]

[190] Talarico TL, Casas IA, Chung TC, Dobrogosz WJ. Production and isolation of reuterin, a growth inhibitor produced by Lactobacillus reuteri. Antimicrob Agents Chemother 1988; 32(12): 1854-8. [http://dx.doi.org/10.1128/AAC.32.12.1854] [PMID: 3245697]

[191] Jay JM. Antimicrobial properties of diacetyl. Appl Environ Microbiol 1982; 44(3): 525-32. [PMID: 7137998]

[192] Chung TC, Axelsson L, Lindgren SE, Dobrogosz WJ. *In vitro* studies on reuterin synthesis by Lactobacillus reuteri. Microb Ecol Health Dis 1989; 2: 137-44. [http://dx.doi.org/10.3109/08910608909140211]

[193] Ohshima T, Kojima Y, Seneviratne CJ, Maeda N. Therapeutic application of synbiotics, a fusion of probiotics and prebiotics, and biogenics as a new concept for oral Candida infections: A mini review. Front Microbiol 2016; 7(10): 10. [PMID: 26834728]

[194] Walter J, Britton RA, Roos S. Host-microbial symbiosis in the vertebrate gastrointestinal tract and the Lactobacillus reuteri paradigm. Proc Natl Acad Sci USA 2011; 108(1) (Suppl. 1): 4645-52. [http://dx.doi.org/10.1073/pnas.1000099107] [PMID: 20615995]

[195] Gänzle MG. Reutericyclin: biological activity, mode of action, and potential applications. Appl Microbiol Biotechnol 2004; 64(3): 326-32. [http://dx.doi.org/10.1007/s00253-003-1536-8] [PMID: 14735324]

[196] Gänzle MG, Vogel RF. Studies on the mode of action of reutericyclin. Appl Environ Microbiol 2003; 69(2): 1305-7. [http://dx.doi.org/10.1128/AEM.69.2.1305-1307.2003] [PMID: 12571063]

[197] Vollenweider S, Evers S, Zurbriggen K, Lacroix C. Unraveling the hydroxypropionaldehyde (HPA) system: an active antimicrobial agent against human pathogens. J Agric Food Chem 2010; 58(19): 10315-22. [http://dx.doi.org/10.1021/jf1010897] [PMID: 20845960]

[198] Hurdle JG, Yendapally R, Sun D, Lee RE. Evaluation of analogs of reutericyclin as prospective candidates for treatment of staphylococcal skin infections. Antimicrob Agents Chemother 2009; 53(9): 4028-31. [http://dx.doi.org/10.1128/AAC.00457-09] [PMID: 19581456]

[199] Hurdle JG, Heathcott AE, Yang L, Yan B, Lee RE. Reutericyclin and related analogues kill stationary phase Clostridium difficile at achievable colonic concentrations. J Antimicrob Chemother 2011; 66(8): 1773-6. [http://dx.doi.org/10.1093/jac/dkr201] [PMID: 21632577]

[200] Yang Z, Suomalainen T, Mäyrä-Mäkinen A, Huttenen E. Antimicrobial activity of 2- pyrrolidone---carboxylic acid produced by lactic acid bacteria. J Food Prot 1997; 60(7): 786-90.

[http://dx.doi.org/10.4315/0362-028X-60.7.786]

[201] Worthington RJ, Melander C. Combination approaches to combat multidrug-resistant bacteria. Trends Biotechnol 2013; 31(3): 177-84.
[http://dx.doi.org/10.1016/j.tibtech.2012.12.006] [PMID: 23333434]

[202] Davies J, Davies D. Origins and evolution of antibiotic resistance. Microbiol Mol Biol Rev 2010; 74(3): 417-33.
[http://dx.doi.org/10.1128/MMBR.00016-10] [PMID: 20805405]

[203] Bush K, Macielag MJ. New β-lactam antibiotics and β-lactamase inhibitors. Expert Opin Ther Pat 2010; 20(10): 1277-93.
[http://dx.doi.org/10.1517/13543776.2010.515588] [PMID: 20839927]

[204] Bhardwaj AK, Mohanty P. Bacterial efflux pumps involved in multidrug resistance and their inhibitors: rejuvenating the antimicrobial chemotherapy. Recent Pat Antiinfect Drug Discov 2012; 7(1): 73-89.
[http://dx.doi.org/10.2174/157489112799829710] [PMID: 22353004]

[205] van Soolingen D, Hernandez-Pando R, Orozco H, *et al.* The antipsychotic thioridazine shows promising therapeutic activity in a mouse model of multidrug-resistant tuberculosis. PLoS One 2010; 5(9): e12640.
[http://dx.doi.org/10.1371/journal.pone.0012640] [PMID: 20844587]

[206] Amaral L, Boeree MJ, Gillespie SH, Udwadia ZF, van Soolingen D. Thioridazine cures extensively drug-resistant tuberculosis (XDR-TB) and the need for global trials is now! Int J Antimicrob Agents 2010; 35(6): 524-6.
[http://dx.doi.org/10.1016/j.ijantimicag.2009.12.019] [PMID: 20188526]

[207] Lin J, Martinez A. Effect of efflux pump inhibitors on bile resistance and *in vivo* colonization of Campylobacter jejuni. J Antimicrob Chemother 2006; 58(5): 966-72.
[http://dx.doi.org/10.1093/jac/dkl374] [PMID: 16963459]

[208] van Panhuis WG, Grefenstette J, Jung SY, *et al.* Contagious diseases in the United States from 1888 to the present. N Engl J Med 2013; 369(22): 2152-8.
[http://dx.doi.org/10.1056/NEJMms1215400] [PMID: 24283231]

[209] Mishra RP, Oviedo-Orta E, Prachi P, Rappuoli R, Bagnoli F. Vaccines and antibiotic resistance. Curr Opin Microbiol 2012; 15(5): 596-602.
[http://dx.doi.org/10.1016/j.mib.2012.08.002] [PMID: 22981392]

CHAPTER 4

Nanoantibiotics: Recent Developments and Future Prospects

Muhammad Ovais[1,2], **Nashmia Zia**[3], **Ali Talha Khalil**[4], **Muhammad Ayaz**[5], **Amjad Khalil**[6] and **Irshad Ahmad**[6,*]

[1] *CAS Center for Excellence in Nanoscience, CAS Key Laboratory for Biomedical Effects of Nanomaterials and Nanosafety, National Center for Nanoscience and Technology (NCNST), Beijing100190, P.R. China*

[2] *University of Chinese Academy of Sciences, Beijing100049, P.R. China*

[3] *Department of Pharmacy, University of Peshawar, Peshawar25000, Pakistan*

[4] *Department of Eastern Medicine and Surgery, Qarshi University, Lahore54000, Pakistan*

[5] *Department of Pharmacy, University of Malakand, Khyber Pakhtunkhwa 18800, Pakistan*

[6] *Department of Life Sciences, King Fahd University of Petroleum and Minerals (KFUPM), Dhahran31261, Saudi Arabia*

Abstract: The currently available antimicrobial drugs against pathogenic microorganisms are inadequate to cope with the challenges of the newly emerging multidrug-resistant (MDR) pathogens. Other shortcomings include their partial absorption in the blood system accompanied with gastrointestinal problems such as vomiting, nausea and diarrhea. To overcome these limitations, an alternate therapeutic approach has to be developed through a combinatorial approach of antimicrobial agents with nanomaterials called "nanoantibiotics". Several research groups have already focused on the development of novel nanoantibiotics (nAbts) that will be efficient with better pharmacokinetic profiles and fast absorption in the blood system. The nAbts will open up new avenues in drug research, but ample work is needed on its development for industrial applications. In the current chapter, we will explain the current status of different nAbts loaded systems, their release mechanisms, key targets, formulations and mode of action. We will also describe the important features of nAbts such as size, surface charge, hydrophobicity/philicity, biofilm formation, stimuli-receptive and functionalization against MDR pathogens.

Keywords: Antimicrobial Resistance, Bacterial Infections, Drug Delivery, Functionalization, Loaded Systems, MDR Pathogens, Mode of Action, Microbial Factors, Nanoantibiotics, Nanoparticles.

[*] **Corresponding author to Irshad Ahmad:** Department of Life Sciences, King Fahd University of Petroleum and Minerals (KFUPM), Dhahran 31261, Saudi Arabia, Tel: +966038608393, Fax: +966038607031; E-mail: irshad@kfupm.edu.sa

Atta-ur-Rahman (Ed.)

BACKGROUND OF ANTIBIOTICS: GLOBAL THREATS, MARKET VALUE, AND STATISTICS

Antibiotics are generally known as antimicrobial agents that have the potential to effectively control bacterial infections. After the discovery of penicillin by Alexander Fleming in 1928, a chain of different types of antibiotics was developed to control the deadly bacterial pathogens in order to improve the quality and life expectancy of humans. However, due to the lack of awareness, mismanagement and overuse of antibiotics in the developing countries, resistant strains of bacteria have been developed with the passage of time. Antimicrobial resistance (AMR) is considered as an alarming threat which may affect not only the poor but also the developed nations of the world. A current report shows that the annual human casualties are estimated to be 10 million deaths by 2050 with a predictable financial burden of 100 trillion US dollars due to the fast emergence rate of AMR [1, 2].

Grand View Research, Inc. has reported the market value of antibiotics as $39.8 billion in 2015 that is predicted to reach $57.0 billion by 2024. The increasing trend was observed due to the upsurge of infectious diseases particularly in the developing counties of Asia Pacific, Middle East and Africa [3]. Recently the World Health Organization has established a Global AMR Surveillance System (GLASS) aimed to strengthen and endorse a scientific method to collect and explore the worldwide AMR data that will be used to combat the widespread problem. GLASS reported *Escherichia coli, Klebsiella pneumoniae, Staphylococcus aureus, Streptococcus pneumoniae,* and *Salmonella* spp as the most resistant bacteria among 500,000 isolates collected from 22 countries [4]. In a current report, the trends and drivers of antibiotic consumption were analyzed from 2000-2015 among 76 countries and the total global antibiotic consumption was estimated till 2030. During this period, a 65% increase was observed in the global antibiotic intake from 21.1 to 34.8 billion defined daily doses (DDDs). The high-income countries (HIC) including United States, France and Italy have consumed antibiotics at a higher rate than the low and middle-income countries (LMIC) predominantly India, China and Pakistan. An increasing trend was observed in the commonly prescribed broad-spectrum antibiotics (penicillins, cephalosporins, quinolones, and macrolides) along with the newer antibiotic groups (carbapenems, glycylcyclines, and oxazolidinones) [5].

Recently, a worldwide survey was conducted in 53 countries in order to determine the antibiotic consumption rate and AMR incidence against the consumed antimicrobial agents in hospitalized patients. The data was collected from patients in 303 hospitals located in eight developing and seventeen developed countries. According to the data, only 34·4% of the patients took one dose of antibiotics for their infections. Among the different classes of antibiotics, the majority of patients were prescribed with penicillins plus β-lactamase inhibitors, cephalosporins and fluoroquinolones. The majority of patients in Latin America and Asian countries are using the carbapenem class of antibiotics that are assumed to develop AMR [6].

In this chapter, we will provide the readers novel perspectives on the development of nAbts that may replace conventional antibiotics due to their profound drawbacks including AMR against MDR pathogens. We believe that this innovative approach will contribute to shift the paradigm towards the development of nAbts that will prove to be an asset in the technological revolution of this millennium.

LIMITATIONS TO CONVENTIONAL ANTIBIOTICS

The development of new antibiotic has slowed down noticeably in the 21st century. Only four new antibiotics are developed and approved for the US market from 2008-2012, compared to 16 antibiotics during 1983-1987. Actually, no novel antibiotics have been developed for the Gram-negative bacteria in the last 40 years. The reasons for that could be scientific, commercial and regulatory. Unfortunately, antibiotic resistance has become a worldwide risk accountable for an extraordinary death rate and deadly infections. The significance of these infections is serious in unstable circumstances such as natural catastrophe, violence, and starvation. The World Health Organization (WHO) has declared that the post-antibiotic period may cause regular infections and very minute infections to become fatal if we are unable to take quick actions alongside antibiotic resistance. The United States government declared that 63,000 patients die every year from bacterial infections acquired in the hospital. Each year, approximately 25,000 patients die due to multiple drug resistance (MDR) bacterial infections in Europe [7].

The infections linked to antibiotic resistance usually arise in patients grouped together as highly susceptible, which are given medical facilities and using antibiotics in such settings along with aggressive actions. According to a report, the pathogens causing communal diseases like sepsis and pneumonia have developed antibiotic resistance responsible for the death of 50,000 Americans in 2006, which costs the United States government approximately $8 billion [8]. Recent studies have raised apprehension regarding the disaster of antibiotic resistance. Investigators have spoken loudly and expressed their worry about the rise in the number of patients who are developing fatal aggressive infections because of MDR pathogens every year, as well as the increasing trend of deaths as a result of these conditions. Thus, it is important to develop novel antibiotics on a priority basis because of the existing bacterial strains which are continuously developing resistance against the currently available antibiotics worldwide [9]. Unfortunately, the pharmaceutical sector is coping with this situation *via* introducing novel antibiotics for the treatment of resistant bacterial strains, whereas the antibiotics are still prescribed haphazardly and misused all over the world. The problem seems to become a severe threat not only for the healthcare and pharmaceutical sector, but also becomes an economic problem for industrialized countries of the world [8].

In the above mentioned circumstances, over dosage and self-medications are among the main reasons for an increasing trend of antibiotic resistance. Other reasons include improper health education, low socio economic status of the people and the unavailability of physicians on time. As a result, the patients are compelled to pursue guidance from a nearest pharmacist or to self-medicate themselves that may contribute to increasing antibiotic resistance [10]. The currently available antibiotics are considered to be a temporary solution to cure the deadly infections. The treatments will not be enough to fight the danger of AMR, therefore novel approaches need to be adopted in countering the menace of antibiotics resistance.

DEVELOPMENT OF NANOANTIBIOTICS: AN ALTERNATIVE PARADIGM

Because of the multidisciplinary nature of nanoscience, the nanotechnology based interventions have entered diverse disciplines from engineering and materials sciences to health and biological sciences. Materials of which are at least one-dimension size ranging from (1-100 nm) are typically termed as nanomaterials. They can have different shapes and sizes which impart them novel properties for a variety of applications. When reduced to such a small size, the properties of the material can show dramatic changes. Among the various fields, health sciences have received tremendous attention from the nanotechnology perspective. The nAbts are a recent feat in health and medicine which represent such nanomaterials that either possess the antimicrobial potential or have the potential to enhance the efficacy of the antimicrobial drug on different diseases [11, 12]. The nAbts offer several advantages as compared to conventional antibiotics such as durability, absorption, controlled release, circulation and targeted delivery. In addition, they are cost effective and economical as well as adaptable for the resistant antimicrobial world [13 - 16]. The nAbts have emerged as being more effective in the treatment of pathogenic bacteria and their related infections. Antibiotic resistance is considered to be an emerging threat in which about 10 million deaths are estimated by 2050 [17]. These nano based systems also provide an alternative and long term solution to the antimicrobial drug resistance. These nAbts are usually emulsion based, metals, metal oxides or their composites and carbon based.

TYPES OF NANOANTIBIOTICS

Metals, their Oxides or Composites

Among the nanoantibiotics, the metal based nAbts mostly possess inherent antimicrobial properties. For example, various metals like Ag and Au or oxides of Zn, Ni, Fe, Ti, Cu *etc.*, possess antimicrobial potential. Metals have been known to have antimicrobial potential since ancient times. Moreover, these metals have the tendency to differentiate the eukaryotic and prokaryotic cells through the metalloproteins and metal transport systems [17]. Fig. (**1**) shows the mechanisms by which the metal-based nanoparticles (NPs) interfere with the bacteria. The nanoscaled metal-based NPs are highly potent due to their surface area to volume ratio which allows a strong antimicrobial action [17]. Here it is pertinent to mention the recent trends in the synthesis and antimicrobial applications of these metal based NPs like green synthesis, doping and nanocomposites. To achieve good antimicrobial potential, metals are frequently doped with metal oxides. The Ag-doped ZnO (Ag-ZnO) indicated enhanced antimicrobial activities against pathogenic microbes like *E. coli* and *S. aureus* [18]. Similarly, in Ag-doped TiO, (Ag-TiO), it was noted that increasing the concentration of Ag has contributed to the enhanced antimicrobial activity of Ag-TiO NPs against *E. coli, P. aeruginosa, K. pneumonia* and *E. cloacae* [19]. Novel materials have been fabricated for enhanced photo induced biocidal activity. TiO_2 and TiO_2:In_2O_3 films modified with Ag and Ag-Ni bimetallic NPs were also investigated [20]. Metal-based TiO_2 hybrid nanomaterials like Ag-TiO_2, Pt-TiO_2, Pd-TiO_2 and Au-TiO_2, have also been checked comparatively for their antibacterial properties. Among them, Ag-TiO_2 appeared most effective in inhibiting *E. coli* [21]. Novel antimicrobial metal based nano fabrications have also been used with materials like chitosan [22], cellulose [23], and zeolites [24]. Herein, it is crucial to signify the versatile nature of these metal-metal oxide based NP systems which is evidenced by different hybrid nanomaterials.

It is noteworthy to mention the recent progress made on the interface of phyto-nanotechnology in which plant based pharmacologically active extracts have been used for the biosynthesis of metal or metal oxide based NPs [25 - 31]. Phyto-synthesis not only possesses advantages such as being cost effective, one step, and an entire green route of fabrication, but also the synthesized NPs are reportedly having higher degrees of biocompatibility [32, 33]. Iron NPs synthesized using tea extracts were found to be nontoxic to human keratinocyte cells as compared to the iron NPs synthesized using sodium borohydride as a reducing agent [34]. The biocompatible nature of phyto-synthesized Ag, CuO, ZnO NPs is also established [35 - 38]. Moreover, plant extracts have also been used to fabricate novel nano hybrid materials and composites. Various hybrid materials like Ag-TiO_2/*Euphorbia heterophylla* [39], Ni-Fe_3O_4/*Moringa oliefera* [40], Cu-Fe_3O_4/*Morinda morindoides* [41], and Zn-TiO_2/*Aspalathus linearis* [42] have also been demonstrated which emphasize the need for further studies on the phyto-inspired hybrid nanomaterials for antimicrobial applications.

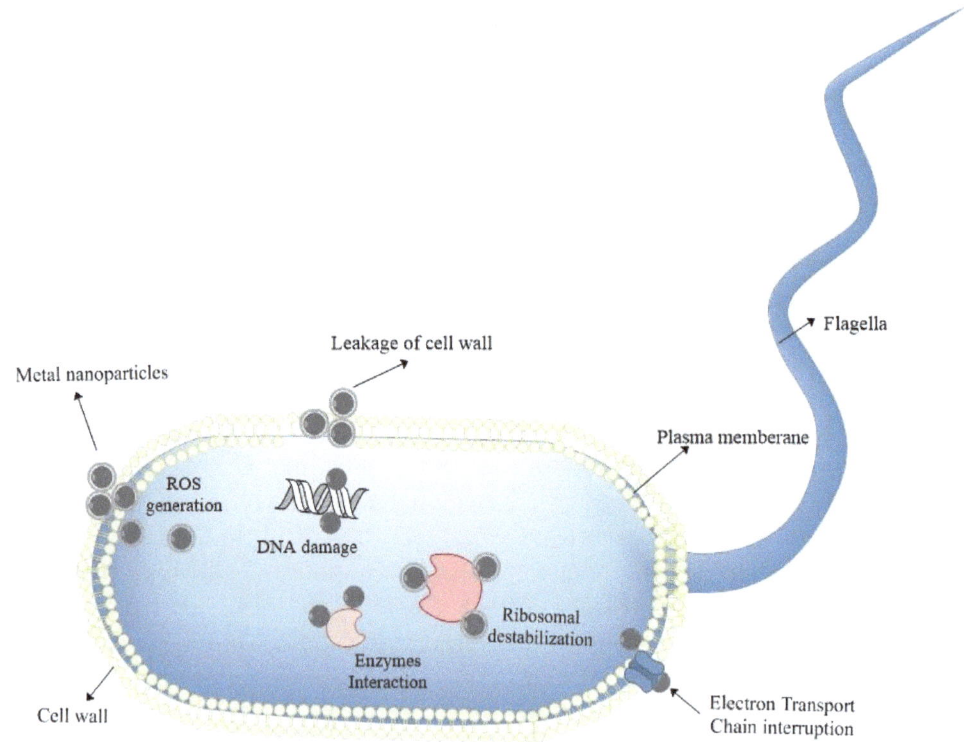

Fig. (1). Proposed anti-bacterial mechanism of metal nanoparticles.

Polymeric Nanoparticles

The polymeric NPs based antibiotic systems are considered to be an exciting prospect to guided delivery and controlled release of the therapeutic drugs. They are synthesized *via* encapsulating or conjugating the therapeutics in the polymer core or backbone [43, 44]. These polymeric NPs systems are flexible in terms of synthesis and functionalization. Initially, polymeric NPs started with the use of non-biodegradable polymers (polystyrene, polyacrylates, polyacrylamide *etc.*) which possessed certain toxicity issues due to accumulation in the body. The science is now more focused on the use of biodegradable, extractable polymers like poly-lactides, poly-amino acids, gelatin, chitosan, albumin and alginates [45]. In comparison to conventional antibiotics as free drugs, the polymeric NP based antibiotics yielded enhanced results in clearing the pathogens from the body and also indicated improved bioavailability, antibiotic protection from hydrolytic and enzymatic degradation, and versatile routes for administration. Additional advantages include their stability and size-shape controlled synthesis [46].

The polymeric based nAbts is a popular area of research and many polymeric systems are being prepared for the treatment of microbial infections. For the chlamydial infections, the polymeric system (rifampin-azithromycin-PLGA (poly(lactic-co-glycolic acid)) was more effective as compared to the antibiotics without PLGA (Fig. **2**) [47]. Protein (Gliadin)-based polymeric systems with omeprazole and clarithromycin co-encapsulation indicated efficacious results against *H. pylori* in rats. The efficiency was further increased by conjugation with lectin and adding amoxicillin in addition to the omeprazole and clarithromycin [48, 49]. Improved anti-tuberculosis potential of the various alginate polymer based NPs is also reported [50]. Tobramycin based polymeric NPs indicated impressive results against *P. aeruginosa* [51]. Various polymeric NPs have also been used against the biofilms successfully [52]. Detailed reviews have already been done which are recommended to study for the various polymeric based nAbts and their results against various infections [46].

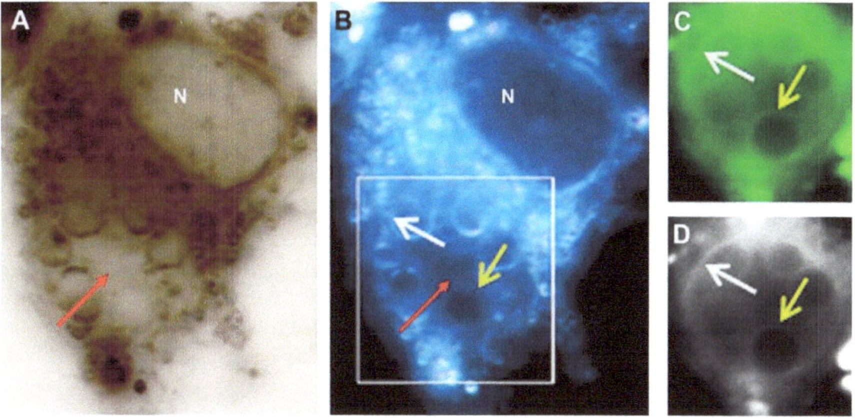

Fig. (2). Images showing the homing of labeled nanoparticles into inclusions of persistently infected cells. HEp2 cells were unstained and captured by real time microscopy using the RTM-3 microscope with Riveal Contrast (Quorum Technologies) (A), fluorescence mode (B), epifluorescence mode using FITC filter (C), and a black and white image of same inclusion (D). 6C-nanoparticles are associated with both the inclusion membrane (white arrow) and RB (yellow arrow). Red arrows in A, B indicate the inclusion which is viewed in C and D. N - Nucleus 100 original magnification. Adopted with permission from reference [47].

Fullerene and Carbon Based

Carbon-60 and other carbon-based materials like graphene have recently been in focus because of their antimicrobial features. The nano carbons like single-walled carbon nanotubes (SWCNTs), multi-walled carbon nanotubes (MWCNTs), and graphene materials (GMs) can inhibit the growth of bacteria through multiple routes, thereby interfering with organisms [53, 54]. The knowledge and exact mechanism by which this carbon based nanomaterials manifest the antibacterial potential is yet to be explored [55]. Some researchers indicated the photocatalytic generation of ROS and lipid peroxidation as the antimicrobial mechanism of nano carbon action [13]. The graphene materials can be used as graphene oxide (GO), reduced graphene oxide (rGO) and graphene quantum dots (GQD). Currently methicillin-resistant *S. aureus* (MRSA) was potentially controlled by physical and chemical mutilation through a nanocomposite system of reduced graphene oxide-iron oxide NPs (rGO-IONP). The physical damage of the bacteria is due to the efficient absorption of near infrared (NIR) energy by rGO which is changed to heat. The chemical damage done by IONP enhances H2O2 degradation by increasing its level during inflammation and forms highly reactive OH radicals through Fenton reactions. During *in vivo* results, the combined action of rGO-IONP and NIR has a profound effect by controlling the bacteria in subcutaneous inflammations and significantly enhances wound healing (Fig. **3**). The GO and rGO indicate toxicity towards *S. aureus* and *E. coli* [56, 57]. Similarly, C-60 indicated impressive antibacterial activity against a drug resistant strain of *Salmonella typhi* [58]. The graphene based nanomaterials provide versatile ways of colloidal use such as only graphene [59], polymer stabilized graphene material [60] and colloidal nano hybrids [61, 62]. The Nano hybrids of graphene are complex systems in which usually metal NPs like Ag are grown on the surfaces of graphene material. The decorated fabrication usually possesses synergistic antimicrobial action as evidenced in the case of Ag-rGO nanocomposites which was even more effective against *E. coli* than ampicillin [63]. Likewise, the Ag-GO nanocomposites indicated good inhibitory potential against *E. coli* than *S. aureus* [64]. Besides metals, the placement of proteins on GMs has also been demonstrated [65].

MECHANISM OF ACTION: HOW DO NANOANTIBIOTICS ACT?

Developing nAbts is relatively a new paradigm that promises a potential solution to the global problem of antibiotic resistance. It is because of the multiple target/action sites of nAbts that theoretically it will be impossible for the microbes to develop resistance genes against multiple killing mechanisms utilized by nAbts. Different nAbts employ different mechanisms to combat microbial infections. These include i) inhibition of enzyme and DNA synthesis, ii) production of reactive oxygen species and reactive NO that damage the cellular components, iii) adsorption to and damaging of the cell membrane/wall, iv) disruption of energy transduction by interrupting transmembrane electron transport chain reaction, v) and release of heavy metal ions with deleterious effects [13].

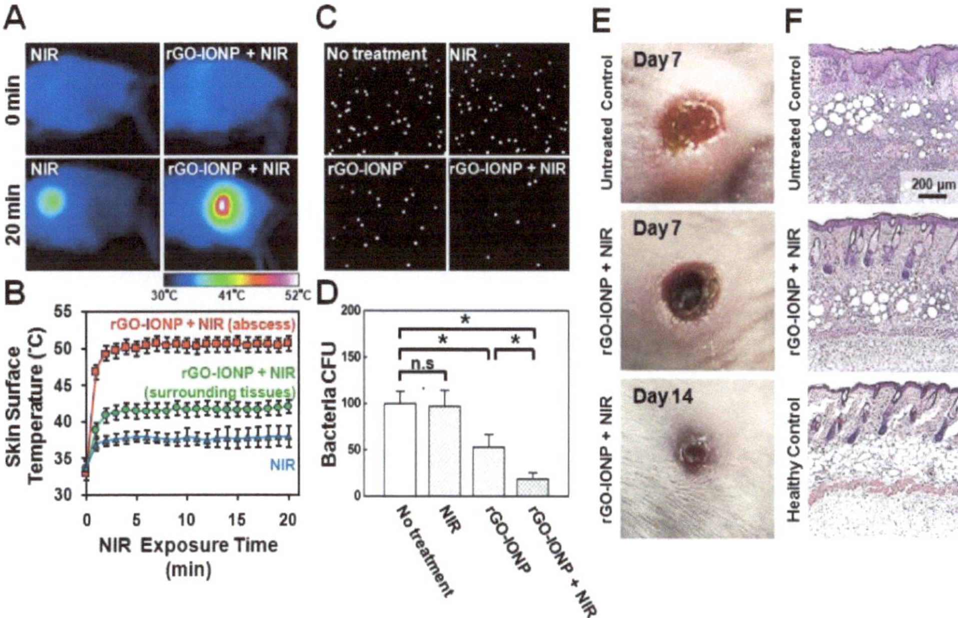

Fig. (3). *In vivo* antibacterial efficacy of rGO-IONP (400 µg/abscess) upon exposure to NIR in a mouse model with MRSA-infected subcutaneous abscesses: (A) thermographic images, (B) temperature evolution curves, and (B) photographs of bacterial CFUs and (D) quantitative results (n = 6). n.s.: not significant; *: statistically significant ($P < 0.05$). (E) Photographs and (F) histological photomicrographs of infected tissues that were harvested at specified times following treatment. Adopted with permission from reference [56].

Many studies have evaluated the mechanism of action for metal-based nAbts. Silver and silver oxide NPs are known for their effective antibacterial action against a large variety of drug resistant organisms. They exert their antibacterial action by multiple mechanisms, which make them highly effective. Silver NPs act by releasing large quantities of silver ions that compromise cell membrane integrity and help in the disruption of energy transduction through the electron transport chain. Furthermore, they are also reported to be involved in the DNA damage of microbial cells [66 - 68]. Zinc oxide NPs are another class of nAbts that exert their action by accumulating inside the cells, and there, they release Zn^{+2} ions. They are involved in the generation of hydrogen peroxide and cell membrane disruption [69, 70]. Titanium dioxide NPs act by generating reactive oxygen species and hence affecting cell membrane integrity [71 - 73].

In addition to metal NPs, different polymeric, liposomal or carbon based nanomaterials are also used as nAbts, all having their own modes of action. Chitosan-based nAbts act by increasing the permeability and rupturing of cell membranes. They are also involved in the enzyme inactivation of microbial machinery [74 - 76]. Carbon nanotubes act by generating reactive oxygen species and resulting in the oxidative degradation of cell membranes, lipid and proteins [77 - 79]. A relatively new class of nAbts comprising fullerenes act by increasing the infiltration of neutrophils and are also involved in the disruption of cell membranes [80 - 82]. These multiple mechanisms of action make the NPs promising agents for targeting the microbial machinery.

DRUG RELEASE KINETICS OF NANOANTIBIOTICS

Efficacy and effectiveness of a drug loaded nano-formulation largely depend on its drug release kinetics. It is essential for optimum efficacy that drug release from nanocarrier should be slow enough to avoid drug loss before reaching the site of action [83 - 85]. Simultaneously, it should be controlled and timed to ensure maximum and sustained effect at the site of action and to prevent off-site action and hence side effects. These requirements make the drug release kinetics a major determining factor for efficacy of nAbts consisting of antibiotic loaded nanocarriers [86, 87].

Commonly used methods for determination of drug release from NPs include dynamic dialysis, ultra-filtration and ultra-centrifugation *etc.*, with dynamic dialysis being the preferable method, because of the elimination of additional steps of separation of free drug from the NPs at various time points of kinetic study. Moreover, the external pressure applied during separation in other techniques also decreases the reliability of those methods [88]. A large variety of nanocarriers have been employed for the targeted, controlled or timely delivery of antibiotics such as liposomes, polymeric NPs, solid lipid NPs, nano-emulsion, micellar systems and dendrimers [89]. Each type shows different characteristics and mechanisms when it comes to drug release kinetics. Moreover, the factors that affect the drug release from these nanocarriers also vary.

For polymeric NPs, the drug release kinetics majorly depends on the i) desorption of surface bound or adsorbed drug, ii) diffusion of drug from within polymeric NPs, and iii) erosion of polymeric NPs and a combined erosion/diffusion effect. If the rate of erosion of matrix is slower as compared to diffusion of drug from the polymer matrix, than drug release occurs by diffusion. Usually drug release from polymeric NPs includes an initial "burst release" effect that indicates the release of the surface bound or adsorbed drug as compared to the drug incorporated within the matrix [90]. In comparison to polymeric NPs, drug release from liposomes largely depends on i) the composition of lipid membrane, ii) the type of drug incorporated, iii) its permeable species percentage, and iv) environmental factors such as pH, temperature, or external stimuli such as ultrasound or enzymatic degradation or protein interaction [91 - 94].

The development of nAbts with optimized drug delivery properties is a new focus. Many strategies have been employed to control the drug release kinetics such as using anionic Gemini surfactant [95], or covalent bonds that can be degraded in the acidic environment [96] to make polymeric NPs pH responsive. Furthermore, loading the drug directly with polyelectrolytes to generate new nAbts with enhanced drug delivery and antibacterial activity is also an emerging trend [97]. The use of aragonite-based nanomaterial for the development of sustained release of a local antibiotic delivery system (LADS) [96] or the use of smart polymers to make the resultant nAbts responsive towards temperature, pH, and ionic strength has been reported [98]. Different studies have evaluated the drug loading and release kinetics of nAbts for guided and controlled delivery. The streptomycin-loaded chitosan-magnetic iron oxide NPs (Strep-CS-MNP) showed an initial burst release followed by controlled drug release kinetics as compared to the physical mixture of chitosan-streptomycin and magnetic iron oxide NPs, indicating the potential for the development of controlled release drug delivery system. The study showed that particles followed pseudo 2nd order release kinetics with a correlation coefficient (R^2) value of 0.9863. The study also indicated that Strep-CS-MNP exhibited better antibacterial activity against MRSA as compared to streptomycin alone [99]. In another study, the pH responsive vancomycin loaded chitosan NPs showed an increased release of vancomycin at pH 6.5 than at pH 7.4. The formulation also showed that MRSA burden in a mouse skin infection model was decreased 8-fold compared to those treated with pure vancomycin [95].

In a recent study, single-walled carbon nanotubes (SWCNTs) covalently functionalized with ciprofloxacin demonstrated the release of 90% loaded drug within 2.5 hrs, following first order release kinetics. Moreover, the formulation showed a 16-fold increase in activity against *Staphylococcus aureus* and *Pseudomonas aeruginosa* and 8-fold against *Escherichia coli* in comparison to free ciprofloxacin [100]. Different complexes have also been utilized to combine the antibacterial effects of metal and antibiotics simultaneously. Recently, a study showed that metal-carbenicillin framework-coated mesoporous silica NPs (MSN) have the potential of a highly efficient and enhanced upshot against MRSA, with better penetration depth into biofilms [101]. In another study, vancomycin-loaded aragonite NPs (VANPs) displayed 120 hrs of drug release profile of vancomycin with a high antibacterial effect against MRSA. These NPs with such an extended release profile may act as a good local antibiotic delivery system for the treatment of osteomyelitis [96].

NANO-BASED STRATEGIES TO OVERCOME MULTI-DRUG RESISTANT (MDR) PATHOGENS

The emergence of multidrug resistant (MDR) pathogens to almost all classes of antibiotics is a global phenomenon and has made the therapy of several diseases impossible [102]. MDR is the result of inappropriate and indiscriminate utilization of antimicrobial agents, and as a result, the globe is leading towards a pre-antibiotic era [103, 104]. This prompts the development of new antimicrobial agents which are i) effective on more than one bacterial target, ii) can counteract the resistance mechanisms of pathogens, and iii) reduce the emergence of MDR pathogens [105, 106]. The antimicrobial efficacy of several antibiotics is linked with their ability to interfere in essential microbial processes including DNA replication, cell wall synthesis and expression of vital proteins. However, microbes have adopted various resistant mechanisms like alteration of the drug target, fabrication of drug metabolizing enzymes and extracellular extrusion of the drugs *via* efflux pumps.

The nAbts are antimicrobial agents, which in nano size ensure efficient drug delivery to the target site, can act as a carrier in the form of nano-carriers and ensure overall efficacy and safety of antibiotics with specialized pharmacokinetic and dynamic profiles [107]. nAbts can overcome multidrug resistance *via* their unique physicochemical properties, thus enabling nAbts to perform their function by applying novel antimicrobial pathways. For instance, nAbts have the affinity to bind with microbial cellular membranes leading to disruption of their integrity and leakage of cytoplasmic components [108]. nAbts subsequent to membrane permeation, bind with essential cellular components including enzymes, DNA, ribosome's thus disrupting cellular functions. Hindrance in these imperative cellular functions leads to enzymes' inhibition, imbalance of electrolytes' oxidative stress causing cell death. The anti-bacterial mechanisms for nAbts is inherently reliant on surface fictionalization, core material, size and shape. In some studies, it was reported that variation in the inherent core of the nAbts leads to variation in the antimicrobial mechanism of these agents. For instance, silver based nAbts utilize Ag^+ as an antimicrobial agent which disrupts microbial membrane, electron transport chain and causes DNA damage leading to cell death [109]. AgNPs exhibit more effective antibacterial spectrum owing to their large surface area, resulting in better contact at the surface of bacteria [110]. The nAbts interfere with the proton-motive forces leading to the inhibition of oxidative phosphorylation and also increase oxidative stress *via* liberation of free radicals resulting in loss of cell membrane integrity and cell death [111, 112]. Copper based nAbts cause the release of Cu^{+2} liberate free radicals including ROS (reactive oxygen species) that impede the synthesis of essential amino acids and DNA leading to cell death. Further, TiO_2 and ZnO based nAbts generate ROS, damage cellular membranes and cause cell death [113]. Metal based nAbts have a strong binding affinity with phosphorus and sulfur containing biogenic substances like DNA bases, which are present in bacterial cells. These nAbts are reported to interact with these bases and cause their destruction, thus leading to cell death [114]. nAbts also inhibit microbial signal transduction *via* de-phosphorylation of peptide substrates on tyrosine residues which results in bacterial cells growth inhibition [115].Various nAbts are proposed to be effective against MDR pathogens however, they are usually narrow spectrum and have low therapeutic selectivity against normal cells. The modification of nAbts surface chemistry is vital for their selective interactions with microbial cells to reduce their toxicity against normal cells and improve their antimicrobial spectrum.

MICROBIAL FACTORS AFFECTING THE ACTION OF NANOANTIBIOTICS

Several factors affect the effectiveness of nAbts, which need to be carefully considered in order to rationally design nAbts against MDR pathogens.

Variation in Microbial Growth Rate

Generally, actively dividing cells are susceptible to the action of conventional antibiotics as well as nAbts. This is a vital contributing factor to the efficacy of antimicrobial agents. Due to lack of rapid division in non-spore forming bacteria, they are resistant to several antibiotics. Owing to the unavailability of sufficient nutrients, and or antibiotic stress, these bacteria exhibit several conformational changes during stationary phase which make them resistant to the action of nAbts [116]. The emergence of stress response genes in slowly diving cells might contribute to their AMR capacity [117].

Bacterial Structural Variations Contributing to MDR

Some bacterial structural features including the lipopolysaccharide (LPS) rich outer membrane of gram negative bacteria contribute to their resistance against several antibiotics [118]. Water soluble drugs can cross this membrane more effectively using porins whereas hydrophobic agents cannot traverse the membrane effectively, providing an intrinsic resistance against these drugs [119]. Additionally, antibiotic degrading enzymes including nucleases, proteases and phosphatases present in the periplasmic space of these bacteria also contribute to the intrinsic resistance of gram-negative bacteria [118]. These bacteria contain beta-lactam antibiotics degrading enzymes including pencillinase and cephalosporinase which inactivate these important antibiotics. By contrast, the LPS of gram-negative bacteria significantly contributes to bacterial membrane negativity and thus facilitates the attachment of positively charged nAbts to the bacterial membrane and subsequent penetration inside the cell [117]. Moreover, the negatively charged membrane repels purely negative charged nAbts, so an ideal carrier for nAbts must contain hydrophilic as well as hydrophobic properties in order for the drugs to be effective [120].

Swarming Phenomenon

Swarming is a condition whereby bacterial cells get transformed as polynucleated, elongated and hyperflagellated cells. During the process, bacteria migrate collectively or spread to the surface as a single unit using flagella [121]. Flagella help bacteria to move in the required direction. This behavior is the most prominent in Coliforms, Pseudomonas, Bacillus, Salmonella and Proteus species [122]. Swarming makes bacteria resistant to several antibiotics. Nevertheless, sub-culturing these swarmed cells reverts them to planktonic cells and they lose their antibiotic tolerance [107].

Biofilm Formation in Bacteria

Biofilm formation is a phenomena whereby pathogenic bacteria get attached to some solid surfaces and form biofilms covered by extracellular polymeric substances [123]. These polymeric substances organize and safeguard the film and aid in cellular communication [124]. Biofilms contain extracellular DNA, proteins and polysaccharides [125]. Biofilm formation makes the bacteria resistant to antibiotics *via* formation of a poly-layered defense. Extracellular polymeric substances make them impermeable to several antimicrobial agents [126]. Moreover, the growth rate of biofilm forming bacteria is slow, rendering more resistance to the action of antibiotics. Biofilm forming bacteria also exhibit withstanding capacity against stress responses which helps them resist the action of antibiotics. Biofilms are reported to assist in lateral and horizontal gene transfer among various species and thus help in the development of acquired resistant [127]. These biofilms possess large amount of antibiotic hydrolyzing enzymes, which is another aspect of biofilm mediated antibiotic resistance [128].

Among the strategies to overcome biofilm-mediated MDR is the use of lipids or polymeric drug delivery systems. Several researchers reported the inhibition of biofilm mediated MDR *via* the use of nAbts [129 - 131]. Polymeric based nAbts are preferred over the conventional dosage form owing to unmatchable characteristics including biocompatibility, capacity to encapsulate hydrophobic and hydrophilic materials, surface fictionalization, release control and versatility [132]. Several researchers suggest encapsulation of antibiotics in NPs to enhance their anti-biofilm efficiency and protect the drug against enzymatic degradation [133]. Liposomes loaded with antibiotics possess low MICs and ensure prolonged contact with biofilm forming bacteria [134]. Furthermore, some polymeric substances like chitosan possess intrinsic antibacterial properties and can act synergistically in combination with antibiotics against biofilm forming bacteria [135].

Persister and Intracellular Microorganisms

Persisters represent metabolically inactive bacterial cells in the culture of metabolically active cells [136]. These bacteria are intrinsically tolerant to the action of antibiotics due to their sluggish growth rate. But unlike resistance, this tolerance is not genetic as it is not mediated *via* mutation [136]. During antibiotic therapy, all pathogenic bacteria are removed but persisters remain at the site infection and are a major cause of re-infection [137]. Pathogens residing intracellularly comprise another scientific threat in the management of infections caused by these pathogens [138]. Liposome-based antibiotic therapy can be effectively used to eradicate intracellular pathogens in comparison to conventional antibiotic therapy [139]. Liposomes after attachment with cellular membrane effectively deliver the antibiotics inside the cell [140]. Several investigators reported the effectiveness of nAbts against intracellular pathogens with promising results [141].

CONCLUSION

Recently, AMR has raised an alarming global threat to human beings. Conventional antibiotics are losing their efficacy against the emerging multiple drug-resistant pathogens. Keeping in view these problems, the scientific community is in search of an alternative approach. In this regard, a nanotechnology-based drug delivery system of potential nanoantibiotic (nAbts) development is considered to be an asset in the technological revolution for the 21st century. The newly emerging area of nAbts is still in its infancy and needs significant focus in order to develop novel therapeutic drugs with the passage of time. The nAbts will have several advantages as compared to the conventional antibiotics such as durability, absorption, controlled release, circulation, and delivery. In addition, nAbts will be cost-effective and economical as well as adaptable for the resistant antimicrobial world. In this chapter, we have discussed different nAbt loaded systems, their release mechanisms, key targets, formulations and mode of action against MDR pathogens. However, further research is required to develop novel nAbts in order to study their efficacy against MDR pathogens. We believe that nAbts will prevail over the traditional antibiotics and will shift the paradigm towards a profound and effective solution of the concerned problems.

CONSENT FOR PUBLICATION

Not applicable.

CONFLICT OF INTEREST

The author (editor) declares no conflict of interest, financial or otherwise.

ACKNOWLEDGMENT

The personal assistance of I.A. by KFUPM is acknowledged.

REFERENCES

[1] Simpkin VL, Renwick MJ, Kelly R, Mossialos E. Incentivising innovation in antibiotic drug discovery and development: progress, challenges and next steps. J Antibiot (Tokyo) 2017; 70(12): 1087-96. [http://dx.doi.org/10.1038/ja.2017.124] [PMID: 29089600]

[2] O'Neill J. Review on Antimicrobial Resistance 2014.http://amr-review.org/Publications

[3] Market Research Report. Antibiotics Market Analysis By Drug Class (Cephalosporins, Penicillins, Fluoroquinolones, Macrolides, Carbapenems, Aminoglycosides, Sulfonamides), By Mechanism Of Action, And Segment Forecasts, 2018 - 2024. Grand view research, inc 2016; Report ID: GVR---68038-149-8

[4] Global antimicrobial resistance surveillance system (GLASS) report: early implementation 2016-2017. Geneva: World Health Organization; Licence: CC BY-NC-SA 2017; 3.0 IGO

[5] Klein EY, Van Boeckel TP, Martinez EM, *et al.* Global increase and geographic convergence in antibiotic consumption between 2000 and 2015. Proc Natl Acad Sci USA 2018; 115(15): E3463-70. [http://dx.doi.org/10.1073/pnas.1717295115] [PMID: 29581252]

[6] Versporten A, Zarb P, Caniaux I, *et al.* Antimicrobial consumption and resistance in adult hospital inpatients in 53 countries: results of an internet-based global point prevalence survey. Lancet Glob Health 2018; 6(6): e619-29. [http://dx.doi.org/10.1016/S2214-109X(18)30186-4] [PMID: 29681513]

[7] Zaman SB, Hussain MA, Nye R, Mehta V, Mamun KT, Hossain N. A review on antibiotic resistance: alarm bells are ringing. Cureus 2017; 9(6): e1403. [PMID: 28852600]

[8] Ventola CL. The antibiotic resistance crisis: part 1: causes and threats. P&T 2015; 40(4): 277-83. [PMID: 25859123]

[9] Rather IA, Kim BC, Bajpai VK, Park YH. Self-medication and antibiotic resistance: Crisis, current challenges, and prevention. Saudi J Biol Sci 2017; 24(4): 808-12. [http://dx.doi.org/10.1016/j.sjbs.2017.01.004] [PMID: 28490950]

[10] Piddock L, Garneau-Tsodikova S, Garner C. Ask the experts: how to curb antibiotic resistance and plug the antibiotics gap? Future Med Chem 2016; 8(10): 1027-32. [http://dx.doi.org/10.4155/fmc-2014-0032] [PMID: 27327784]

[11] Jamil B, Imran M. Factors pivotal for designing of nanoantimicrobials: an exposition. Crit Rev Microbiol 2018; 44(1): 79-94. [http://dx.doi.org/10.1080/1040841X.2017.1313813] [PMID: 28421881]

[12] Edson JA, Kwon YJ. Design, challenge, and promise of stimuli-responsive nanoantibiotics. Nano Converg 2016; 3(1): 26.
[http://dx.doi.org/10.1186/s40580-016-0085-7] [PMID: 28191436]

[13] Huh AJ, Kwon YJ. "Nanoantibiotics": a new paradigm for treating infectious diseases using nanomaterials in the antibiotics resistant era. J Control Release 2011; 156(2): 128-45.
[http://dx.doi.org/10.1016/j.jconrel.2011.07.002] [PMID: 21763369]

[14] Zazo H, Colino CI, Lanao JM. Current applications of nanoparticles in infectious diseases. J Control Release 2016; 224: 86-102.
[http://dx.doi.org/10.1016/j.jconrel.2016.01.008] [PMID: 26772877]

[15] Jamil B, Habib H, Abbasi SA, Ihsan A, Nasir H, Imran M. Development of cefotaxime impregnated chitosan as nano-antibiotics: De novo strategy to combat biofilm forming multi-drug resistant pathogens. Front Microbiol 2016; 7: 330.
[http://dx.doi.org/10.3389/fmicb.2016.00330] [PMID: 27047457]

[16] Saúde AC, Cherobim MD, Amaral AC, Dias SC, Franco OL. Nanoformulated antibiotics: the next step for pathogenic bacteria control. Curr Med Chem 2013; 20(10): 1232-40.
[http://dx.doi.org/10.2174/0929867311320100004] [PMID: 23432585]

[17] Gold K, Slay B, Knackstedt M, Gaharwar AK. Antimicrobial activity of metal and metal-oxide based nanoparticles. Adv Ther 2018; 1700033.
[http://dx.doi.org/10.1002/adtp.201700033]

[18] Nagaraju G, Prashanth S, Shastri M, Yathish K, Anupama C, Rangappa D. Electrochemical heavy metal detection, photocatalytic, photoluminescence, biodiesel production and antibacterial activities of Ag–ZnO nanomaterial. Mater Res Bull 2017; 94: 54-63.
[http://dx.doi.org/10.1016/j.materresbull.2017.05.043]

[19] Ali T, Ahmed A, Alam U, Uddin I, Tripathi P, Muneer M. Enhanced photocatalytic and antibacterial activities of Ag-doped TiO2 nanoparticles under visible light. Mater Chem Phys 2018; 212: 325-35.
[http://dx.doi.org/10.1016/j.matchemphys.2018.03.052]

[20] Skorb E, Antonouskaya L, Belyasova N, Shchukin D, Möhwald H, Sviridov D. Antibacterial activity of thin-film photocatalysts based on metal-modified TiO2 and TiO2: In2O3 nanocomposite. Appl Catal B 2008; 84(1-2): 94-9.
[http://dx.doi.org/10.1016/j.apcatb.2008.03.007]

[21] Chen SF, Li JP, Qian K, Xu WP, Lu Y, Huang WX, *et al.* Large scale photochemical synthesis of M@ TiO 2 nanocomposites (M= Ag, Pd, Au, Pt) and their optical properties, CO oxidation performance, and antibacterial effect. Nano Res 2010; 3(4): 244-55.
[http://dx.doi.org/10.1007/s12274-010-1027-z]

[22] Potara M, Jakab E, Damert A, Popescu O, Canpean V, Astilean S. Synergistic antibacterial activity of chitosan-silver nanocomposites on Staphylococcus aureus. Nanotechnology 2011; 22(13): 135101.
[http://dx.doi.org/10.1088/0957-4484/22/13/135101] [PMID: 21343644]

[23] Pinto RJ, Marques PA, Neto CP, Trindade T, Daina S, Sadocco P. Antibacterial activity of nanocomposites of silver and bacterial or vegetable cellulosic fibers. Acta Biomater 2009; 5(6): 2279-89.
[http://dx.doi.org/10.1016/j.actbio.2009.02.003] [PMID: 19285455]

[24] Shameli K, Ahmad MB, Zargar M, Yunus WMZW, Ibrahim NA. Fabrication of silver nanoparticles doped in the zeolite framework and antibacterial activity. Int J Nanomedicine 2011; 6: 331-41.
[http://dx.doi.org/10.2147/IJN.S16964] [PMID: 21383858]

[25] Khalil AT, Ovais M, Ullah I, *et al. Sageretia thea* (Osbeck.) modulated biosynthesis of NiO nanoparticles and their *in vitro* pharmacognostic, antioxidant and cytotoxic potential. Artif Cells Nanomed Biotechnol 2018; 46(4): 838-52.
[http://dx.doi.org/10.1080/21691401.2017.1345928] [PMID: 28687045]

[26] Khalil AT, Ovais M, Ullah I, Ali M, Shinwari ZK, Maaza M. Biosynthesis of iron oxide (Fe2O3) nanoparticles *via* aqueous extracts of *Sageretia thea* (Osbeck.) and their pharmacognostic properties. Green Chem Lett Rev 2017; 10(4): 186-201. [http://dx.doi.org/10.1080/17518253.2017.1339831]

[27] Khalil AT, Ovais M, Ullah I, Ali M, Shinwari ZK, Maaza M. Physical properties, biological applications and biocompatibility studies on biosynthesized single phase cobalt oxide (Co3O4) nanoparticles *via Sageretia thea* (Osbeck.). Arab J Chem In Press

[28] Khalil AT, Ovais M, Ullah I, Ali M, Shinwari ZK, Maaza M. Bioinspired synthesis of pure massicot phase lead oxide nanoparticles and assessment of their biocompatibility, cytotoxicity and in-vitro biological properties. Arab J Chem In Press

[29] Ovais M, Khalil AT, Raza A, *et al.* Green synthesis of silver nanoparticles *via* plant extracts: beginning a new era in cancer theranostics. Nanomedicine (Lond) 2016; 11(23): 3157-77. [http://dx.doi.org/10.2217/nnm-2016-0279] [PMID: 27809668]

[30] Ovais M, Raza A, Naz S, *et al.* Current state and prospects of the phytosynthesized colloidal gold nanoparticles and their applications in cancer theranostics. Appl Microbiol Biotechnol 2017; 101(9): 3551-65. [http://dx.doi.org/10.1007/s00253-017-8250-4] [PMID: 28382454]

[31] Ovais M, Khalil AT, Islam NU, *et al.* Role of plant phytochemicals and microbial enzymes in biosynthesis of metallic nanoparticles. Appl Microbiol Biotechnol 2018. Epub ahead of print [http://dx.doi.org/10.1007/s00253-018-9146-7] [PMID: 29882162]

[32] Singh P, Kim YJ, Zhang D, Yang DC. Biological synthesis of nanoparticles from plants and microorganisms. Trends Biotechnol 2016; 34(7): 588-99. [http://dx.doi.org/10.1016/j.tibtech.2016.02.006] [PMID: 26944794]

[33] Barabadi H, Ovais M, Shinwari ZK, Saravanan M. Anticancer Green Bionanomaterials: Present Status and Future Prospects. Green Chem Lett Rev 2017; 10(4): 285-314. [http://dx.doi.org/10.1080/17518253.2017.1385856]

[34] Nadagouda MN, Castle AB, Murdock RC, Hussain SM, Varma RS. *In vitro* biocompatibility of nanoscale zerovalent iron particles (NZVI) synthesized using tea polyphenols. Green Chem 2010; 12(1): 114-22. [http://dx.doi.org/10.1039/B921203P]

[35] Shi LB, Tang PF, Zhang W, Zhao YP, Zhang LC, Zhang H. Green synthesis of CuO nanoparticles using Cassia auriculata leaf extract and *in vitro* evaluation of their biocompatibility with rheumatoid arthritis macrophages (RAW 264.7). Trop J Pharm Res 2017; 16(1): 185-92. [http://dx.doi.org/10.4314/tjpr.v16i1.25]

[36] Khalil AT, Ovais M, Ullah I, *et al.* Sageretia thea (Osbeck.) mediated synthesis of zinc oxide nanoparticles and its biological applications. Nanomedicine (Lond) 2017; 12(15): 1767-89. [http://dx.doi.org/10.2217/nnm-2017-0124] [PMID: 28699838]

[37] Moulton MC, Braydich-Stolle LK, Nadagouda MN, Kunzelman S, Hussain SM, Varma RS. Synthesis, characterization and biocompatibility of "green" synthesized silver nanoparticles using tea polyphenols. Nanoscale 2010; 2(5): 763-70. [http://dx.doi.org/10.1039/c0nr00046a] [PMID: 20648322]

[38] Mukherjee S, Chowdhury D, Kotcherlakota R, *et al.* Potential theranostics application of bio-synthesized silver nanoparticles (4-in-1 system). Theranostics 2014; 4(3): 316-35. [http://dx.doi.org/10.7150/thno.7819] [PMID: 24505239]

[39] Atarod M, Nasrollahzadeh M, Mohammad Sajadi S. *Euphorbia heterophylla* leaf extract mediated green synthesis of Ag/TiO2 nanocomposite and investigation of its excellent catalytic activity for reduction of variety of dyes in water. J Colloid Interface Sci 2016; 462: 272-9. [http://dx.doi.org/10.1016/j.jcis.2015.09.073] [PMID: 26469545]

[40] Prasad C, Sreenivasulu K, Gangadhara S, Venkateswarlu P. Bio inspired green synthesis of Ni/Fe3O4 magnetic nanoparticles using Moringa oleifera leaves extract: A magnetically recoverable catalyst for organic dye degradation in aqueous solution. J Alloys Compd 2017; 700: 252-8. [http://dx.doi.org/10.1016/j.jallcom.2016.12.363]

[41] Nasrollahzadeh M, Atarod M, Sajadi SM. Green synthesis of the Cu/Fe3O4 nanoparticles using Morinda morindoides leaf aqueous extract: a highly efficient magnetically separable catalyst for the reduction of organic dyes in aqueous medium at room temperature. Appl Surf Sci 2016; 364: 636-44. [http://dx.doi.org/10.1016/j.apsusc.2015.12.209]

[42] Mayedwa N, Mongwaketsi N, Khamlich S, Kaviyarasu K, Matinise N, Maaza M. Green synthesis of zin tin oxide (ZnSnO3) nanoparticles using Aspalathus Linearis natural extracts: Structural, morphological, optical and electrochemistry study. Appl Surf Sci 2018; 446: 250-7. [http://dx.doi.org/10.1016/j.apsusc.2017.12.161]

[43] Gao W, Thamphiwatana S, Angsantikul P, Zhang L. Nanoparticle approaches against bacterial infections. Wiley Interdiscip Rev Nanomed Nanobiotechnol 2014; 6(6): 532-47. [http://dx.doi.org/10.1002/wnan.1282] [PMID: 25044325]

[44] El-Say KM, El-Sawy HS. Polymeric nanoparticles: Promising platform for drug delivery. Int J Pharm 2017; 528(1-2): 675-91. [http://dx.doi.org/10.1016/j.ijpharm.2017.06.052] [PMID: 28629982]

[45] Banik BL, Fattahi P, Brown JL. Polymeric nanoparticles: the future of nanomedicine. Wiley Interdiscip Rev Nanomed Nanobiotechnol 2016; 8(2): 271-99. [http://dx.doi.org/10.1002/wnan.1364] [PMID: 26314803]

[46] Jijie R, Barras A, Teodorescu F, Boukherroub R, Szunerits S. Advancements on the molecular design of nanoantibiotics: current level of development and future challenges. Mol Syst Des Eng 2017; 2(4): 349-69. [http://dx.doi.org/10.1039/C7ME00048K]

[47] Toti US, Guru BR, Hali M, *et al.* Targeted delivery of antibiotics to intracellular chlamydial infections using PLGA nanoparticles. Biomaterials 2011; 32(27): 6606-13. [http://dx.doi.org/10.1016/j.biomaterials.2011.05.038] [PMID: 21652065]

[48] Ramteke S, Jain NK. Clarithromycin- and omeprazole-containing gliadin nanoparticles for the treatment of Helicobacter pylori. J Drug Target 2008; 16(1): 65-72. [http://dx.doi.org/10.1080/10611860701733278] [PMID: 18172822]

[49] Ramteke S, Ganesh N, Bhattacharya S, Jain NK. Amoxicillin, clarithromycin, and omeprazole based targeted nanoparticles for the treatment of H. pylori. J Drug Target 2009; 17(3): 225-34. [http://dx.doi.org/10.1080/10611860902718649] [PMID: 19241256]

[50] Ahmad Z, Pandey R, Sharma S, Khuller GK. Alginate nanoparticles as antituberculosis drug carriers: formulation development, pharmacokinetics and therapeutic potential. Indian J Chest Dis Allied Sci 2006; 48(3): 171-6. [PMID: 18610673]

[51] Deacon J, Abdelghany SM, Quinn DJ, *et al.* Antimicrobial efficacy of tobramycin polymeric nanoparticles for Pseudomonas aeruginosa infections in cystic fibrosis: formulation, characterisation and functionalisation with dornase alfa (DNase). J Control Release 2015; 198: 55-61. [http://dx.doi.org/10.1016/j.jconrel.2014.11.022] [PMID: 25481442]

[52] Baelo A, Levato R, Julián E, *et al.* Disassembling bacterial extracellular matrix with DNase-coated nanoparticles to enhance antibiotic delivery in biofilm infections. J Control Release 2015; 209: 150-8. [http://dx.doi.org/10.1016/j.jconrel.2015.04.028] [PMID: 25913364]

[53] Zou X, Zhang L, Wang Z, Luo Y. Mechanisms of the antimicrobial activities of graphene materials. J Am Chem Soc 2016; 138(7): 2064-77. [http://dx.doi.org/10.1021/jacs.5b11411] [PMID: 26824139]

[54] Szunerits S, Boukherroub R. Antibacterial activity of graphene-based materials. J Mater Chem B Mater Biol Med 2016; 4(43): 6892-912. [http://dx.doi.org/10.1039/C6TB01647B]

[55] Karahan HE, Wiraja C, Xu C, *et al.* Graphene materials in antimicrobial nanomedicine: Current status and future perspectives. Adv Healthc Mater 2018; 7(13): e1701406. [http://dx.doi.org/10.1002/adhm.201701406] [PMID: 29504283]

[56] Pan WY, Huang CC, Lin TT, *et al.* Synergistic antibacterial effects of localized heat and oxidative stress caused by hydroxyl radicals mediated by graphene/iron oxide-based nanocomposites. Nanomedicine (Lond) 2016; 12(2): 431-8. [http://dx.doi.org/10.1016/j.nano.2015.11.014] [PMID: 26711965]

[57] Akhavan O, Ghaderi E. Toxicity of graphene and graphene oxide nanowalls against bacteria. ACS Nano 2010; 4(10): 5731-6. [http://dx.doi.org/10.1021/nn101390x] [PMID: 20925398]

[58] Skariyachan S, Parveen A, Garka S. Nanoparticle Fullerene (C60) demonstrated stable binding with antibacterial potential towards probable targets of drug resistant Salmonella typhi - a computational perspective and *in vitro* investigation. J Biomol Struct Dyn 2017; 35(16): 3449-68. [http://dx.doi.org/10.1080/07391102.2016.1257441] [PMID: 27817242]

[59] Liu S, Zeng TH, Hofmann M, *et al.* Antibacterial activity of graphite, graphite oxide, graphene oxide, and reduced graphene oxide: membrane and oxidative stress. ACS Nano 2011; 5(9): 6971-80. [http://dx.doi.org/10.1021/nn202451x] [PMID: 21851105]

[60] Mejías Carpio IE, Santos CM, Wei X, Rodrigues DF. Toxicity of a polymer-graphene oxide composite against bacterial planktonic cells, biofilms, and mammalian cells. Nanoscale 2012; 4(15): 4746-56. [http://dx.doi.org/10.1039/c2nr30774j] [PMID: 22751735]

[61] Cai X, Tan S, Lin M, *et al.* Synergistic antibacterial brilliant blue/reduced graphene oxide/quaternary phosphonium salt composite with excellent water solubility and specific targeting capability. Langmuir 2011; 27(12): 7828-35. [http://dx.doi.org/10.1021/la201499s] [PMID: 21585214]

[62] Zhao R, Lv M, Li Y, *et al.* Stable nanocomposite based on PEGylated and silver nanoparticles loaded graphene oxide for long-term antibacterial activity. ACS Appl Mater Interfaces 2017; 9(18): 15328-41. [http://dx.doi.org/10.1021/acsami.7b03987] [PMID: 28422486]

[63] Xu WP, Zhang LC, Li JP, Lu Y, Li HH, Ma YN, *et al.* Facile synthesis of silver@ graphene oxide nanocomposites and their enhanced antibacterial properties. J Mater Chem 2011; 21(12): 4593-7. [http://dx.doi.org/10.1039/c0jm03376f]

[64] Tang J, Chen Q, Xu L, *et al.* Graphene oxide-silver nanocomposite as a highly effective antibacterial agent with species-specific mechanisms. ACS Appl Mater Interfaces 2013; 5(9): 3867-74. [http://dx.doi.org/10.1021/am4005495] [PMID: 23586616]

[65] Lu F, Zhang S, Gao H, Jia H, Zheng L. Protein-decorated reduced oxide graphene composite and its application to SERS. ACS Appl Mater Interfaces 2012; 4(6): 3278-84. [http://dx.doi.org/10.1021/am300634n] [PMID: 22692825]

[66] Klasen HJ. Historical review of the use of silver in the treatment of burns. I. Early uses. Burns 2000; 26(2): 117-30. [http://dx.doi.org/10.1016/S0305-4179(99)00108-4] [PMID: 10716354]

[67] Raimondi F, Scherer GG, Kötz R, Wokaun A. Nanoparticles in energy technology: examples from electrochemistry and catalysis. Angew Chem Int Ed Engl 2005; 44(15): 2190-209. [http://dx.doi.org/10.1002/anie.200460466] [PMID: 15776488]

[68] Sondi I, Salopek-Sondi B. Silver nanoparticles as antimicrobial agent: a case study on E. coli as a model for Gram-negative bacteria. J Colloid Interface Sci 2004; 275(1): 177-82. [http://dx.doi.org/10.1016/j.jcis.2004.02.012] [PMID: 15158396]

[69] Dastjerdi R, Montazer M. A review on the application of inorganic nano-structured materials in the modification of textiles: focus on anti-microbial properties. Colloids Surf B Biointerfaces 2010; 79(1): 5-18.
[http://dx.doi.org/10.1016/j.colsurfb.2010.03.029] [PMID: 20417070]

[70] Uğur ŞS, Sarıışık M, Aktaş AH, Uçar MÇ, Erden E. Modifying of cotton fabric surface with nano-ZnO multilayer films by layer-by-layer deposition method. Nanoscale Res Lett 2010; 5(7): 1204-10.
[http://dx.doi.org/10.1007/s11671-010-9627-9] [PMID: 20596450]

[71] Pratap Reddy M, Venugopal A, Subrahmanyam M. Hydroxyapatite-supported Ag-TiO2 as Escherichia coli disinfection photocatalyst. Water Res 2007; 41(2): 379-86.
[http://dx.doi.org/10.1016/j.watres.2006.09.018] [PMID: 17137613]

[72] Kühn KP, Chaberny IF, Massholder K, *et al.* Disinfection of surfaces by photocatalytic oxidation with titanium dioxide and UVA light. Chemosphere 2003; 53(1): 71-7.
[http://dx.doi.org/10.1016/S0045-6535(03)00362-X] [PMID: 12892668]

[73] Johnston HJ, Hutchison G, Christensen FM, Peters S, Hankin S, Stone V. A review of the *in vivo* and *in vitro* toxicity of silver and gold particulates: particle attributes and biological mechanisms responsible for the observed toxicity. Crit Rev Toxicol 2010; 40(4): 328-46.
[http://dx.doi.org/10.3109/10408440903453074] [PMID: 20128631]

[74] Li Q, Mahendra S, Lyon DY, *et al.* Antimicrobial nanomaterials for water disinfection and microbial control: potential applications and implications. Water Res 2008; 42(18): 4591-602.
[http://dx.doi.org/10.1016/j.watres.2008.08.015] [PMID: 18804836]

[75] Qi L, Xu Z, Jiang X, Hu C, Zou X. Preparation and antibacterial activity of chitosan nanoparticles. Carbohydr Res 2004; 339(16): 2693-700.
[http://dx.doi.org/10.1016/j.carres.2004.09.007] [PMID: 15519328]

[76] Rabea EI, Badawy MET, Stevens CV, Smagghe G, Steurbaut W. Chitosan as antimicrobial agent: applications and mode of action. Biomacromolecules 2003; 4(6): 1457-65.
[http://dx.doi.org/10.1021/bm034130m] [PMID: 14606868]

[77] Kang S, Pinault M, Pfefferle LD, Elimelech M. Single-walled carbon nanotubes exhibit strong antimicrobial activity. Langmuir 2007; 23(17): 8670-3.
[http://dx.doi.org/10.1021/la701067r] [PMID: 17658863]

[78] Aslan S, Loebick CZ, Kang S, Elimelech M, Pfefferle LD, Van Tassel PR. Antimicrobial biomaterials based on carbon nanotubes dispersed in poly(lactic-co-glycolic acid). Nanoscale 2010; 2(9): 1789-94.
[http://dx.doi.org/10.1039/c0nr00329h] [PMID: 20680202]

[79] Vecitis CD, Zodrow KR, Kang S, Elimelech M. Electronic-structure-dependent bacterial cytotoxicity of single-walled carbon nanotubes. ACS Nano 2010; 4(9): 5471-9.
[http://dx.doi.org/10.1021/nn101558x] [PMID: 20812689]

[80] Sayes CM, Gobin AM, Ausman KD, Mendez J, West JL, Colvin VL. Nano-C60 cytotoxicity is due to lipid peroxidation. Biomaterials 2005; 26(36): 7587-95.
[http://dx.doi.org/10.1016/j.biomaterials.2005.05.027] [PMID: 16005959]

[81] Markovic Z, Todorovic-Markovic B, Kleut D, *et al.* The mechanism of cell-damaging reactive oxygen generation by colloidal fullerenes. Biomaterials 2007; 28(36): 5437-48.
[http://dx.doi.org/10.1016/j.biomaterials.2007.09.002] [PMID: 17884160]

[82] Lyon DY, Brunet L, Hinkal GW, Wiesner MR, Alvarez PJ. Antibacterial activity of fullerene water suspensions (nC60) is not due to ROS-mediated damage. Nano Lett 2008; 8(5): 1539-43.
[http://dx.doi.org/10.1021/nl0726398] [PMID: 18410152]

[83] Loew S, Fahr A, May S. Modeling the release kinetics of poorly water-soluble drug molecules from liposomal nanocarriers. J Drug Deliv 2011; 2011: 376548.
[http://dx.doi.org/10.1155/2011/376548] [PMID: 21773045]

[84] Zeng L, An L, Wu X. Modeling drug-carrier interaction in the drug release from nanocarriers. J Drug Deliv 2011; 2011: 370308.
[http://dx.doi.org/10.1155/2011/370308] [PMID: 21845225]

[85] Drummond DC, Noble CO, Guo Z, Hong K, Park JW, Kirpotin DB. Development of a highly active nanoliposomal irinotecan using a novel intraliposomal stabilization strategy. Cancer Res 2006; 66(6): 3271-7.
[http://dx.doi.org/10.1158/0008-5472.CAN-05-4007] [PMID: 16540680]

[86] Johnston MJ, Semple SC, Klimuk SK, *et al.* Therapeutically optimized rates of drug release can be achieved by varying the drug-to-lipid ratio in liposomal vincristine formulations. Biochim Biophys Acta 2006; 1758(1): 55-64.
[http://dx.doi.org/10.1016/j.bbamem.2006.01.009] [PMID: 16487476]

[87] Joguparthi V, Anderson BD. Liposomal delivery of hydrophobic weak acids: enhancement of drug retention using a high intraliposomal pH. J Pharm Sci 2008; 97(1): 433-54.
[http://dx.doi.org/10.1002/jps.21135] [PMID: 17918731]

[88] Modi S, Anderson BD. Determination of drug release kinetics from nanoparticles: overcoming pitfalls of the dynamic dialysis method. Mol Pharm 2013; 10(8): 3076-89.
[http://dx.doi.org/10.1021/mp400154a] [PMID: 23758289]

[89] Kalhapure RS, Suleman N, Mocktar C, Seedat N, Govender T. Nanoengineered drug delivery systems for enhancing antibiotic therapy. J Pharm Sci 2015; 104(3): 872-905.
[http://dx.doi.org/10.1002/jps.24298] [PMID: 25546108]

[90] Lu XY, Wu DC, Li ZJ, Chen GQ. Polymer nanoparticles. Progress in molecular biology and translational science. 104: Elsevier 2011; 299-323
[http://dx.doi.org/10.1016/B978-0-12-416020-0.00007-3]

[91] Modi S, Anderson BD. Bilayer composition, temperature, speciation effects and the role of bilayer chain ordering on partitioning of dexamethasone and its 21-phosphate. Pharm Res 2013; 30(12): 3154-69.
[http://dx.doi.org/10.1007/s11095-013-1143-z] [PMID: 23884570]

[92] Joguparthi V, Xiang TX, Anderson BD. Liposome transport of hydrophobic drugs: gel phase lipid bilayer permeability and partitioning of the lactone form of a hydrophobic camptothecin, DB-67. J Pharm Sci 2008; 97(1): 400-20.
[http://dx.doi.org/10.1002/jps.21125] [PMID: 17879989]

[93] Xiang TX, Anderson BD. Liposomal drug transport: a molecular perspective from molecular dynamics simulations in lipid bilayers. Adv Drug Deliv Rev 2006; 58(12-13): 1357-78.
[http://dx.doi.org/10.1016/j.addr.2006.09.002] [PMID: 17092601]

[94] Lindner LH, Hossann M. Factors affecting drug release from liposomes. Curr Opin Drug Discov Devel 2010; 13(1): 111-23.
[PMID: 20047152]

[95] Kalhapure RS, Jadhav M, Rambharose S, *et al.* pH-responsive chitosan nanoparticles from a novel twin-chain anionic amphiphile for controlled and targeted delivery of vancomycin. Colloids Surf B Biointerfaces 2017; 158: 650-7.
[http://dx.doi.org/10.1016/j.colsurfb.2017.07.049] [PMID: 28763772]

[96] Saidykhan L, Abu Bakar MZ, Rukayadi Y, Kura AU, Latifah SY. Development of nanoantibiotic delivery system using cockle shell-derived aragonite nanoparticles for treatment of osteomyelitis. Int J Nanomedicine 2016; 11: 661-73.
[http://dx.doi.org/10.2147/IJN.S95885] [PMID: 26929622]

[97] Sikwal DR, Kalhapure RS, Rambharose S, *et al.* Polyelectrolyte complex of vancomycin as a nanoantibiotic: Preparation, *in vitro* and in silico studies. Mater Sci Eng C 2016; 63: 489-98.
[http://dx.doi.org/10.1016/j.msec.2016.03.019] [PMID: 27040243]

[98] Wang YL, He M, Miron RJ, Chen AY, Zhao YB, Zhang YF. Temperature/pH-sensitive nanoantibiotics and their sequential assembly for optimal collaborations between antibacterial and immunoregulation. ACS appl Mater Interfaces 2017; 9(37): 31589-99. [http://dx.doi.org/10.1021/acsami.7b10384] [PMID: 28856893]

[99] Hussein-Al-Ali SH, El Zowalaty ME, Hussein MZ, Ismail M, Webster TJ. Synthesis, characterization, controlled release, and antibacterial studies of a novel streptomycin chitosan magnetic nanoantibiotic. Int J Nanomedicine 2014; 9: 549-57. [PMID: 24549109]

[100] Assali M, Zaid AN, Abdallah F, Almasri M, Khayyat R. Single-walled carbon nanotubes-ciprofloxacin nanoantibiotic: strategy to improve ciprofloxacin antibacterial activity. Int J Nanomedicine 2017; 12: 6647-59. [http://dx.doi.org/10.2147/IJN.S140625] [PMID: 28924348]

[101] Duan F, Feng X, Jin Y, *et al.* Metal-carbenicillin framework-based nanoantibiotics with enhanced penetration and highly efficient inhibition of MRSA. Biomaterials 2017; 144: 155-65. [http://dx.doi.org/10.1016/j.biomaterials.2017.08.024] [PMID: 28834764]

[102] Ayaz M, Subhan F, Sadiq A, Ullah F, Ahmed J, Sewell RD. Cellular efflux transporters and the potential role of natural products in combating efflux mediated drug resistance. Front Biosci 2017; 22: 732-56. [http://dx.doi.org/10.2741/4513] [PMID: 27814643]

[103] Ayaz M, Subhan F, Ahmed J, *et al.* Sertraline enhances the activity of antimicrobial agents against pathogens of clinical relevance. J Biol Res (Thessalon) 2015; 22(1): 4. [http://dx.doi.org/10.1186/s40709-015-0028-1] [PMID: 26029671]

[104] Ayaz M, Subhan F, Ahmed J, *et al.* Citalopram and venlafaxine differentially augments antimicrobial properties Of antibiotics. Acta Pol Pharm Drug Res 2015; 72(6): 1269-78.

[105] Ovais M, Ahmad I, Khalil AT, *et al.* Wound healing applications of biogenic colloidal silver and gold nanoparticles: recent trends and future prospects. Appl Microbiol Biotechnol 2018; 102(10): 4305-18. [http://dx.doi.org/10.1007/s00253-018-8939-z] [PMID: 29589095]

[106] Ovais M, Khalil AT, Raza A, *et al.* Multifunctional theranostic applications of biocompatible green-synthesized colloidal nanoparticles. Appl Microbiol Biotechnol 2018; 102(10): 4393-408. [http://dx.doi.org/10.1007/s00253-018-8928-2] [PMID: 29594356]

[107] Pelgrift RY, Friedman AJ. Nanotechnology as a therapeutic tool to combat microbial resistance. Adv Drug Deliv Rev 2013; 65(13-14): 1803-15. [http://dx.doi.org/10.1016/j.addr.2013.07.011] [PMID: 23892192]

[108] Gupta A, Landis RF, Rotello VM. Nanoparticle-Based Antimicrobials: Surface Functionality is Critical. F1000Research 2016, 5(F1000 Faculty Rev):364

[109] Matsumura Y, Yoshikata K, Kunisaki S, Tsuchido T. Mode of bactericidal action of silver zeolite and its comparison with that of silver nitrate. Appl Environ Microbiol 2003; 69(7): 4278-81. [http://dx.doi.org/10.1128/AEM.69.7.4278-4281.2003] [PMID: 12839814]

[110] Rai M, Yadav A, Gade A. Silver nanoparticles as a new generation of antimicrobials. Biotechnol Adv 2009; 27(1): 76-83. [http://dx.doi.org/10.1016/j.biotechadv.2008.09.002] [PMID: 18854209]

[111] Lok CN, Ho CM, Chen R, *et al.* Proteomic analysis of the mode of antibacterial action of silver nanoparticles. J Proteome Res 2006; 5(4): 916-24. [http://dx.doi.org/10.1021/pr0504079] [PMID: 16602699]

[112] Kim JS, Kuk E, Yu KN, *et al.* Antimicrobial effects of silver nanoparticles. Nanomedicine (Lond) 2007; 3(1): 95-101. [http://dx.doi.org/10.1016/j.nano.2006.12.001] [PMID: 17379174]

[113] Wang L, Hu C, Shao L. The antimicrobial activity of nanoparticles: present situation and prospects for the future. Int J Nanomedicine 2017; 12: 1227-49. [http://dx.doi.org/10.2147/IJN.S121956] [PMID: 28243086]

[114] Feng QL, Wu J, Chen GQ, Cui FZ, Kim TN, Kim JO. A mechanistic study of the antibacterial effect of silver ions on *Escherichia coli* and *Staphylococcus aureus*. J Biomed Mater Res 2000; 52(4): 662-8. [http://dx.doi.org/10.1002/1097-4636(20001215)52:4<662::AID-JBM10>3.0.CO;2-3] [PMID: 11033548]

[115] Shrivastava S, Bera T, Roy A, Singh G, Ramachandrarao P, Dash D. Characterization of enhanced antibacterial effects of novel silver nanoparticles. Nanotechnology 2007; 18(22): 225103. [http://dx.doi.org/10.1088/0957-4484/18/22/225103]

[116] Gilbert P, Collier PJ, Brown MR. Influence of growth rate on susceptibility to antimicrobial agents: biofilms, cell cycle, dormancy, and stringent response. Antimicrob Agents Chemother 1990; 34(10): 1865-8. [http://dx.doi.org/10.1128/AAC.34.10.1865] [PMID: 2291653]

[117] Hajipour MJ, Fromm KM, Ashkarran AA, *et al.* Antibacterial properties of nanoparticles. Trends Biotechnol 2012; 30(10): 499-511. [http://dx.doi.org/10.1016/j.tibtech.2012.06.004] [PMID: 22884769]

[118] Silhavy TJ, Kahne D, Walker S. The bacterial cell envelope. Cold spring harb Perspect Biol 2010; 2(5): a000414. [http://dx.doi.org/10.1101/cshperspect.a000414] [PMID: 20452953]

[119] Delcour AH. Outer membrane permeability and antibiotic resistance. Biochim Biophys Acta 2009; 1794(5): 808-16. [http://dx.doi.org/10.1016/j.bbapap.2008.11.005] [PMID: 19100346]

[120] Habimana O, Le Goff C, Juillard V, *et al.* Positive role of cell wall anchored proteinase PrtP in adhesion of lactococci. BMC Microbiol 2007; 7(1): 36. [http://dx.doi.org/10.1186/1471-2180-7-36] [PMID: 17474995]

[121] Kearns DB. A field guide to bacterial swarming motility. Nat Rev Microbiol 2010; 8(9): 634-44. [http://dx.doi.org/10.1038/nrmicro2405] [PMID: 20694026]

[122] Flemming HC, Neu TR, Wozniak DJ. The EPS matrix: the "house of biofilm cells". J Bacteriol 2007; 189(22): 7945-7. [http://dx.doi.org/10.1128/JB.00858-07] [PMID: 17675377]

[123] Davies DG, Parsek MR, Pearson JP, Iglewski BH, Costerton JW, Greenberg EP. The involvement of cell-to-cell signals in the development of a bacterial biofilm. Science 1998; 280(5361): 295-8. [http://dx.doi.org/10.1126/science.280.5361.295] [PMID: 9535661]

[124] Simões M, Cleto S, Pereira MO, Vieira MJ. Influence of biofilm composition on the resistance to detachment. Water Sci Technol 2007; 55(8-9): 473-80. [http://dx.doi.org/10.2166/wst.2007.293] [PMID: 17547019]

[125] Høiby N, Bjarnsholt T, Givskov M, Molin S, Ciofu O. Antibiotic resistance of bacterial biofilms. Int J Antimicrob Agents 2010; 35(4): 322-32. [http://dx.doi.org/10.1016/j.ijantimicag.2009.12.011] [PMID: 20149602]

[126] Madsen JS, Burmølle M, Hansen LH, Sørensen SJ. The interconnection between biofilm formation and horizontal gene transfer. FEMS Immunol Med Microbiol 2012; 65(2): 183-95. [http://dx.doi.org/10.1111/j.1574-695X.2012.00960.x] [PMID: 22444301]

[127] Smith AW. Biofilms and antibiotic therapy: is there a role for combating bacterial resistance by the use of novel drug delivery systems? Adv Drug Deliv Rev 2005; 57(10): 1539-50. [http://dx.doi.org/10.1016/j.addr.2005.04.007] [PMID: 15950314]

[128] Sheng Z, Liu Y. Effects of silver nanoparticles on wastewater biofilms. Water Res 2011; 45(18):

6039-50.
[http://dx.doi.org/10.1016/j.watres.2011.08.065] [PMID: 21940033]

[129] Radzig MA, Nadtochenko VA, Koksharova OA, Kiwi J, Lipasova VA, Khmel IA. Antibacterial effects of silver nanoparticles on gram-negative bacteria: influence on the growth and biofilms formation, mechanisms of action. Colloids Surf B Biointerfaces 2013; 102: 300-6.
[http://dx.doi.org/10.1016/j.colsurfb.2012.07.039] [PMID: 23006569]

[130] Hou J, Miao L, Wang C, *et al.* Inhibitory effects of ZnO nanoparticles on aerobic wastewater biofilms from oxygen concentration profiles determined by microelectrodes. J Hazard Mater 2014; 276: 164-70.
[http://dx.doi.org/10.1016/j.jhazmat.2014.04.048] [PMID: 24880618]

[131] Sahoo SK, Misra R, Parveen S. Nanoparticles: A boon to drug delivery, therapeutics, diagnostics and imaging. In: Nanomedicine in Cancer. edn.: Pan Stanford; 2017: 73-124

[132] Forier K, Raemdonck K, De Smedt SC, Demeester J, Coenye T, Braeckmans K. Lipid and polymer nanoparticles for drug delivery to bacterial biofilms. J Control Release 2014; 190: 607-23.
[http://dx.doi.org/10.1016/j.jconrel.2014.03.055] [PMID: 24794896]

[133] Meers P, Neville M, Malinin V, *et al.* Biofilm penetration, triggered release and *in vivo* activity of inhaled liposomal amikacin in chronic Pseudomonas aeruginosa lung infections. J Antimicrob Chemother 2008; 61(4): 859-68.
[http://dx.doi.org/10.1093/jac/dkn059] [PMID: 18305202]

[134] Chakraborty SP, Sahu SK, Mahapatra SK, *et al.* Nanoconjugated vancomycin: new opportunities for the development of anti-VRSA agents. Nanotechnology 2010; 21(10): 105103.
[http://dx.doi.org/10.1088/0957-4484/21/10/105103] [PMID: 20154376]

[135] Wood TK, Knabel SJ, Kwan BW. Bacterial persister cell formation and dormancy. Appl Environ Microbiol 2013; 79(23): 7116-21.
[http://dx.doi.org/10.1128/AEM.02636-13] [PMID: 24038684]

[136] Bald D, Koul A. Advances and strategies in discovery of new antibacterials for combating metabolically resting bacteria. Drug Discov Today 2013; 18(5-6): 250-5.
[http://dx.doi.org/10.1016/j.drudis.2012.09.007] [PMID: 23032727]

[137] Maurin M, Raoult D. Use of aminoglycosides in treatment of infections due to intracellular bacteria. Antimicrob Agents Chemother 2001; 45(11): 2977-86.
[http://dx.doi.org/10.1128/AAC.45.11.2977-2986.2001] [PMID: 11600345]

[138] Briones E, Colino CI, Lanao JM. Delivery systems to increase the selectivity of antibiotics in phagocytic cells. J Control Release 2008; 125(3): 210-27.
[http://dx.doi.org/10.1016/j.jconrel.2007.10.027] [PMID: 18077047]

[139] Sachetelli S, Khalil H, Chen T, Beaulac C, Sénéchal S, Lagacé J. Demonstration of a fusion mechanism between a fluid bactericidal liposomal formulation and bacterial cells. Biochim Biophys Acta 2000; 1463(2): 254-66.
[http://dx.doi.org/10.1016/S0005-2736(99)00217-5] [PMID: 10675504]

[140] Oh YK, Nix DE, Straubinger RM. Formulation and efficacy of liposome-encapsulated antibiotics for therapy of intracellular Mycobacterium avium infection. Antimicrob Agents Chemother 1995; 39(9): 2104-11.
[http://dx.doi.org/10.1128/AAC.39.9.2104] [PMID: 8540724]

[141] Lutwyche P, Cordeiro C, Wiseman DJ, *et al.* Intracellular delivery and antibacterial activity of gentamicin encapsulated in pH-sensitive liposomes. Antimicrob Agents Chemother 1998; 42(10): 2511-20.
[http://dx.doi.org/10.1128/AAC.42.10.2511] [PMID: 9756749]

CHAPTER 5

Cranberry Juice and Other Functional Foods in Urinary Tract Infections in Women: A Review of Actual Evidence and Main Challenges

Rebeca Monroy-Torres[*,1] and **Ana Karen Medina-Jiménez**[2]

[1] *Environmental Nutrition and Food Security Laboratory, Medicine and Nutrition Department, Health and Science Division, University of Guanajuato, Guanajuato, Mexico*

[2] *University Observatory of Food and Nutritional Security of the State of Guanajuato/ Observatorio Universitario de Seguridad Alimentaria y Nutricional del Estado de Guanajuato (OUSANEG), Mexico*

Abstract: An urinary tract infection (UTI) is defined as the presence of signs and symptoms caused by a pathogenic external agent, which can be of viral or bacterial origin, and the most common bacterial agent is E. coli. An UTI can affect upper or lower urinary tract. Signs and symptoms could include, but are not limited to: dysuria, urinary urgency and frequency, sensation of bladder fullness, lower abdominal discomfort, suprapubic tenderness, flank pain, costovertebral angle tenderness, bloody urine, fevers, chills, and malaise. UTIs are the most common infections seen in medical practice in women of reproductive age and adults in general. Most UTI are caused by *enterobacteriaceae* such as *Escherichia Coli,* which is a gram-negative bacterium present in 80% of these infections. The presence of fimbriae in some strains determines its ability to colonize the urethra and migrate to the bladder. The antibiotic resistant genes of this bacterium have resulted in the need to search for alternatives to antibiotic therapy, which has become less effective due to misuse and abuse, and it is now a major worldwide problem. Several nutrients have been associated with better urinary health, mainly vitamin C, probiotics and flavonoids. Some studies have shown that frequent consumption of yogurt with probiotics and fruit juices rich in vitamin C decrease the recurrence of UTI in women. Vitamin C and flavonoids are considered non-enzymatic antioxidants that slow down the production of free radicals and oxidation, thus strengthening the immune system. The flavonoids in cranberries have a deleterious effect on *E. Coli* fimbriae, causing the bacteria to lose its grip. Diet and nutritional status, especially regarding the consumption of certain antioxidants, have a particularly strong relationship with urinary function and health. These findings may lead to defining programs for nutritional and dietary monitoring which could improve health and nutritional prognosis for women and their newborns, as well as prevent obstetric complications from Urinary Tract Infection. The primary objective of this

[*] **Corresponding author Rebeca Monroy Torres:** Environmental Nutrition and Food Security Laboratory, Medicine and Nutrition Department, Health and Science Division, University of Guanajuato, Guanajuato, Mexico; Tel: +52(477)2674900, Ext: 3677; Fax: +52(477)2674900, Ext: 3677; E-mail: rmonroy79@gmail.com.

Atta-ur-Rahman (Ed.)

review is to analyze the most conclusive evidence about the effectiveness of these options and to integrate other alternatives in future treatment that can fight against antimicrobial resistance and impact on morbidity, mortality and health cost.

Keywords: Antibiotic Resistance, Bacteria, Cramberry Juice, Eschecrichia Coli, Environment, Functional Foods, Flavonoids, Immunonutrition, Nutritional Status, Prevention, Urinary Tract Infection.

SELECTION CRITERIA FOR HIS CHAPTER REVIEW

We searched databases such as PubMed, MEDLINE, Cochrane Central Library of Controlled Trials (Cochrane Library), Research Gate, Google Academic Research conducted in 2003 to 2015 to present experimental designs of intervention over prophylaxis, prevention, treatment of UTI with natural alternatives, as nutrients or preparations in different presentations other than antibiotics.

INTRODUCTION

Epidemiology of UTI

An urinary tract infection is defined as the presence of signs and symptoms caused by a pathogenic external agent, wich can be viral or bacterial, and the most common bacterial agent is E. coli. An UTI can affect upper or lower urinary tract. Signs and symptoms could include, but are not limited to: dysuria, urinary urgency and frequency, sensation of bladder fullness, lower abdominal discomfort, suprapubic tenderness, flank pain, costovertebral angle tenderness, bloody urine, fevers, chills, and malaise. Concerning diagnostic studies for UTI, options include dipstick, urinalysis, and culture. Currently, for the diagnosis of UTI emphasis is done on the detection of pyuria. In most situations a positive leukocyte esterase dipstick test would suffice to support the diagnosis, more than 10 white blood cells (WBCs)/mL in unspun fresh urine by counting leukocytes using a hemocytometer chamber also contributes to diagnosis. It is very important to mention that this and other laboratory findings should be correlated with clinical findings in patients [1].

Adult Population in General

UTIs continue to be a global public health problem, they are associated with nearly 150 million deaths per year [2]. Frequency is higher in women (40-50%) than in men, who can develop them just by 5% [3]. According to the US Centers for Disease Control and Prevention (CDCP), in 2005, 4 million medical consultations were due to UTI, with annual high costs of $ 1.6 billion [4].

In Mexico, the prevalence of UTI has been increasing since 2003 [5]. In 2013, UTI remained as one of the leading causes of morbidity, with *Escherichia coli* being the main causative agent, followed by other bacteria of the genus *Klebsiella, Proteus* and *Staphylococcus* [6]. According to the National Epidemiological Surveillance System reported that by 2014, in Mexico, UTI ranked third among the main causes of national morbidity with 4,244,053 cases, as in the state of Guanajuato, where a prevalence of 171, 429 cases, and mainly affected the population group aged 25-44 [5].

Antibiotics are the best alternative to treat UTI, but the increase in bacterial resistance has generated the need to search for alternatives for preventing or treating UTI as the cranberry juice and other functional foods.

It is necessary to clarify that manifestations, causes, treatment and prognosis are different in non-pregnant women from pregnant women. Therefore, it is important to consider this review in giving an individualized treatment according to each clinical history.

Pregnant Women

The population group most likely to be affected by UTI are women, mainly in a physiological state such as pregnancy, due to the increased concentrations of steroid hormones and the pressure exerted by the uterus on ureters and bladder, favoring hypotonia and congestion that generate a greater predisposition to vesicoureteral reflux and urinary stasis [7]. Therefore, this disease could increase susceptibility and risks of adverse events to the fetus. Children and elderlies are also a vulnerable group. In case of women, the causal factor is anatomic because the female urethra is shorter and exposed to more bacteria [3]. There are treatment regimens for UTI in this age group, as presented in Table (**1**) [8]. Fosfomycin is the first-line treatment regimen due to its high sensitivity and its safety to use in pregnancy [9].

Table 1. Treatment schedules in pregnancy for UTIs.

UTI	Antibiotic	Doses	Duration (days)
Asymptomatic bacteriuria	Nitrofurantoin	100 mg, oral use, c/6 hours	7
	Amoxicillin / clavulanic acid	250/125 mg, 2 times/d	7
Acute cystitis	Nitrofurantoin	100 mg, oral use, c/6 hours	10
	Amoxicillin / clavulanic acid	250/125 mg, 2 times/d	10

(Table 1) cont.....

UTI	Antibiotic	Doses	Duration (days)
Acute pyelonephritis	Cefotaxime	1 g, IV, c/8 hours	10
	Ceftriaxone	1 g, IV, c/24 hours	10
	Amikacin	15 mg/kg/día,IV, 1 time/d	10
	Gentamicin	3.5 a 5 mg/kg/día, IV, 1 time/d	10

Reference: With permission of Pacheco-Gahbler C, *et al.* [8].

The UTI's are considered the second most common comorbidity in pregnant women, after anemia and diabetes with a prevalence of 19.7% and 13%, respectively. UTI may have two symptomatic and asymptomatic manifestations [10], the most common being asymptomatic (asymptomatic bacteriuria in lower pathways); while acute pyelonephritis is generated in 25 to 40% of treated cases treated as an asymptomatic bacteriuria, which is one of the main causes of hospitalization before delivery [11]. Other data show that 3 to 12% of pregnant women will present a urinary tract infection, of which 3 to 10% will present asymptomatic bacteriuria, with a risk of 20% to 30% of developing asymptomatic urinary tract infection if they are not treated appropriately [7].

Nutritional Status: Determinants of UTI

It is essential to consider the relationship between the nutritional status and the immunological state during pregnancy. In a study conducted by Posadas *et al.*, in 2014, found that a poor nutritional status during pregnancy increased up to 50% in patients with a poor diet (consumptions below the recommendations, for example, 67 to 77% of 100% of energy and nutrients requirement) low intake of fruits, proteins and vegetables with increased or decreased of weight. The quality of the diet measured mainly in form of nutrients rich in antioxidants, fiber and proteins play an important role in the prevention of infectious diseases [12].

Diabetes is a pathology that can increase the occurrence of UTI, mostly in the female population. It is relevant to take into consideration gestational diabetes as a comorbidity, since there is a high prevalence in México; almost 6 to 7% of pregnant women are afflicted by this pathology [13]. It has been studied that the mechanisms and pathogens that promote the presence of UTI in women with diabetes and women without diabetes is similar. However, one of the conditions that may be related to a greater susceptibility is glucosuria, whose large extended may facilitate the decrease of the phagocytic activity of leukocytes [14, 15]. Glycemic control, measured through glycosylated hemoglobin, has remained as a controversial risk factor, since no relationship was found between the two variables, so an adequate glycemic control clinic should be integrated [16 - 18]. It is opportune to mention that cardiometabolic diseases such as diabetes and hypertension are worldwide public Health problem; Mexico has one of the highest rates in the aforementioned diseases in addition to obesity and overweight. Obesity is the main risk factor associated with the development of metabolic and cardiovascular diseases [8]. According to the 2012 National Health and Nutrition Survey (NHNS), the diagnosis of diabetes was higher in women than in men, at national level (10.3% *vs.* 8.4%). Also, Mexico presents high rates of adolescent pregnancy and these women will present at least one case of UTI during their pregnancy; besides, 35% of this population is overweight and obese. High consumption of products with added sugar is also common, along with poor ingestion of fruits and vegetables [19].

Obesity, diabetes and hypertension generate and inflammatory process as well as immune system alterations, which could explain the increased prevalence of UTI in these patients, this is why the continuous search for alternate treatment is fundamental. It is necessary to create a nutritional scheme to approach this pathology, depending on UTI stage, as follows: a) prevention of UTI, b) treatment and c) symptom management.

Resistance to Antibiotics and Therapeutic Alternatives

Bacterial antibiotic resistance is generated by misuse and deficient practices in antibiotic prescriptions. Most evidence of this problem was between the 70's and 80's, and despite the advances in research and development, the progress in the generation of new antimicrobial drugs has been slow. In view of this, the World Health Organization's strategy [20] has generated postures and guidelines to contain antibiotic resistance by the promotion and design of new alternatives. Something interesting is that within these non-therapeutic alternatives, they have shown that maintaining a proper nutritional status that allows the immune system to cooperate as a first pathway or physiological mechanism that functions as prevention (Fig. **1**). Bacteria presents several mechanisms of resistance to antibiotics, such as *E. coli* and *Salmonell*a, that generate defense mechanisms against antibiotics, so they present genes that are responsible for the synthesis of membrane-carrying proteins, known as pumps of effluent. This is the foundation for high resistance to antibiotics since these pumps will only be activated in the presence of antibiotics. Bacterial genes coding for Efflux bombs are in the chromosome or present themselves in the form of plasmids [21].

E. coli and Klebsiella are the most common pathogenic microorganisms isolated in UTI, and are resistant to three of the antibiotics indicated in the clinical practice guidelines (ampicillin, trimethoprim-sulfamethoxazole and cephalosporins) this indicates a clear health problem since they are widely used in public health institutions [22].

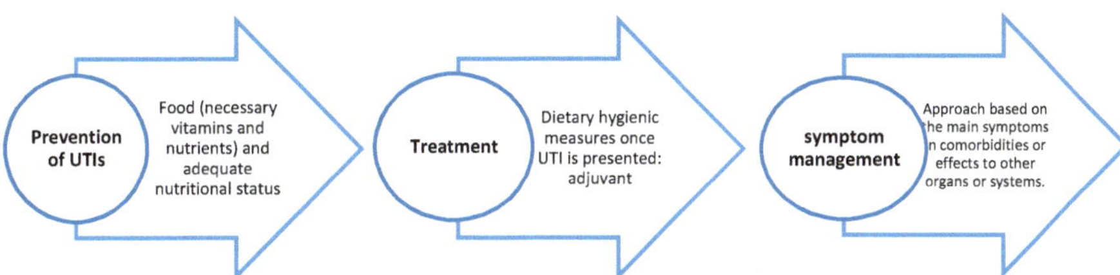

Fig. (1). Stages of the intervention of the nutritional approach depending on the clinical picture or status of the UTI.

A recent study has shown that *E. Coli* bacterial resistance against oflaxacin and cefixime has not been significant, however, resistance tends to increase in patients with uncontrolled diabetes [23]. In this sense, the maintenance of a glycemic control, mainly during gestational diabetes, could be considered as a factor that reduces the risk of occurrence of UTI.

Another recent study has revealed that *E. coli* bacterial resistance against oflaxacin and cefixime has not been significant, however, resistance tends to increase in patients with uncontrolled diabetes [24]. Again, evidence suggests that glycemic control is key to reduce the risk of occurrence of UTI.

Prevention should be considered in addition to exploring new and different alternatives for treatment, derived from the risks of antibiotic resistance. One of the approaches has been from the consumption of foods with nutrients or substances that possess (or show) antimicrobial action or properties without inducing the risk of resistance by their mechanisms. Foods or nutrients with more scientific evidence are cranberry, vitamin C and A, sugars such as mannose, vitamin A, probiotics as well as acidification or alkalinization of urinary pH" [24].

Although the most common treatment for UTI is antibiotics, its association to adverse effects and induction of antibiotic resistance Table **3**, has led to the search for effective therapeutic alternatives and preventive strategies. Such is the case with cranberry juice. It is now known that the most dangerous bacteria due to their high resistance is *E. coli*, which causes 90% of urinary tract infections, followed by the bacterial genera *Klebsiella, Pseudomona, Proteus* and *Acinetobacter* [6].

The list of the main antibiotics and their resistance percentage is observed in (Table **3**). A Health regulation to restrict the use and prescription of antibiotics was approved; this was done as a preventive measure, despite regulations, antibiotic resistance continues to arise, so institutions and health care providers have turned to foods and drugs that are capable of acidifying urine, like consumption of vitamin C, cranberry juice, among others. The following Tables **1** and **2** show the antimicrobial treatment regimens for UTI according to the order in which they should be prescribed and their availability [9].

Table 2. Treatment for UTI in adults.

UTI	Drug	Doses	Duration (days)
Acute non-complicated cystitis	Nitrofurantoin	100 mg, 3 a 4 times /d	5 a 7
	Ciproflaxacin	250 mg, 2 times/d	3
	Ciproflaxacin extended release	500 mg, 1 time/d	3
	Fosmfomycin trometamol	3g, the Single Dose Treatment	
	Trimetropim sulfamethoxazole	160/800 mg, 2 times/d	3
Acute pyelonephritis uncomplicated	Ciproflaxacin	500 mg, 2 times/d	14
	Ciproflaxacin extended release	1 g, 1 time/d	
	Levoflaxacin	500 mg, 1 time/d	
	Ceftibuten	400 mg, 1 time/d	
	Cefixime	400 mg, 1 time/d	
Complicated acute pyelonephritis	Ciproflaxacin	400 mg, IV, 2 times/d	
	Levoflaxacin	500 mg, IV, 1 time/d	
	Cefriaxone	1 g, IV, 2 times/d	
	Cefotaxime	1 g, IV, titrating doses	
	Amikacin	15 mg/kg/d, IV, 1 time/d	
	Gentamicin	3.5 a 5 mg/kg/d, IV, 1 time/d	
Bacterial Prostatitis	Ciproflaxacin	500 mg, 2 times/d	4 a 8 weeks
	Ciproflaxacin extended release	1 g, 1 time/d	
	Levoflaxacin	500 mg, 1 time/d	
	Oflaxacino	400 mg, 1 time/d	
	Cefriaxone *	1g, IV, 2 times/d	
*Alone or in association with aminoglycoside during acute systemic inflammatory response syndrome.			

Reference: With permission of Pacheco-Gahbler C, *et al.* Diagnosis and Treatment of Urinary Tract Infections (UTI). Academia Nacional de Medicina. Guías Mappa. 2010 [8].

Table 3. Prevalence of antibiotic resistance of 119 isolates of uropathogenic *Escherichia Coli* [6].

Antibiotic	Resistence	
	N	%
Ampicilin	98	83.7
Carbenicillin	74	63.2

(Table 3) cont.....

Antibiotic	Resistence	
	N	%
Piperacillin	63	53.8
Meropenem	1	0.85
Amikacine	2	1.7
Gentamicin	28	23.9
Tobramycin	36	30.7
Nalidixic acid	66	56.4
Ofloxacin	71	60.6
Norfloxacin	71	60.6
Ciprofloxacin	65	55.5
Cefuroxime	17	14.5
Ceftriaxone	12	10.2
Ceftazidime	10	8.5
Cefepime	9	7.6
Cefazolin	24	20.5
Nitrofurantoin	6	5.1
Amoxicillin / clavulanate	23	19.6
Ticarcillin / clavulanate	30	25.6
Trimethoprim / sulfamethoxazole	66	56.4

Reference: With permisión of Molina J & Manjarrez A; 2015 [6].

Cranberry Juice: Main Scientific Evidence

A search was conducted as strongest evidence to address main evidence, and it is so in table **4** it was integrated 33 articles, published from 2003 to 2015. Of these, five were interventions with cranberry juice, two experimental trials, one in mice and three clinical trials in elderly, kidney transplant patient and children. Five studies are interventions with non-juice blueberry products, one experimental in cells and four clinical trials, one in patients with low urinary tract infection, two in women with recurrent UTI and one in women diagnosed with uncomplicated acute cystitis. Seven studies were interventions with other foods such as mandarin (citrus reticulata), garlic, rice vinegar, sea berry, and vitamin A or *Lactobacillus Rhamnosus* supplements whose research suggests a beneficial effect in the prophylaxis or treatment of UTI. Four studies were clinical trials in cells and murine models and three clinical trials in humans, one in patients with long-term urinary catheterization, one in healthy men and women, and finally one in patients with recurrent UTI.

Table 4. Main evidence related to the beneficial effects of blueberry on UTI as well as other substrates

Cranberry Juice						
Reference	**Objective**	**Population**	**Results**	**Biological marker**	**Treatment (dose)**	**Sugar content of the product**
Experimental studies in cells						
Hotchkiss A., Nuñez A., *et al.* 2015 [29].	To determine the role of the xyloglucan structure in cranberry over inhibiting the adhesion of *E. Coli* to epithelial cells.	*E. Coli* cultures O157:H7 (ATCC BAA-1883) and HT29 human colonic epithelial cells (ATCC HTB-38)	The blueberry xyloglucan structure was first characterized as an arabino-xyloglucan with structure SSGG. In addition to proanthocyanidins, SSGG type xyloglucan oligosaccharides are novel bioactive cranberry juice ingredients with potential for the control of UTIs.	*E. Coli* *Epithelial cells*	ND	ND
González D., Esteban A., Sánchez F., *et al* . 2015 [22].	To determine the anti-adhesive activity of cranberry phenolic compounds and their metabolite derived from bacteria against uropathogenic *E. Coli* in cultures of bladder epithelial cells.	Epithelial cell cultures	The anti-adhesive activity of some phenolic metabolites derived from cranberry against UPEC was demonstrated for the first time *in vitro*, suggesting that its presence in the urine could reduce bacterial colonization and progression of UTI	*E. Coli* (UPEC) ATCC®53503	16 different phenolic compounds at concentrations of 100, 250 and 500 micrograms.	

(Table 4) cont.....

Cranberry Juice						
Reference	**Objective**	**Population**	**Results**	**Biological marker**	**Treatment (dose)**	**Sugar content of the product**
Experimental animal studies						
Harjai K., Kumar G., and Sehgal H. 2014 [49].	To evaluate the effectiveness of cranberry in attenuating the virulence of P. aeruginosa in UTIs in experimental mouse model and to explore the action of cranberry towards the virulence of P. aeruginosa through the detection of quorum inhibition.	Mice 6-8 weeks old without bacteria in urine. Division into 2 groups (experimental and control). The control group was subdivided into 2 groups. 10 mice for each experimental group.	Cranberry inhibits QS of P. aeruginosa and affects its ability to stick. This approach may lead to the discovery of the new category of safe antibacterial drugs from dietary sources such as blueberry with reduced toxicity without the risk of antibiotic resistance.	*P. Aeruginosa* (PAO1)	Cranberry Juice (Ocean Spray). 25% blueberry, 75% water, 100 g / L glucose added with trace amounts of vitamin C. Experimental group: prophylactic treatment of mice, 14 days with 1 mL of sub-inhibitory concentration of cranberry through oral, followed by infection (50 µl inoculum of bacteria (108 CFU / mL)) on day 15. Control group 1: Day 15 without treatment. Control group 2: prophylactic treatment with oral sterile saline, followed by infection at day 15.	100 gr sugar (glucose) / L

(Table 4) cont.....

Cranberry Juice						
Reference	**Objective**	**Population**	**Results**	**Biological marker**	**Treatment (dose)**	**Sugar content of the product**
Controlled Clinical Trials						
Handelanda M, *et al.*, 2014 [50].	To demonstrate that black chokeberry juice consumption decreases the incidence of UTI in nursing home residents.	6 nursing homes, two groups: A (n = 110) and B (126).	The incidence of UTI was reduced by 55% in group A and 38% in group B. This fruit is not considered potentially commercial because of its astringency.	*Staphylococcus aureus, E coli,* and *Bacillus cereus*) and antiviral activity (*influenza A virus*) *in vitro*.	Juice high in phenolic compounds (715 mg of gallic acid equivalent, 100 mL), including B-type cyanidines, anthocyanins and chlorogenic acids. Dose: 156 mL per resident (Group A); 89mL per resident (Group B).	19.6 gr and 12 gr. The added sugar to the drink was to improve the flavor and to diminish the astringency of the fruit.
Pagonas N, *et al.* 2012 [51].	To determine the effect of cranberry juice and L-Methionine in UTI after renal transplantation, as prophylaxis.	Experimental group: 82 patients with kidney transplantation and UTI treated with cranberry juice prophylaxis. Control group: 30 patients without prophylaxis.	Prophylaxis significantly reduced the incidence of UTI and UTI episodes. Cranberry juice and L-Methionine reduced the incidence of UTI after renal transplantation.	*E. Coli, Enteroccocus fecalis, Klebsiella pneumoniae, Klebsiella oxitoca, Proteus mirabilis*	Two dose doses of 50 mL / day (39 patients) 3 doses of 500 mg / day (25 patients) Both treatments (18 patients)	

(Table 4) cont.....

Cranberry Juice						
Reference	**Objective**	**Population**	**Results**	**Biological marker**	**Treatment (dose)**	**Sugar content of the product**
Kontiokari T., Salo J., Eerola E., *et al.* 2005 [26].	To evaluate the effect of cranberry juice on nasopharyngeal and colonic microbiota.	Children (mean 4.3 years) n = 171 (experimental group) n = 170 (control group)	Cranberry juice was well tolerated by children and did not alter the bacterial load on nasopharynx or the fatty acid composition of bacteria from feces.	ND	5 mL / kg cranberry juice Ocean spray, divided into 3 doses per day, for 3 months	30gr/240 mL

Interventions with cxranberry Products						
Reference	**Objective**	**Population**	**Results**	**Biological marker**	**Treatment (dose)**	**Sugar content of the product**
Basic experimental studies						
Feliciano R, Meudt J, *et al.* 2014 [52].	To determine the proportion of type A PACs of cranberry and type B of the apple based on its effect on ExPEC.	Enterocytes: Caco-2 cells (HTB-37)	Cranberry "type A" cap has greater bioactivity than apple type "B" cap, to increase agglutination of *E. Coli* extra intestinal pathogen (ExPEC) and decrease ExPEC by invasion of epithelial cells.	*E. Coli* strain 5011	ND	ND

(Table 4) cont.....

Interventions with cxranberry Products						
Reference	**Objective**	**Population**	**Results**	**Biological marker**	**Treatment (dose)**	**Sugar content of the product**
Clinical trials						
Ledda A., Bottari A., Luzzi R., *et al.* 2015 [53].	To evaluate the prophylactic effects of oral supplementation with a new well-standardized blueberry extract in patients with R-UTI with a two-month follow-up.	N = 22 control group and 22 experimental group. Subjects with a history of R-UTI (at least 3 episodes of IVUs in the year prior to the inclusion of two infections in the last six months). Subjects with ECD, compromised immune factors, concomitant infections, use of antibiotics, blood in urine were excluded.	Cranberry extract may have a role in prevention to decrease UI episodes in subjects with a history of I-UI.	ND	Dose: one capsule of Anthocran TM blueberry extract for 60 consecutive days and lifestyle recommendations for control group; For experimental group, only recommendations..	ND

(Table 4) cont.....

Interventions with cxranberry Products						
Reference	**Objective**	**Population**	**Results**	**Biological marker**	**Treatment (dose)**	**Sugar content of the product**
Vicariotto F. 2014 [54].	To determine the efficacy of a combination of dry cranberry extract, D-mannose in women unaffected by uncomplicated acute cystitis.	33 women without menopause, not pregnant women diagnosed with uncomplicated acute cystitis	Long-term use of a combination of cranberry, D-mannose, an innovative gelling complex and the 2 microorganisms tested is suggested to significantly improve the symptoms reported by women with acute cystitis.	*E. Coli, Enterobacter, Citrobacter, Klebsiella.*	Two envelopes a day the first month, then one envelope per day for six days.	ND
Burleigh A., Reed *et al.* 2013 [55].	To determine if consumption of sweetened dry blueberries reduces UTIs and whether this intervention can alter the heterogeneity, virulence factor profiles or amount of *E. Coli* in the intestine.	20 women with recurrent UTI	Beneficial effect of blueberry consumption to reduce the number of UTIs in susceptible women. No changes were found in the *E. Coli* profiles.	*E. Coli*	One serving of blueberries for two weeks	ND
Bailey D, Dalton C, (<bold><xref ref-type="table" rid="T1">1</xref></bold>). 2007 [56].	Examine the ability of a cranberry concentrate to prevent UTIs in women with a history of recurrent infections.	Women aged 25 to 70 years with a history of at least 6 UTIs in the preceding year.	Cranberry concentrate can prevent UTIs in women exposed to recurrent infections.		One capsule twice a day for 12 weeks with 200 mg of 30% blueberry concentrate of phenolic compounds.	ND

(Table 4) cont.....

Interventions with other products						
Reference	Objective	Population	Results	Biological Marker	Treatment (dose)	Sugar Content of the Product
Basic experimental studies						
Vollmerhausena T., Ramosa N., Ngoc D., and Braunera. 2013 [43].	To investigate the effect of the seeds of Citrus reticulata on the uroephitelium and to determine the mechanisms responsible for its protection against UTI.	Human bladder cells (T24 y 5637)	Treatment with Citrus reticulata decreased the expression of the β1 integrin and reduced bacterial invasion while adhesion of the uroepithelial cells was not affected. The expression of Caveolina-1 was unaffected and Citrus reticulata did not demonstrate antimicrobial effect nor interfere with fimbria type 1 binding.	*Escherichia coli* NU14	Seed infusion: 100 g of seeds in 600 mL of water, boiled and reduced for 8 hours to the third part, which was stored at 4 ° C, performed in triplicate and subsequently reduced the total volume (600 mL) to 100 mL.	ND
Karlsson M, Scherbak N, Olsson P, and Jass J. 2012 [57].	Identify and characterize the components of L. Rhamnosus GR-1 that influence the modulation of NF-Î°B activity in urothelial cells.	T24 cells (TCC HTB-4), a human bladder epithelial cell line, purchased from ATCC.	Several immunomodulatory proteins produced by lactobacilli, such as elongation factor Tu, PNL / P60 and GroEL, proved to elicit immune responses in human cells.	*Lactobacillus Rhamnosus* GR-1, *E. Coli* GR-12		ND

(Table 4) cont.....

Harjai K, Kumar R, and Singh S. 2010 [58]	To determine the effect of garlic on the blockade of quorum perception and the attenuation of virulence of P. aureginosa.	Experimental model with mice with UTI	Oral treatment with garlic significantly lowered renal bacterial counts and protected the mouse kidney from tissue destruction. The results suggest that decreased virulence of P. aeruginosa in mice fed with garlic can be attributed to the inhibitory property of garlic quorum detection.	*P. Aeruginosa*	Garlic: 40mgmL-1	ND
Sohn D, Hann C, Jung Y, *et al.* 2009 [59].	To evaluate the anti-inflammatory and antimicrobial effects of garlic and its synergy with ciprofloxacin in the treatment of CBP in an animal model.	60 male mice	Garlic treatment for chronic bacterial prostatitis demonstrated efficacy as an anti-inflammatory and antimicrobial. Its synergy with ciprofloxacin showed results with greater success.	0.2 ml of bacterial suspension of *E. Coli* Z17.02: K1: H-(1x1018)	9 mg / kg of garlic concentrate diluted in 1 mL of distilled water, twice daily.	NA
Clinical Trials						
Chung Y, Chen H, and Yeh M. 2012 [60].	To assess the effect of rice vinegar administered *via* nasogastric tube feeding on catheter-associated bacteriuria in patients with long-term urinary catheterization	n= 60 Experimental group and control group	Rice vinegar may decrease bacteriuria and the risk of symptomatic UTI, but more studies are needed to determine the effects of vinegar intake over a longer period and with a larger sample size.	*E. Coli* y *Proteus spp*	Diluted rice vinegar (30%): 100 mL / day for 4 weeks in the experimental group. 100 mL of water to control group.	ND
Larmo P., Alin J., *et al.* 2008 [61].	To study the effect of sea berry on the duration of common cold infection, in ITD and UTI, as well as the concentration of CRP	233 volunteers without diagnosed disease, men and women aged 19-50.	Sea berries did not demonstrate CC or DTI, but in UTI, further research is suggested. A reductive effect was detected in the PCR.		28gr frozen sea berry puree	ND

(Table 4) cont.....

Yilmaz A., Batah E., Gulay G., *et al.* 2007 [62].	To investigate the effect of vitamin A supplementation on RUTI	24 patients with RUTI. Cases and controls	Supplementation with vitamin A may have an adjuvant effect in the treatment of RUTI.		200 000 IU of vitamin A for each individual in the experimental group.	ND
Reviews and meta analysis						
Reference	Objective	Study	Results			
Lasekan O. 2014 [63].	Address the issue of berry intake and its potential functionality and highlight an idea of what the future holds for berries.	Review	It is suggested that berries such as blueberries and goji berries are effective for the treatment of UTI and seasonal influenza, respectively. Cranberry (blueberry) has been shown to have a protective effect against CVD. Strawberries, lingonberries, bilberries, blackcurrant, and chokeberry (aronia), significantly reduce posttransplant insulin response in non-diseased women (a meal of 50 grams of white or rye bread and 150 grams of berry puree).			
Foxman B., and Buxton M. 2013 [30].	To perform an updated literature review on the effects of functional foods, probiotics, vaccines, and alternative treatments for the treatment and prevention of UTI.	Review	American cranberry is the main functional food recommended for UTI prevention, however; Other foods (with less scientific evidence) are garlic (against E. Coli), a preparation containing horseradish and cappuccino, rice vinegar, and a sage herb found in Asia (Salvia plebeia). On the other hand, the consumption of preparations with lactobacilli limits colonization by uropathogens.			
Vasileiou I., Katsargyris A., and Teocharis S. 2013 [41].	To summarize the proposed mechanisms of cranberry against IVU and clinical trials evaluating the efficacy of cranberry supplementation in different populations.	Review	The benefit of blueberry is based on the prevention of bacterial adherence to uroepithelial cells. There is a decrease in symptoms due to UTI derivatives through the suppression of inflammatory cascades and an increase in the immune response. Cranberry juice is mostly recommended for symptomatic women. However, its prophylactic use is not clear as a treatment. No dose has been determined. PACs (proanthocyanidin polymers) are compounds responsible for their benefits. The low bioavailability, could be cause of little results in clinical tests.			

(Table 4) cont.....

Krueger C., Reed J., Feliciano R., and Howel A. 2013 [64].	To discuss quantitative and qualitative analyzes of "type A" PACs interflavane bonds in relation to their biological activity for the prevention of UTI.	Revisión	The structural heterogeneity of "type A" PAC makes its exact quantification difficult, requiring a high level of knowledge of how its structure affects its bioactivity; however, as a result of recent advances in MALDI-TOF and MALDI-FT-ICR mass spectrometry, generation of c-PAC standards, and a better understanding of DMAC reaction kinetics, integrity (including authenticity, normalization, efficacy and safety) of cranberry fruit, juice and dietary supplements can now be determined.
Wang C., *et al.* 2012 [42].	Evaluate products containing blueberry for the prevention of UTI and examine the factors that influence its effectiveness.	Systematic review and meta-analysis of randomized controlled trials	Consumption of blueberry-containing products may protect against UTIs in certain populations. However, due to heterogeneity between trials, this conclusion should be interpreted with caution. Cranberry-containing products tend to be more effective in women with recurrent UTIs, children, blueberry juice drinkers, and people who use cranberry-containing products more than twice a day.
Yang S., Chiang I., Lin C., and Chang S. 2012 [65].	Review progress in non-surgical treatment of pediatric UTIs.	Review: 2 systematic reviews, 5 randomized controlled trials and 3 randomized crossover studies.	The proposed non-surgical treatments for pediatrics are: behavioral modification (timed emptying and adequate fluid intake), topical steroids for phimosis, nutritional supplements (breast milk, blueberry, probiotics and vitamin A), biofeedback training for dysfunctional micturition, Anticholinergics to reduce intravesical pressure, alpha blockers in mictional dysfunction and neurogenic bladder, and intermittent catheterization for children with large PVR. Moreover, published reports usually include small numbers of patients and lack randomized and controlled group. Further well-designed studies are warranted to support the concepts of non-surgical treatment for pediatric UTI.
Rossi R., Porta S., and Canovi B. 2010 [66].	To review the relationship between blueberry and UTIs in women	Review	Cranberry products may play a role in older women with recurrent UTIs. Still in relation to antibiotic treatment in women, according to a recently published study also investigated the potential interaction of cranberry juice with b-lactam antibiotics that support the hypothesis that cranberry juice in normal amounts as prophylaxis for urinary infection is not likely to alter the pharmacokinetics of these by oral antibiotics. Positive results have been found in pregnant women. Consider the high content of oxalic acid by the potential nephrolithiasis.

(Table 4) cont.....

Nowack R., and Schmitt W. 2008 [40].	To review existing evidence for the effect of cranberry juice as prophylaxis for UTI	Review	Cramberry juice is effective in women although there is no defined dose. The effective mechanism of cranberry may be because of its anti-adhesive ability of bacteria to intestinal epithelium. The use of cranberry products appears to be safe and provide additional benefits because of its antioxidant activity and cholesterol lowering.
Jepson R., and Craig J. 2007 [67].	Review the effectiveness of blueberry and blueberry products for the prevention of UTI symptoms	Review	Cranberry juice is effective in reducing the symptoms of RUTIs in women. However, in other populations its effectiveness is uncertain. For blueberry, no relevant studies were found.
Heionen M. 2007 [68].	Carry out a review of antioxidant and antimicrobial activity of phenolic compounds of berries.	Review	The antioxidant effect depends on the environment, *i.e.* the structure of the food, whether it is oil in bulk, emulsion, or containing particles of liposomes. Ingestion of cranberry juice, blackcurrants as well as red wine have been shown to increase plasma antioxidant activity or to decrease the formation of lipid oxidation products. Cranberry proanthocyanidins, at a dose of 36 mg (less than a fist of cranberries) help reduce *E. Coli* adhesion in the urinary tract.
Puupponen R., Nohynek L., *et al.* 2005 [69].	Review the baya bioactive compounds novel tools against human pathogens.	Mini- Review	Several studies show that berry compounds inhibit the growth of human pathogenic bacteria, such as *Salmonella, Staphylococcus, Helicobacter and E. Coli* O157: H7. The use of antimicrobial activity of berry phenolic compounds as natural antimicrobial agents can offer many new applications for the food and medicine industry.
Descriptive Studies			
Reference	Objective	Results	Biological marker
Leahu A., Oroian N M., and Ropciuc S. 2014 [70].	Analyze the antioxidant potential of berries of minor importance and biological abundance (blueberries, cranberries) based on their ascorbic acid content, total polyphenols, antiradical activities, crude protein, moisture and dry matter in fresh and frozen fruit.	The content of total phenolic compounds was measured in extracts of fruit methanol as well as their radical scavenging activity. At the end of the frozen storage period, the ascorbic acid content and the total polyphenols remained significantly unchanged compared to the measured values just after the freezing process.	ND

(Table 4) cont.....

Chrubasik-Hausmann S., Vlachojannis C., and Zimmermann B. 2014 [25].	To analyze the content of proanthocyanidins of four cranberry products by photometry and by high performance liquid chromatography, and compare them with the recommended daily doses for the products by their manufacturers with reference doses obtained from the literature.	The PAC content of all products was below that stated. For two of the products, even the consumption of the maximum dose would not provide the health benefits declared by the product. And only for one of them, the maximum dose could have an opportunity to generate some effect. Products containing less than 100 mg per day should not be traded, as this dose has been shown to be ineffective.	ND
Sánchez F., Bartolomé B., *et al.* 2012 [71].	To evaluate the quality of commercial blueberry products, phenolic characterization and *in vitro* bioactivity. Antioxidant and antimicrobial capacity. 19 different products containing cranberry (syrups, powders, capsules)	The products analyzed differ widely in their phenolic content. The product presentation form and the polyphenolic profile affected the bacterial anti-adhesion activity (MIC: 0.5 mg / ML for powders at MIC: 112 mL / mL for gel capsules). Only 4 products provided the recommended dose of 36 mg PACs / day.	*E. Coli* P-fimbria
Kontiokari T., Latinen J., *et al.* 2003 [72].	To study dietary factors and other factors for UTIs in fertile women in a case-control design. 139 women with UTIs, and 185 women without UTI registered in the last 5 years.	Women's dietary habits were an important risk factor for UTI. Consumption of fresh juices and dairy products rich in probiotics decreased the risk of UTIs.	*E.Coli*
UTI: Urinary Tract Infection **CFU: Colony-forming units.** **PACs: Proanthocyanidins** **ND: Not disponible**			

In addition, nine reviews, a meta analysis and a minireview were analyzed. Four descriptive studies were integrated, three of them focused mainly on analyzing the content of phenolic compounds and proanthocyanidins and their antioxidant capacity. It is considered important to note that *Chrubasik-Hausmann S, Vlachojannis C, and Zimmermann B* in 2014 [25] found that proanthocyanidins content was below what was stated for four cranberry products (considering a recommendation that products containing less than 100 mg of these substances should not be marketed as products that provide health benefits, like a reduction in the incidence of UTI). Therefore, the prevention and treatment of these pathologies should be based on different approaches, and it is considered that special attention should be given to integrate proper (or healthy) eating habits into the risk analysis for presenting UTI, as already suggested by *Kontiokari T* and *Latinen J* in 2003 [26].

The 2016 study titled "*Cranberry Products or Topical Estrogen-Based Therapy for the Prevention of Urinary Tract infection*" integrated the best evidence about effectiveness and safety guidelines for cranberry juice as UTI prophylaxis [27]. Cranberry juice has been established as an option for UTI prevention. Several studies have analyzed the efficacy of cranberry juice against grape juice, and although there is some controversy, the evidence suggests that the beneficial effect of this functional food does not resides on its bacteriostatic effect, but on the inhibition of bacterial adhesion that has been attributed to tannins [28], and more recently to polysaccharides such as arabino-xyloglucans [29]. Similarly, a controlled clinical trial in a population of 359 pregnant women showed that consumption of cranberry juice may reduce the time of onset of UTI, although it does not prevent it [30].

Nutritional status and feeding are indicators of prevention or protection with bacterial susceptibility. Obesity and diabetes decrease immunocompetence, thus creating an ideal environment to development of UTI mainly associated with concomitant diseases. Since a nutritional state of obesity alone implies an alteration in the metabolism that leads to the appearance of other pathologies, the loss of metabolic homeostasis is related to an increased risk of suffering from infections. Likewise, the pharmacokinetics of antibiotics may differ in individuals with increased obesity that are susceptible to infection by microbial resistance [31, 32]. Other factors implicated are reduced immunogenicity in response to vaccination, probably due to alteration in the generation or function of antibody-generating plasma cells, or reduced absorption of the vaccine at the injection site due to excess adiposity [33, 34].

Although there is great evidence regarding the subject in which cranberry could play an important role in the prevention of UTI, the authors conclude that further research on evaluation of its therapeutic effects should be continued. One of the causes is the great heterogeneity found in the studies that complicates an integral analysis. However, the evidence found allows us to identify areas of opportunity such as integrating other variables in the studies that are based on the support and maintenance of the immune system, such as adequate nutrition, lifestyle, and food quality and quantity. As will be reviewed below.

Inmunonutrition

Immunonutrition is the use of specific nutrients to increase immune response and modify response to trauma in periods of critical illness and stress. The main nutrients used are glutamine, arginine and omega-3 fatty acids [35]. Glutamine, being a precursor of glutathion antioxidant promotes the integrity of the gastrointestinal system by providing energy for colonocytes and enterocytes, thereby decreasing the resistance to insulin, cellular catabolism and bacterial translocation [36]. The importance of an adequate nutritional status lies in the contribution of key nutrients that help maintain the proper functioning of T lymphocytes and other defense mechanisms [37 - 39].

Their relationships to the presence of infections should not be underestimated. Therefore the constant search for effective nutritional strategies for the intervention or prevention of infectious diseases such as UTI.

Evidence-Based Treatment Alternatives: Functional Foods

Evidence exists to support that some foods such as blueberries have the property of reducing UTI incidence; their phenolic compounds and proanthocyanidins can explain this. Some mechanisms of functional foods are to avoid adherence through bacterial fimbriae, specifically with *E. coli*, which prevents adhesion in uroepithelial cells [40, 41]. One of the main mechanisms for the colonization of the *E. coli* bacteria is the adhesins produced by the fimbriae found on its surface. Two cranberry compounds have been shown to be effective in inhibiting such adhesins: fructose and proanthocyanidins. A study by González *et al.* in 2015 [22], demonstrated the anti-adhesive activity of some phenolic metabolites derived from cranberry against uropathogenic *E. coli* of bladder epithelial cells. Recently, it was found in a study that arabino-xyloglucans, by their structure, may be another functional compound contained in cranberry with functional effects on the prevention or reduction of UTI [29].

Despite various studies and systematic reviews on cranberry juice, efficacy and utility in the treatment of UTI in different populations, have not been conclusive and designs and dosages tested have been heterogeneous. However, the use of cranberry products has shown more effectiveness for the prevention of UTI in women with recurrent UTI, in children and in people whose consumption of this type of product occurs more than twice a day, although it was not possible to specify a dose (95% CI): 0.28-0.8 (0.12, 1.22) in 17 studies, and one study in pregnant women with RR= 4.57 (95% CI: 0.25, 83.60) [42, 43].

Although no dose has been established for the preventive effect of cranberry products in UTI, the main presentations are cranberry juice, cranberry extract capsules and cranberry or dry blueberry with different results, which focus mainly on the amount or dose of antioxidants that can contain this type of product. According to a study by Chrubasik-Hausmann S. *et al.*, in 2014, where the content of proanthocyanidins of some cranberry products was analyzed and it was concluded that they should have more than 100 mg of this compound to be marketed with a “health benefit” since it has been shown that a lower dose is not effective [25].

Another type of functional food has been studied for its possible protective effect against UTI, Vollmerhausena T. *et al.*, studied in 2013 the effect of mandarin (*citrus reticulata blanco*) seeds on the uroepithelium of human bladder cells, demonstrating a decrease in bacterial invasion without changes in adhesion [43].

Concerning vitamin C, in the study of *Posadas et al.,* 50% of the participants reported adequate intake of 70mg/d [12]; this amount is controversial, as some authors consider it insufficient to saturate 80% of the vitamin C reserves in neutrophils, monocytes and lymphocytes without a significant loss of urine. This suggests that a suitable dose would be 100mg/d [44, 45], and concluded, according to the results, that none of the participants would fulfill the adequate intake of vitamin C, which could influence the development of UTI [46].

Recent studies and systematic reviews have concluded that cranberry juice has not been effective for the prevention and treatment of UTI, due to the lack of studies with consistency and methodological quality [47, 48].

Finally, pertinent to pregnant women, a single-blind clinical trial was carried out in pregnant women and the main conclusions were that a daily intake of 100 mg of ascorbic acid and other micronutrients (ferrous sulphate, folic acid) over the prophylaxis of UTI during pregnancy for 3 months reduced the urinary infections compared with a control group (Only ferrous sulphate, folic acid). The global frequency of urinary infections was 25% (The urinary infections in group case (12.7%) was significantly lower than control group (29.1%), (p=0.03, OR=0.35, CI 95%=0.13-0.91). The suggestion for pregnant women with a high incidence of bacteriuria and UTI is to ingest additional vitamin C [46].

CONCLUSIONS

Most studied foods or preparations include cranberry, which, despite being the most studied, remains controversial. The basis of the study of infectious diseases remains based on the ability of the immune system to respond to different bacterial agents, so a first protective mechanism will be the preservation of a good nutritional status, a situation that in current times is a problem of Public Health with the high prevalence of obesity, malnutrition, diabetes, cancer. Besides, exposure to various contaminants, that are commonly accumulated within the food ,during production chain; sexually transmitted infectious diseases mainly, probably exacerbate the problem and the prognosis. Therefore, these variables should be considered at the time of studying the infectious diseases of the urinary tract and thus would have a better analysis of the impacts of maintaining an adequate function of the immune system through diet, nutritional status and these foods specific.

Hypothesis to be Studied

The following approaches would help to redirect or generate new hypothesis for different and safer treatments for UTI in pregnant women and non-pregnant women:

1. Is the effect of cranberry juice on UTI equal in participants with or without obesity?
2. Will the exposure to various environmental pollutants (aromatic hydrocarbons, arsenic, persistent organic compounds, *etc.*) change the effectiveness of these alternatives?
3. Does the consumption of adequate food (rich in fruits and vegetables, low in saturated fats and additives), improve the effectiveness of the alternative treatments studied?
4. Will the pharmacokinetics of these alternatives be affected by exposure to other dietary nutrients or is there any interaction of different treatment alternatives with dietary nutrients?

CONSENT FOR PUBLICATION

Not applicable.

CONFLICT OF INTEREST

The authors confirm no conflict of interest in the publication of this chapter.

ACKNOWLEDGMENTS

To the teacher Gabriela Ramírez Tavares for her valuable help in reviewing the English version of this chapter.

REFERENCES

[1] Brusch JL. Urinary Tract Infection (UTI) and Cystitis (Bladder Infection) in Females [Internet]. Updated: Jul 19, 2018. Medscape. [Cited May 2018]. Available from https://emedicine. medscape. com/article/233101-overview

[2] Bono MJ, Reygaert WC. Urinary Tract Infection [Internet] nd StatPearls StatPearls Publishing 2018-2017 Dec 12; Available from: http://www.ncbi.nlm.nih.gov/pubmed/29261874 [Cited 2018 May 29].

[3] Amiri M, Lavasani Z, Norouzirad R, *et al.* Prevalence of Urinary Tract Infection Among Pregnant Women and its Complications in Their Newborns During the Birth in the Hospitals of Dezful City, Iran, 2012 - 2013. Iran Red Crescent Med J 2015; 17(8): e26946.
[http://dx.doi.org/10.5812/ircmj.26946] [PMID: 26430526]

[4] Murillo-Rojas OA, Leal-Castro AL, Eslava-Schmalbach JH. [Using antibiotics in urinary tract infection in a first level of attention health care unit in Bogotá, Colombia]. Rev Salud Publica (Bogota) 2006; 8(2): 170-81.
[PMID: 17191601]

[5] Secretaría de Salud. Anuario de Morbilidad 1984–2016. Sistema Integral para la Vigilancia Epidemiológica. México. 2014. [Internet]. [cited 2018 Jan] Available from: http://www.epidemiologia. Secretaría de salud.gob.mx/anuario/html/morbilidad_nacional.html

[6] Molina J, Manjarrez A. Urinary tract infections-Escherichia coli. [Internet]. National Autonomous University of Mexico. Department of Public Health. School of Medicine. 2015. [Cited 2017 Nov]. Available from http://www.facmed.unam.mx/deptos/microbiologia/bacteriologia/ enfermedades-via--urinarias.html

[7] Reséndiz J. Prevalence of urinary tract infections pregnant women threatening preterm labor. [internet]. Revista Electrónica de PortalesMedicos.com. Ginecología y Obstetricia. 2012 [Cited 2017 Nov]; 6 (18) Available from: http://www. portalesmedicos. com/ publicaciones/ articles/ 4190/1/Prevalencia-de-infecciones-del-tracto-urinario-en-mujeres-embarazadas-sint-maticas-o-asintomaticas-con-amenaza-de-parto-pretermino

[8] Pacheco-Gahbler C, Aragón-Tovar AR, Cantellano-Orozco M, *et al.* Diagnosis and Treatment of Urinary Tract Infections (UTI). [Internet]. National Academy of Medicine. Mappa Guides. First actualization: 2010 [Cited 2017 Nov]. Available from: https://cmu.org.mx/media/ cms_page_media/57/GUIAS_MAPPA_IVU.pdf

[9] Unlu BS, Yildiz Y, Keles I, *et al.* Urinary tract infection in pregnant population, which empirical antimicrobial agent should be specified in each of the three trimesters? Ginekol Pol 2014; 85(5): 371-6.
[http://dx.doi.org/10.17772/gp/1744] [PMID: 25011219]

[10] Yasemi M, Peyman H, Asadollahi K, *et al.* Frequency of bacteria causing urinary tract infections and their antimicrobial resistance patterns among pediatric patients in Western Iran from 2007-2009. J Biol Regul Homeost Agents 2014; 28(3): 443-8.
[PMID: 25316131]

[11] Dawkins JC, Fletcher HM, Rattray CA, Reid M, Gordon-Strachan G. Acute pyelonephritis in Pregnancy: A Retrospective Descriptive Hospital Based-Study. ISRN Obstet Gynecol 2012; 6

[12] Posadas P, Monroy-Torres R, Naves-Sanchez J. Intake of vitamin C, probiotics, flavonoids and nutritional status in pregnant women with urinary tract infection. Immunol Endocr Metab Agents Med Chem 2014; 14(1): 40-5.

[13] Committe on practice Bulletins Obstetrics. Gestational diabetes mellitus. Practice Bulletin No. 137. American College of Obstetricians and Gynecologists. Obstet Gynecol 2013; 122: 406-16. [http://dx.doi.org/10.1097/01.AOG.0000433006.09219.f1] [PMID: 23969827]

[14] Jijón S, Mondragón L, Morales X, Barrios C, Muñoz M. Frequency of urinary tract infections in women with type 2 diabetes mellitus of Chilpancingo, Guerrero 2009; 34: 105

[15] Pallarés J, López A, Cano A, Fábrega J, Mendive J. [Urinary infection in the diabetic patient]. Aten Primaria 1998; 21(9): 630-7. [PMID: 9677749]

[16] Boroumand MA, Sam L, Abbasi SH, Salarifar M, Kassaian E, Forghani S. Asymptomatic bacteriuria in type 2 Iranian diabetic women: a cross sectional study. BMC Womens Health 2006; 6: 4. [http://dx.doi.org/10.1186/1472-6874-6-4] [PMID: 16504076]

[17] Al-Rubeaan KA, Moharram O, Al-Naqeb D, Hassan A, Rafiullah MR. Prevalence of urinary tract infection and risk factors among Saudi patients with diabetes. World J Urol 2013; 31(3): 573-8. [http://dx.doi.org/10.1007/s00345-012-0934-x] [PMID: 22956119]

[18] Renko M, Tapanainen P, Tossavainen P, Pokka T, Uhari M. Meta-analysis of the significance of asymptomatic bacteriuria in diabetes. Diabetes Care 2011; 34(1): 230-5. [http://dx.doi.org/10.2337/dc10-0421] [PMID: 20937688]

[19] Instituto Nacional de Salud Pública (INSP). Instituto Nacional de Salud Publica (INSP). Encuesta Nacional de Salud y Nutrición 2012, Resultados Nacionales.Primera edición.[internet]. 2012 [cited 2018 Jan] Available from: http://ensanut.insp.mx/informes/ENSANUT2012ResultadosNacionales.pdf

[20] WHO. [homepage on the Internet] WHO Global Strategy to Contain Antimicrobial Resistance. [updated 2002; cited: 29th August 2017] Available from: http://www. antibioticos. msssi. gob.es/PDF/resist_OMS_estrategia_mundial_contra_resistencias.pdf

[21] Marchetti ML, Errecalde JO, Mestorino ON. Bacterial resistance to antimicrobials caused by efflux pumps. Analecta Vet 2011; 31(2): 40-53.

[22] de Llano DG, Esteban-Fernández A, Sánchez-Patán F, Martínlvarez PJ, Moreno-Arribas MV, Bartolomé B. Anti-adhesive activity of cranberry phenolic compounds and their microbial-derived metabolites against uropathogenic escherichia coli in bladder epithelial cell cultures. Int J Mol Sci 2015; 16(6): 12119-30. [http://dx.doi.org/10.3390/ijms160612119] [PMID: 26023719]

[23] Malmartel A, Ghasarossian C. Bacterial resistance in urinary tract infections in patients with diabetes matched with patients without diabetes. J Diabetes Complications 2016; 30(4): 705-9. [http://dx.doi.org/10.1016/j.jdiacomp.2016.01.005] [PMID: 26851821]

[24] Head KA. Natural approaches to prevention and treatment of infections of the lower urinary tract. Alternative Medicine Review [internet]. 2008, 13 (3): 227 – 244. [cited 2017 Nov] Available from: http://go.galegroup.com/ps/anonymous?id=GALE%7CA187494439&sid=googleScholar&v=2.1&it=r &linkaccess=fulltext&issn=10895159&p=AONE&sw=w&authCount=1&isAnonymousEntry=true

[25] Chrubasik-Hausmann S, Vlachojannis C, Zimmermann BF. Proanthocyanin content in cranberry CE medicinal products. Phytother Res 2014; 28(11): 1612-4. [http://dx.doi.org/10.1002/ptr.5172] [PMID: 24849530]

[26] Kontiokari T, Salo J, Eerola E, Uhari M. Cranberry juice and bacterial colonization in children--a placebo-controlled randomized trial. Clin Nutr 2005; 24(6): 1065-72. [http://dx.doi.org/10.1016/j.clnu.2005.08.009] [PMID: 16194582]

[27] Canadian Agency for Drugs and Technologies in Health. Cranberry Products or Topical Estrogen-Based Therapy for the Prevention of Urinary Tract Infections: A Review of Clinical Effectiveness and Guidelines 2016. [PMID: 27929627]

[28] Monroy-Torres R, Macías AE. [Does cranberry juice have bacteriostatic activity?]. Rev Invest Clin 2005; 57(3): 442-6.
[PMID: 16187705]

[29] Hotchkiss AT Jr, Nuñez A, Strahan GD, *et al.* Cranberry Xyloglucan Structure and Inhibition of Escherichia coli Adhesion to Epithelial Cells. J Agric Food Chem 2015; 63(23): 5622-33.
[http://dx.doi.org/10.1021/acs.jafc.5b00730] [PMID: 25973733]

[30] Foxman B, Buxton M. Alternative approaches to conventional treatment of acute uncomplicated urinary tract infection in women. Curr Infect Dis Rep 2013; 15(2): 124-9.
[http://dx.doi.org/10.1007/s11908-013-0317-5] [PMID: 23378124]

[31] Falagas ME, Kompoti M. Obesity and infection. Lancet Infect Dis 2006; 6(7): 438-46.
[http://dx.doi.org/10.1016/S1473-3099(06)70523-0] [PMID: 16790384]

[32] Hanley MJ, Abernethy DR, Greenblatt DJ. Effect of obesity on the pharmacokinetics of drugs in humans. Clin Pharmacokinet 2010; 49(2): 71-87.
[http://dx.doi.org/10.2165/11318100-000000000-00000] [PMID: 20067334]

[33] Milner JJ, Beck MA. The impact of obesity on the immune response to infection. Proc Nutr Soc 2012; 71(2): 298-306.
[http://dx.doi.org/10.1017/S0029665112000158] [PMID: 22414338]

[34] Eliakim A, Schwindt C, Zaldivar F, Casali P, Cooper DM. Reduced tetanus antibody titers in overweight children. Autoimmunity 2006; 39(2): 137-41.
[http://dx.doi.org/10.1080/08916930600597326] [PMID: 16698670]

[35] Kurmis R, Parker A, Greenwood J. The use of immunonutrition in burn injury care: where are we? J Burn Care Res 2010; 31(5): 677-91.
[http://dx.doi.org/10.1097/BCR.0b013e3181eebf01] [PMID: 20671563]

[36] Vincent JL. Metabolic support in sepsis and multiple organ failure: more questions than answers .... Crit Care Med 2007; 35(9) (Suppl.): S436-40.
[http://dx.doi.org/10.1097/01.CCM.0000278601.93369.72] [PMID: 17713390]

[37] Drover JW, Dhaliwal R, Weitzel L, Wischmeyer PE, Ochoa JB, Heyland DK. Perioperative use of arginine-supplemented diets: a systematic review of the evidence. J Am Coll Surg 2011; 212(3): 385-399, 399.e1.
[http://dx.doi.org/10.1016/j.jamcollsurg.2010.10.016] [PMID: 21247782]

[38] Zapatera B, Prados A, Gómez-Martínez S, Marcos A. Immunonutrition: methodology and applications. Nutr Hosp. 2015; 26; 31(Suppl 3): 145-54

[39] Monroy-Torres R. Naves- Sánchez J. The Role of the Healthy Dietary Intake in women with Human Papilloma Virus. Immunol Endocr Metab Agents Med Chem 2014; 14.

[40] Nowack R, Schmitt W. Cranberry juice for prophylaxis of UTIs. Conclusions from clinical experience and research. Phytomedicine 2008; 15: 653-67.
[http://dx.doi.org/10.1016/j.phymed.2008.07.009] [PMID: 18691859]

[41] Vasileiou I, Katsargyris A, Teocharis S. Current clinical status on the preventive effects of cranberry consumption against UTI. Nutr Res J. 2013; 33: 595-607

[42] Wang CH, Fang CC, Chen NC, *et al.* Cranberry-containing products for prevention of urinary tract infections in susceptible populations: a systematic review and meta-analysis of randomized controlled trials. Arch Intern Med 2012; 172(13): 988-96.
[http://dx.doi.org/10.1001/archinternmed.2012.3004] [PMID: 22777630]

[43] Vollmerhausen TL, Ramos NL, Dzung DT, Brauner A. Decoctions from Citrus reticulata Blanco seeds protect the uroepithelium against Escherichia coli invasion. J Ethnopharmacol 2013; 150(2): 770-4.
[http://dx.doi.org/10.1016/j.jep.2013.09.050] [PMID: 24120518]

[44] Food and Nutrition Board. Dietary reference intakes for vitamin C, vitamin E, selenium and

carotenoids. Washington: National Academy Press 2000; pp. 95-185.

[45] Casanueva E, Angulo ME, Goidberg S, *et al.* Bases to determine the dose of vitamin C in pregnancy. Gac Med Mex 2005; 141(4): 273-7.
[PMID: 16164121]

[46] Ochoa-Brust GJ, Fernández AR, Villanueva-Ruiz GJ, Velasco R, Trujillo-Hernández B, Vásquez C. Daily intake of 100 mg ascorbic acid as urinary tract infection prophylactic agent during pregnancy. Acta Obstet Gynecol Scand 2007; 86(7): 783-7.
[http://dx.doi.org/10.1080/00016340701273189] [PMID: 17611821]

[47] Cayley WE Jr. Are cranberry products effective for the prevention of urinary tract infections? Am Fam Physician 2013; 88(11): 745-6.
[PMID: 24364520]

[48] Jepson RG, Williams G, Craig JC. Cranberries for preventing urinary tract infections. Cochrane Database Syst Rev 2012; 10(10): CD001321.
[PMID: 23076891]

[49] Harjai K, Gupta RK, Sehgal H. Attenuation of quorum sensing controlled virulence of Pseudomonas aeruginosa by cranberry. Indian J Med Res 2014; 139(3): 446-53.
[PMID: 24820840]

[50] Handeland M, Grude N, Torp T, Slimestad R. Black chokeberry juice (Aronia melanocarpa) reduces incidences of urinary tract infection among nursing home residents in the long term--a pilot study. Nutr Res 2014; 34(6): 518-25.
[http://dx.doi.org/10.1016/j.nutres.2014.05.005] [PMID: 25026919]

[51] Pagonas N, Hörstrup J, Schmidt D, *et al.* Prophylaxis of recurrent urinary tract infection after renal transplantation by cranberry juice and L-methionine. Transplant Proc 2012; 44(10): 3017-21.
[http://dx.doi.org/10.1016/j.transproceed.2012.06.071] [PMID: 23195017]

[52] Feliciano RP, Meudt JJ, Shanmuganayagam D, Krueger CG, Reed JD. Ratio of "A-type" to "B-type" proanthocyanidin interflavan bonds affects extra-intestinal pathogenic Escherichia coli invasion of gut epithelial cells. J Agric Food Chem 2014; 62(18): 3919-25.
[http://dx.doi.org/10.1021/jf403839a] [PMID: 24215458]

[53] Ledda A, Bottari A, Luzzi R, *et al.* Cranberry supplementation in the prevention of non-severe lower urinary tract infections: a pilot study. Eur Rev Med Pharmacol Sci 2015; 19(1): 77-80.
[PMID: 25635978]

[54] Vicariotto F. Effectiveness of an association of a cranberry dry extract, D-mannose, and the two microorganisms Lactobacillus plantarum LP01 and Lactobacillus paracasei LPC09 in women affected by cystitis: a pilot study. J Clin Gastroenterol 2014; 48(1) (Suppl. 1): S96-S101.
[http://dx.doi.org/10.1097/MCG.0000000000000224] [PMID: 25291140]

[55] Burleigh AE, Benck SM, McAchran SE, Reed JD, Krueger CG, Hopkins WJ. Consumption of sweetened, dried cranberries may reduce urinary tract infection incidence in susceptible women--a modified observational study. Nutr J 2013; 12(1): 139.
[http://dx.doi.org/10.1186/1475-2891-12-139] [PMID: 24139545]

[56] Bailey DT, Dalton C, Joseph Daugherty F, Tempesta MS. Can a concentrated cranberry extract prevent recurrent urinary tract infections in women? A pilot study. Phytomedicine 2007; 14(4): 237-41.
[http://dx.doi.org/10.1016/j.phymed.2007.01.004] [PMID: 17296290]

[57] Karlsson M, Scherbak N, Khalaf H, Olsson PE, Jass J. Substances released from probiotic Lactobacillus rhamnosus GR-1 potentiate NF-κB activity in Escherichia coli-stimulated urinary bladder cells. FEMS Immunol Med Microbiol 2012; 66(2): 147-56.
[http://dx.doi.org/10.1111/j.1574-695X.2012.00994.x] [PMID: 22620976]

[58] Harjai K, Kumar R, Singh S. Garlic blocks quorum sensing and attenuates the virulence of

Pseudomonas aeruginosa. FEMS Immunol Med Microbiol 2010; 58(2): 161-8.
[http://dx.doi.org/10.1111/j.1574-695X.2009.00614.x] [PMID: 19878318]

[59] Sohn DW, Han CH, Jung YS, Kim SI, Kim SW, Cho YH. Anti-inflammatory and antimicrobial effects of garlic and synergistic effect between garlic and ciprofloxacin in a chronic bacterial prostatitis rat model. Int J Antimicrob Agents 2009; 34(3): 215-9.
[http://dx.doi.org/10.1016/j.ijantimicag.2009.02.012] [PMID: 19375896]

[60] Chung YC, Chen HH, Yeh ML. Vinegar for decreasing catheter-associated bacteriuria in long-term catheterized patients: a randomized controlled trial. Biol Res Nurs 2012; 14(3): 294-301.
[http://dx.doi.org/10.1177/1099800411412767] [PMID: 21708892]

[61] Larmo P, Alin J, Salminen E, Kallio H, Tahvonen R. Effects of sea buckthorn berries on infections and inflammation: a double-blind, randomized, placebo-controlled trial. Eur J Clin Nutr 2008; 62(9): 1123-30.
[http://dx.doi.org/10.1038/sj.ejcn.1602831] [PMID: 17593932]

[62] Yilmaz A, Bahat E, Yilmaz GG, Hasanoglu A, Akman S, Guven AG. Adjuvant effect of vitamin A on recurrent lower urinary tract infections. Pediatr Int 2007; 49(3): 310-3.
[http://dx.doi.org/10.1111/j.1442-200X.2007.02370.x] [PMID: 17532826]

[63] Lasekan O. Exotic berries as a functional food. Curr Opin Clin Nutr Metab Care 2014; 17(6): 589-95.
[http://dx.doi.org/10.1097/MCO.0000000000000109] [PMID: 25159559]

[64] Krueger CG, Reed JD, Feliciano RP, Howell AB. Quantifying and characterizing proanthocyanidins in cranberries in relation to urinary tract health. Anal Bioanal Chem 2013; 405(13): 4385-95.
[http://dx.doi.org/10.1007/s00216-013-6750-3] [PMID: 23397091]

[65] Yang SS, Chiang IN, Lin CD, Chang SJ. Advances in non-surgical treatments for urinary tract infections in children. World J Urol 2012; 30(1): 69-75.
[http://dx.doi.org/10.1007/s00345-011-0700-5] [PMID: 21614468]

[66] Rossi R, Porta S, Canovi B. Overview on cranberry and urinary tract infections in females. J Clin Gastroenterol 2010; 44 (Suppl. 1): S61-2.
[http://dx.doi.org/10.1097/MCG.0b013e3181d2dc8e] [PMID: 20495471]

[67] Jepson RG, Craig JC. A systematic review of the evidence for cranberries and blueberries in UTI prevention. Mol Nutr Food Res 2007; 51(6): 738-45.
[http://dx.doi.org/10.1002/mnfr.200600275] [PMID: 17492798]

[68] Heinonen M. Antioxidant activity and antimicrobial effect of berry phenolics--a Finnish perspective. Mol Nutr Food Res 2007; 51(6): 684-91.
[http://dx.doi.org/10.1002/mnfr.200700006] [PMID: 17492800]

[69] Puupponen-Pimiä R, Nohynek L, Alakomi HL, Oksman-Caldentey KM. Bioactive berry compounds-novel tools against human pathogens. Appl Microbiol Biotechnol 2005; 67(1): 8-18.
[http://dx.doi.org/10.1007/s00253-004-1817-x] [PMID: 15578177]

[70] Leahu A, Oroian M, Ropciuc S. Total phenolics of fresh and frozen minor berries and their antioxidant properties. Journal of Faculty of Food Engineering 2014; 13(1): 87-93.

[71] Sánchez-Patán F, Bartolomé B, Martín-Alvarez PJ, Anderson M, Howell A, Monagas M. Comprehensive assessment of the quality of commercial cranberry products. Phenolic characterization and *in vitro* bioactivity. J Agric Food Chem 2012; 60(13): 3396-408.
[http://dx.doi.org/10.1021/jf204912u] [PMID: 22439747]

[72] Kontiokari T, Laitinen J, Järvi L, Pokka T, Sundqvist K, Uhari M. Dietary factors protecting women from urinary tract infection. Am J Clin Nutr 2003; 77(3): 600-4.
[http://dx.doi.org/10.1093/ajcn/77.3.600] [PMID: 12600849]

CHAPTER 6

Targeting Magnesium Homeostasis as Potential Anti-infective Strategy Against Mycobacteria

Saif Hameed* and **Zeeshan Fatima***

Amity Institute of Biotechnology, Amity University Haryana, Gurugram (Manesar)-122413, India

Abstract: The emergence of Multi-Drug Resistance (MDR) in treating Tuberculosis (TB) has been led by the unnecessary overconsumption of anti tubercular drugs. MDR being a multifactorial phenomenon is governed by many unknown mechanisms, hence dissecting novel mechanisms for combating MDR is urgently needed. The success of pathogenic *Mycobacterium tuberculosis* (*Mtb*) as a dreadful pathogen is due to its capability to persist in chronic infection, despite of the robust immune response by the host. During the course of infection, *Mtb* resides inside the host macrophages which in turn provide a relatively hostile environment to the pathogen. This host pathogen interaction causes the macrophages to undergo maturation, leading to lowering of pH and limiting nutrients, including magnesium (Mg^{2+}). Since Mg^{2+} is the most abundant divalent cation in living cells it is mandatory for all organisms to maintain its physiological levels to an optimum concentration. Mg^{2+} has a critical role in stabilizing membranes, ribosomes, required as a cofactor in various enzymatic reactions and the neutralization of nucleic acids. Thus harnessing the Mg^{2+} dependent pathways in *Mtb* could be targeted as effective anti-TB strategy. This chapter helps in gaining insights into the association of Mg^{2+} in survival of *Mtb* and how it can be exploited as potential anti-mycobacterial drug targets.

Keywords: Anti-TB target, Drug resistance, Magnesium, MgtC, *Mycobacterium*.

INTRODUCTION

Tuberculosis (TB) caused by *Mycobacterium tuberculosis* (*Mtb*) is one of the major cause of death affecting approximately one third of the world's population in latent form [1]. It is known to spread easily through air *via* coughs, sneezes, or spits by a diseased person. Although the diagnosis of TB is not impossible; it is very challenging [2]. This is due to various reasons including: the slow growing nature of the bacteria, detection of false positive results in Bacille Calmette-Guérin (BCG) vaccinated individual, and many others. Although, BCG shows

* **Corresponding Authors Zeeshan Fatima and Saif Hameed:** Amity Institute of Biotechnology, Amity University Haryana, Gurugram (Manesar)-122413, India; Tel: +91-124-2337015, Ext: 1116; Emails: drzeeshanfatima@gmail.com; saifhameed@yahoo.co.in

Atta-ur-Rahman (Ed.)

variability in its efficacy, it is the only licensed vaccine which is currently considered to be effective against active TB [3]. As there is lack of more effective vaccine, chemotherapy comes forward to be a major control tool for TB. During the last few decades, the dramatical increase in the microbial infections with concomitant deployment of antimicrobial drugs has led to the emergence of multidrug resistance (MDR) and extensively drug resistance (XDR) [4]. Tubercle bacilli which are resistant to at least rifampicin and isoniazid, two key drugs in the treatment of the disease, are known to cause MDR-TB; while on the other hand, *Mtb* which in addition to being MDR, are also resistant to any quinolone and to one of the three injectable second-line drugs: kanamycin, capreomycin or amikacin cause XDR-TB [5]. Taking such crucial circumstances into consideration, search for novel strategies to combat MDR, remains a priority area of research.

Micronutrients are one of the requisite elements for the propagation and survival of microorganisms as well as human. Micronutrients stress has been widely exploited as one of the alternative strategies against TB [6]. The commonly utilized micronutrients include iron, magnesium, calcium, zinc, copper and phosphorous. Restricting these micronutrients availability modulates cellular responses and bacterial morphogenesis to drug-induced stress and unravels a novel strategy in which metal homeostasis impacts host-pathogen interactions. Any alterations in the concentrations of these micronutrients inside host environment can affect virulence attributes of the organism. For instance iron deprivation affects cell envelope thickness and cellular membrane integrity, which in turn presumably allows faster entry of drugs leading to enhanced drug sensitivity of the cells in mycobacteria [7]. SodC, a superoxide mutase containing a Cu/Zn pair, is known for its contribution in oxidative stress resistance by detoxification of the toxic free radicals generated by host. Piddington *et al.*, 2001 showed that inactivation of *sodc* in *Mtb* resulted in killing of the cells when superoxide was generated externally; further this mutation also increased the sensitivity toward killing by gamma interferon (IFN-γ) [8]. Although the physiological function of Mn^{2+} is poorly known, but it participates in a wide range of metabolic functions as a cofactor; which includes free radical detoxification, signal transduction, carbon metabolism and growth regulation [9]. Pandey *et al.*, 2015 reported that a deletion of Mn^{2+} transporter gene, *mnth,* led to attenuation of growth intracellularly, indicating significance of Mn^{2+} uptake for survival within macrophages [10]. Likewise, among all essential micronutrients, magnesium (Mg^{2+}) is one of the most significant elements for *Mtb,* which is known for its participation in a wide range of biochemical pathways essential for survival of all organisms [11]. Despite these roles, Mg^{2+} transport and homeostasis are poorly understood at the physiological level in mycobacteria and in further sections we attempt to describe the role of Mg^{2+} in *Mtb*.

SIGNIFICANCE OF MAGNESIUM

Magnesium is believed to be the most abundant divalent cation in living cells and the second most abundant cation after potassium [12]. It is well known for its role as cofactor for several enzymes, regulatory factors for ribosomes, membranes and other cellular structures, which include neutralizing nucleic acids in cytoplasm, and stabilizing phospholipid head groups and surface molecules in the membrane (Fig. **1**). It also regulates cell wall integrity and plays a role in the synthesis of tRNA [11]. Although some of these functions can be performed by other divalent cations in absence of Mg^{2+}, most of them are strictly dependent on Mg^{2+}; because of which the bacteria has to establish various mechanisms in order to maintain the Mg^{2+} homeostasis. Being the fourth most abundant metal in human body Mg^{2+} is present as both free ionized and bound forms. Its existence as bound or free ionized form depends on the pH, ionic strength, temperature as well as the competing ions [13]. Although most divalent cations are maintained at a relatively low concentration in the intracellular pool, a higher level of Mg^{2+} (~0.5-2.0mM) is maintained in the cytoplasm [14]; thus specific transporters are required. Apart from the above mentioned functions, Mg^{2+} also plays a vital role in the virulence of the bacterial kingdom. For example PhoPQ, a two component signal transduction system, has a critical role in activating genes related to bacterial virulence; where PhoP serves a s a response regulator while PhoQ, a sensor kinase [15]. Ford *et al.,* 2014 has shown in a study that Mg^{2+} plays a significant role in the signal controlling gene regulation through PhoPQ system [16]. Therefore, host macrophages have established special mechanisms that eliminate Mg^{2+} after the bacteria enters; similarly the bacteria uses specific systems in order to detect tiny amount of Mg^{2+} in the intracellular pool [17, 18].

After invasion, *Mtb* replicates inside the phagocytic vacuoles of host macrophages. These phagosomes provide *Mtb* a hostile environment, generally mildly acidic and poor in nutrients including Mg^{2+}. This was well observed in various studies which involved *Mtb* mutants which were unable to grow in acidic medium with low Mg^{2+} concentration. Piddington *et al.,* 2000 observed that bacterial growth in low pH is interconnected with requirement of Mg^{2+}; the need for Mg^{2+} could not be replaced by any other divalent cations such as Ca^{2+}, Zn^{2+}, or Mn^{2+} [19]. Goodsmith *et al.,* 2015 showed that an increased level of Mg^{2+} was required for the replication and survival of a *Mtb* mutant, with disrupted of a membrane protein, PerM; which is responsible for *Mtb* persistence in host body, thus resulting in chronic TB. These facts suggest that Mg^{2+} is essential for promoting survival of *Mtb* inside host after invasion as shown in Fig. (**2**) [11].

Cellular mechanisms	Mg^{2+} abundance	Mg^{2+} deficiency
Membrane integrity	Stable	Unstable
ATP	Forms complex with Mg^{2+} to be Biologically active	Cannot form complex with Mg^{2+} so inactive
Cell division	Binds to nucleic acids, hence helps in cell division	Abnormal cell division
Phagosomal fate	Survival	Death
Cord formation	Normal	Diminished
Ribosome	Normal ribosomal function	Rapid breakdown and loss of ribosome

Fig. (1). Comparison between the cellular mechanisms in presence and absence of magnesium.

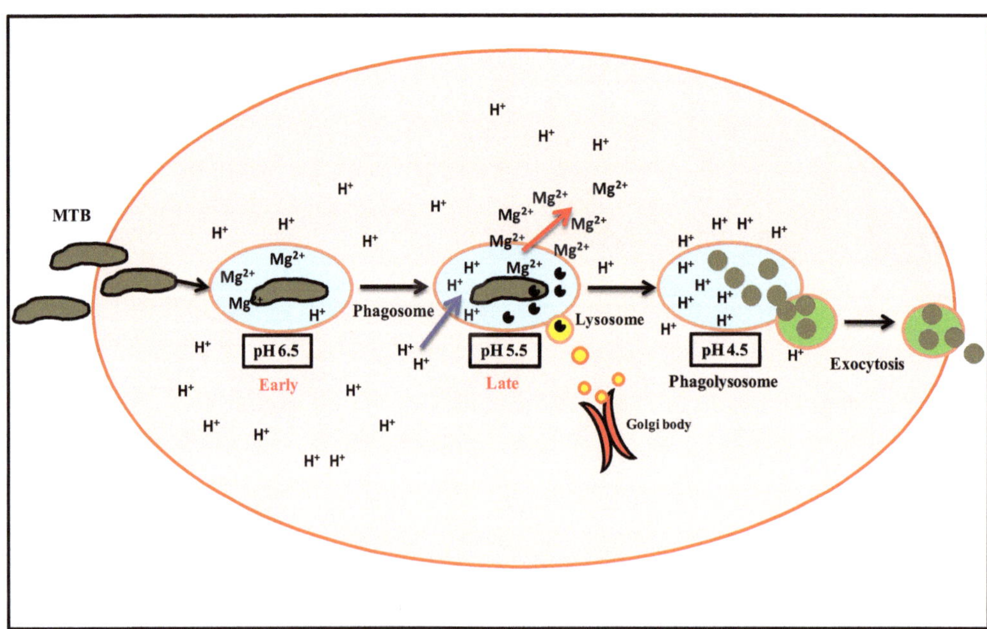

Fig. (2). Phagosomal killing of *Mtb* in macrophage. Figure shows the series of events that take place when *Mtb* enters the macrophage leading to phagosome maturation and acidification causing phagocytosis. Purple arrow shows the entry of H^+ ions inside the phagosome, creating an acidic environment; while red arrow shows the depreciation of Mg^{2+} ions from the phagolysosome. These processes lead to exocytosis and clearance of bacteria.

AVAILABLE MAGNESIUM SOURCES INSIDE HOST

Various studies have revealed that the environment inside the host macrophage is nutrient limited (mostly Mg^{2+}) as well as acidic in pH [20]. In order to overcome these hostile conditions encountered by *Mtb* during infection, it requires Mg^{2+}. Hence, acquisition of Mg^{2+} poses a challenge for *Mtb*; although it is able to accumulate Mg^{2+} even from nutrient limited environment. While Mg^{2+} is the second most abundant divalent cation in living cells, NMR and X-ray microanalysis have revealed that total Mg^{2+} in mammalian cells ranges from 17 to 20 mM. Among these, 15-18 mM is found bound to nucleic acids, phospholipids, proteins *etc.* within structures such as, nucleus, mitochondria and endo-(sarco)-plasmic reticulum; this leaves only 15-22% of free Mg^{2+} in the lumen of these organelles [21]. Although the Mg^{2+} content in both nucleus and endoplasmic reticulum is difficult to detect, the cytoplasmic Mg^{2+} pool is well detectable. Most of this Mg^{2+} is found to be bound to phosphonucleotides and ATP to form a biologically active complex, ATP-Mg^{2+}. Concentration of ATP is approximately more than 5mM within the cytoplasm and mitochondria, so it is the largest metabolic pool which abundantly binds to Mg^{2+}. After binding with ATP, phospholipids, and proteins, there is only less than 5% of the total free cellular Mg^{2+} left in the cytosol [22, 23]. Although very little is known about proteins that bind to Mg^{2+}, one-third of the Mg^{2+} present in plasma and extracellular fluid is known to bind to extracellular proteins like albumin [24].

Magnesium Transporters in *Mtb*

The fact that Mg^{2+} is a unique cation in a biological system is well known because a) it is the most abundant divalent cation in living cells, and b) it acts as a cofactor for numerous enzymatic reactions. Therefore, it can be understood that lack of this particular cation can cause several cellular machinery to fall apart, including ribosomes, nucleic acids and cell membrane; this can lead to cell death [25]. *Mtb*, undergoes a non-replicating persistent (NRP) state when encounters a nutrient-poor and hypoxic environment inside the host. This characteristic feature enables the bacteria to establish chronic infections and acquire drug tolerance [26]. Its growth and virulence is highly regulated by pH and availability of essential cations in the host environment. As already mentioned in previous section, host macrophages have established mechanisms to limit Mg^{2+} from macrophages at the time of infection. Similarly *Mtb* can sense as well as acquire Mg^{2+} even from a nutrient limited environment. The various Mg^{2+} transporters which are found in *Mtb* are CorA, MgtE and MgtC as shown in Fig. (**3**). CorA is a member of 2TM-GxN family of metal ion transporters, which has a significant role in both influx and efflux of Mg^{2+} across the membrane [27]. Another annotated gene which is found in *Mtb* genome to be encoding a potential Mg^{2+} transporter is *MgtE* [28].

The third Mg^{2+} transporter protein is predicted to have a major role, not only in the acquisition of Mg^{2+} but also in the phagosomal survival of *Mtb*. This protein, a virulence factor, shares 38% sequence similarity with MgtC protein of *Salmonella enterica* and therefore, also shares similar functions; it is responsible for virulence, promoting intraphagosomal survival and for growth in low-Mg^{2+} levels [9]. Various research groups created phagosomal environment *in vitro* to prove that *Mtb* lacking *mgtC* was not able to grow and replicate under Mg^{2+} deprivation in combination with decreasing pH [20]. At present, there are no other genes regulating Mg^{2+} homeostasis known apart from these annotated genes.

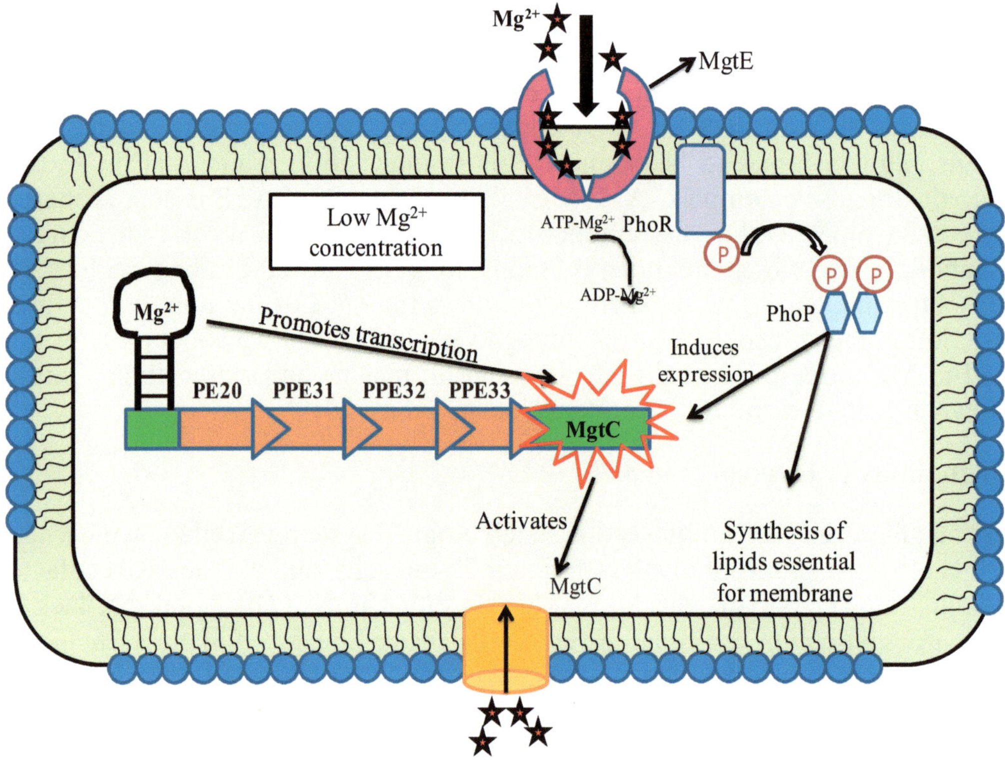

Fig. (3). Magnesium acquisition mechanisms and transcriptional regulation of magnesium homeostasis.

TRANSCRIPTIONAL REGULATION OF MAGNESIUM HOMEOSTASIS

M-Box Riboswitch

As already known, metal ions have very significant role in biology; but on the contrary, a high intracellular concentration of these metal ions could result in cellular toxicity to the pathogen. Therefore, their cellular concentration is subjected to a tight regulation *via* metalloregulatory proteins. Several studies have already shown that metal ion homeostasis is regulated through various posttranscriptional mechanisms; which involves small, non-coding RNAs (ncRNAs) as trans/cis-acting RNA elements, also known as riboswitches. These elements are generally located within the untranslated regions (UTRs) of mRNAs; they specifically sense molecules, and regulate gene expression [29]. There are various riboswitches present in *Mtb*; one of which is represented twice in the *Mtb* genome, *Ykok* leader, also known as Mbox. This is generally found in relation with Mg^{2+} transporters.The first Mbox is found upstream of a conserved protein, Rv1535, which is induced when the cell lacks Mg^{2+}. Similarly a second Mbox is found upstream of an operon consisting of an Mg^{2+} transporter, MgtC, and proteins, belonging to a family containing proline-glutamate (PE) or proline-proline-glutamate (PPE) as N-terminal motifs as shown in Fig. (**3**). As both these operons are induced by Mg^{2+} starvation, it is presumed that these RNA elements are activated by the Mg^{2+} limited environment encountered by *Mtb* in the host macrophages [30].

PhoPR (2-CS) Signal Transduction System Works in a Magnesium Dependent Manner

To survive in the unsuitable environment of the host body, microbes have evolved various other mechanisms in order to overcome the changing conditions. Intracellular pathogens harbor a special sensor system, which is not only essential for their virulence but also has a role in response to host defense system. These are highly conserved prokaryotic signal transduction pathway, a two component system (TCS); which contains a sensor, histidine kinase (HK) and an effector, response regulator (RR) [28, 31]. For example, the *Salmonella* PhoP/PhoQ system is the most widely studied example of a biological system that responds to the low Mg^{2+} environment, encountered by the pathogen inside the host macrophage [32, 28]. With a significant similarity to the *Salmonella* TCS, the PhoP/PhoR systems of *Mtb* also plays a vital role in its pathogenesis. It governs the invasion of epithelial cells, alteration of antigen presentation, resistance to killing by host antimicrobial peptides, and survival inside the acidic and nutrient limited macrophage [33].

PhoR, a 52-kda protein is held by two transmembrane helices onto the inner phospholipid membrane. It gets autophophorylated upon receiving signal in the periplasmic space. It is responsible for signal sensing various concentrations of Mg^{2+}, Cl^- or pH (H^+) during fusion of phagosome and lysosome (*i.e.*, phagosomal maturation), which will cause the PhoP to get phosphorylated. PhoP acts as the response regulator and is encoded by *rv0750* gene of *Mtb*. It autoregulates both the genes of the PhoPR TCS, present in a single operon. This 27.5-kda protein undergoes Mg^{2+}-dependent phosphorylation when PhoR senses the presence of

Mg^{2+}. This further result in PhoP to regulate the transcription of the operon, which could lead to the synthesis of various lipids, such as saccharolipids (SL) and DAT/PAT, essential for *Mtb* virulence as shown in Fig. (**3**). It has been shown in several studies that the intracellular pathogens couple the regulation of the gene expression of their virulence factors with Mg^{2+} concentration in the surroundings. PhoP has been found to induce the expression of a predicted Mg^{2+} transporter, MgtC under low Mg^{2+} conditions [29]. Walters *et al.,* 2006 has shown the fact that disruption of PhoPR in the bacterial system prevents growth in low Mg^{2+} media, further in macrophages [28]. Other studies have also demonstrated that *phoP* mutants showed reduced cord formation leading to altered colony morphology, and also relatively diminished bacillary size than the wild type [31].

TARGETING MAGNESIUM HOMEOSTASIS AGAINST *MTB*

We know that *Mtb* is a slow growing bacteria and this helps in establishing chronic infections and acquiring drug resistance. And like other bacteria, *Mtb* has also figured out a way to cheat killing by the host defense system. Hence it has become a priority to develop efficient therapeutics and strategies in order to stop *Mtb* propagation and persistence. Active uptake of Mg^{2+} across the membrane have been mediated by basically three transport systems, namely CorA, MgtE and MgtC. Although Mg^{2+} homeostasis had been studied only in few bacterial species over the past years, the intraphagosomal survival strategy shared by *Mtb* and *Salmonella enterica* have been well studied. Buchmeier *et al.*, 2000 screened *Mtb* mutants lacking *mgtC* which were unable to grow at a concentration of Mg^{2+} as low as ~20μM. These studies also showed that Mg^{2+} was required in order to survive the mildly acidic environment encountered by the pathogen during infection [20]. This provided new insights into the fact that MgtC protein might function by coupling the exchange of H^+ and Mg^{2+} ions; hence suggesting that MgtC does not mediate Mg^{2+} transport all by itself and that requirement of Mg^{2+} is interlinked with pH conditions in the surrounding.

Mtb can be possibly targeted at molecular level also, because there are other entities which play vital role at the transcription level. These elements are naturally essential for Mg^{2+} homeostasis; hence have potential to fulfill the aim of developing new strategies to prevent *Mtb* replication during infection. Mbox, a riboswitch related to Mg^{2+} transporters, is usually located in the upstream of two proteins namely MgtC and Rv1535, both of which are induced upon Mg^{2+} starvation. These RNA elements sense low Mg^{2+} and induce the transcription of these genes which are predicted to be Mg^{2+} transporters. Therefore it can be assumed that the genes downstream of this riboswitch definitely take part in Mg^{2+} homeostasis; which suggests that not only these proteins but the riboswitch, Mbox, can also pose as potential targets in the development of novel therapeutics against TB. Disruption of these elements might lead to abolition of Mg^{2+} transport which will lead to an Mg^{2+} poor environment for the bacteria inside the macrophage as usual, resulting in its death [30].

Another biochemical pathway in *Mtb* which was studied in relation to Mg^{2+} homeostasis was PhoPR two component transduction signal system. Experiments have revealed that replication of *phoP* mutant *Mtb* was inhibited in mice macrophages. This mutation also caused the size of the bacteria to decrease and reduced cord formation, resulting in altered morphology. PhoPR also functions in an Mg^{2+} dependent manner; PhoR senses the availability of Mg^{2+} in the surrounding and eventually phosphorylates PhoP which activates a series of additional genes. These genes are generally responsible for the synthesis of lipids which are major part of the unique cell membrane. The cell membrane is a unique structure of *Mtb* which differentiates it from other intracellular pathogens and plays a crucial role in its virulence. Hence, it is assumed that PhoPR transduction system can also be a potential element to pose as a possible drug target [31].

CONCLUSION

The growing appreciation that metal homeostasis do play significant role is evident in not only *Mtb* but other pathogens as well. For instance, iron availability is known to influence drug susceptibilities in *Candida albicans* [34, 35]. Even in *Mtb* our group has deciphered that iron availability does play vital roles in governing drug susceptibilities, virulence and efflux mechanisms [7, 36, 37]. As already discussed in previous sections, Mg^{2+} is a critical element playing important role in the growth and survival of *Mtb* inside the host; although research in this area is still at infancy, it can be concluded that Mg^{2+} homeostasis is crucial for *Mtb* to establish infection as well as its pathogenesis.

CONSENT FOR PUBLICATION

None declared.

CONFLICT OF INTEREST

The author(s) confirm that this chapter contents have no conflict of interest.

ACKNOWLEDGEMENT

Declared none.

REFERENCES

[1] Gibson SER, Harrison J, Cox JAG. Modelling a Silent Epidemic: A Review of the *In Vitro* Models of Latent Tuberculosis. Pathogens 2018; 7(4): E88. [http://dx.doi.org/10.3390/pathogens7040088] [PMID: 30445695]

[2] Sarro YD, Kone B, Diarra B, *et al.* Simultaneous diagnosis of tuberculous and non-tuberculous mycobacterial diseases: Time for a better patient management 2018; 3(3): 10. [http://dx.doi.org/10.15761/CMID.1000144] [PMID: 30613797]

[3] Fatima Z, Hameed S, Saibabu V, Sharma S, Hans S. Tuberculosis: Propagation beyond Lungs.Diagnosis and Management of Tuberculosis. 1st ed. Wilmington: Open Access eBooks 2017; pp. 1-21.

[4] World Health Organization (Geneva). Multidrug and extensively drug-resistant TB (M/XDR-TB). 2010 Global Report on Surveillance and Response 2010.

[5] Migliori GB, Loddenkemper R, Blasi F, Raviglione MC. 125 years after Robert Koch's discovery of the tubercle bacillus: the new XDR-TB threat. Is "science" enough to tackle the epidemic? Eur Respir J 2007; 29(3): 423-7. [http://dx.doi.org/10.1183/09031936.00001307] [PMID: 17329486]

[6] Li Y, Sharma MR, Koripella RK, *et al.* Zinc depletion induces ribosome hibernation in mycobacteria. Proc Natl Acad Sci USA 2018; 115(32): 8191-6. [http://dx.doi.org/10.1073/pnas.1804555115] [PMID: 30038002]

[7] Pal R, Hameed S, Fatima Z. Iron deprivation affects drug susceptibilities of mycobacteria targeting membrane integrity. J Pathogens 2015; 2015: 938523. [http://dx.doi.org/10.1155/2015/938523] [PMID: 26779346]

[8] Piddington DL, Fang FC, Laessig T, Cooper AM, Orme IM, Buchmeier NA. Cu,Zn superoxide dismutase of Mycobacterium tuberculosis contributes to survival in activated macrophages that are generating an oxidative burst. Infect Immun 2001; 69(8): 4980-7. [http://dx.doi.org/10.1128/IAI.69.8.4980-4987.2001] [PMID: 11447176]

[9] Agranoff D, Krishna S. Metal ion transport and regulation in *Mycobacterium tuberculosis.* Front Biosci 2004; 9: 2996-3006. [http://dx.doi.org/10.2741/1454] [PMID: 15353332]

[10] Pandey R, Russo R, Ghanny S, Huang X, Helmann J, Rodriguez GM. MntR(Rv2788): a transcriptional regulator that controls manganese homeostasis in *Mycobacterium tuberculosis.* Mol Microbiol 2015; 98(6): 1168-83. [http://dx.doi.org/10.1111/mmi.13207] [PMID: 26337157]

[11] Goodsmith N, Guo XV, Vandal OH, *et al.* Disruption of an *M. tuberculosis* membrane protein causes a magnesium-dependent cell division defect and failure to persist in mice. PLoS Pathog 2015; 11(2): e1004645. [http://dx.doi.org/10.1371/journal.ppat.1004645] [PMID: 25658098]

[12] Laires MJ, Monteiro CP, Bicho M. Role of cellular magnesium in health and human disease. Front Biosci 2004; 9: 262-76. [http://dx.doi.org/10.2741/1223] [PMID: 14766364]

[13] Glasdam SM, Glasdam S, Peters GH. The importance of magnesium in the human body: A systematic literature review. Adv Clin Chem 2016; 73: 169-93. [http://dx.doi.org/10.1016/bs.acc.2015.10.002] [PMID: 26975973]

[14] Shin JH, Wakeman CA, Goodson JR, *et al.* Transport of magnesium by a bacterial Nramp-related gene. PLoS Genet 2014; 10(6): e1004429. [http://dx.doi.org/10.1371/journal.pgen.1004429] [PMID: 24968120]

[15] Bozue J, Mou S, Moody KL, *et al.* The role of the phoPQ operon in the pathogenesis of the fully virulent CO92 strain of Yersinia pestis and the IP32953 strain of Yersinia pseudotuberculosis. Microb

Pathog 2011; 50(6): 314-21.
[http://dx.doi.org/10.1016/j.micpath.2011.02.005] [PMID: 21320584]

[16] Ford DC, Joshua GW, Wren BW, Oyston PC. The importance of the magnesium transporter MgtB for virulence of Yersinia pseudotuberculosis and Yersinia pestis. Microbiology 2014; 160(Pt 12): 2710-7.
[http://dx.doi.org/10.1099/mic.0.080556-0] [PMID: 25234474]

[17] Garcia-del Portillo F, Foster JW, Maguire ME, Finlay BB. Characterization of the micro-environment of Salmonella typhimurium-containing vacuoles within MDCK epithelial cells. Mol Microbiol 1992; 6(22): 3289-97.
[http://dx.doi.org/10.1111/j.1365-2958.1992.tb02197.x] [PMID: 1484485]

[18] Belon C, Gannoun-Zaki L, Lutfalla G, Kremer L, Blanc-Potard AB. Mycobacterium marinum MgtC plays a role in phagocytosis but is dispensable for intracellular multiplication. PLoS One 2014; 9(12): e116052.
[http://dx.doi.org/10.1371/journal.pone.0116052] [PMID: 25545682]

[19] Piddington DL, Kashkouli A, Buchmeier NA. Growth of *Mycobacterium tuberculosis* in a defined medium is very restricted by acid pH and Mg($^{2+}$) levels. Infect Immun 2000; 68(8): 4518-22.
[http://dx.doi.org/10.1128/IAI.68.8.4518-4522.2000] [PMID: 10899850]

[20] Buchmeier N, Blanc-Potard A, Ehrt S, Piddington D, Riley L, Groisman EA. A parallel intraphagosomal survival strategy shared by *mycobacterium tuberculosis* and *Salmonella enterica.* Mol Microbiol 2000; 35(6): 1375-82.
[http://dx.doi.org/10.1046/j.1365-2958.2000.01797.x] [PMID: 10760138]

[21] Romani AMP. Cellular magnesium homeostasis. Arch Biochem Biophys 2011; 512(1): 1-23.
[http://dx.doi.org/10.1016/j.abb.2011.05.010] [PMID: 21640700]

[22] Scarpa A, Brinley FJ. *In situ* measurements of free cytosolic magnesium ions. Fed Proc 1981; 40(12): 2646-52.
[PMID: 7286245]

[23] Lüthi D, Günzel D, McGuigan JA. Mg-ATP binding: its modification by spermine, the relevance to cytosolic Mg^{2+} buffering, changes in the intracellular ionized Mg^{2+} concentration and the estimation of Mg^{2+} by 31P-NMR. Exp Physiol 1999; 84(2): 231-52.
[PMID: 10226168]

[24] Giménez-Mascarell P, Schirrmacher CE, Martínez-Cruz LA, Müller D. Novel Aspects of Renal Magnesium Homeostasis. Front Pediatr 2018; 6: 77.
[http://dx.doi.org/10.3389/fped.2018.00077] [PMID: 29686978]

[25] Smith RL, Maguire ME. Microbial magnesium transport: unusual transporters searching for identity. Mol Microbiol 1998; 28(2): 217-26.
[http://dx.doi.org/10.1046/j.1365-2958.1998.00810.x] [PMID: 9622348]

[26] Baker JJ, Johnson BK, Abramovitch RB. Slow growth of *Mycobacterium tuberculosis* at acidic pH is regulated by *phoPR* and host-associated carbon sources. Mol Microbiol 2014; 94(1): 56-69.
[http://dx.doi.org/10.1111/mmi.12688] [PMID: 24975990]

[27] Hu J, Sharma M, Qin H, Gao FP, Cross TA. Ligand binding in the conserved interhelical loop of CorA, a magnesium transporter from *Mycobacterium tuberculosis.* J Biol Chem 2009; 284(23): 15619-28.
[http://dx.doi.org/10.1074/jbc.M901581200] [PMID: 19346249]

[28] Walters SB, Dubnau E, Kolesnikova I, Laval F, Daffe M, Smith I. The Mycobacterium tuberculosis PhoPR two-component system regulates genes essential for virulence and complex lipid biosynthesis. Mol Microbiol 2006; 60(2): 312-30.
[http://dx.doi.org/10.1111/j.1365-2958.2006.05102.x] [PMID: 16573683]

[29] Ramesh A, Winkler WC. Magnesium-sensing riboswitches in bacteria. RNA Biol 2010; 7(1): 77-83.
[http://dx.doi.org/10.4161/rna.7.1.10490] [PMID: 20023416]

[30] Arnvig K, Young D. Non-coding RNA and its potential role in Mycobacterium tuberculosis pathogenesis. RNA Biol 2012; 9(4): 427-36.
[http://dx.doi.org/10.4161/rna.20105] [PMID: 22546938]

[31] Broset E, Martín C, Gonzalo-Asensio J. Evolutionary landscape of the Mycobacterium tuberculosis complex from the viewpoint of PhoPR: implications for virulence regulation and application to vaccine development. MBio 2015; 6(5): e01289-15.
[http://dx.doi.org/10.1128/mBio.01289-15] [PMID: 26489860]

[32] Groisman EA, Hollands K, Kriner MA, Lee EJ, Park SY, Pontes MH. Bacterial Mg^{2+} homeostasis, transport, and virulence. Annu Rev Genet 2013; 47: 625-46.
[http://dx.doi.org/10.1146/annurev-genet-051313-051025] [PMID: 24079267]

[33] Soncini FC, García Véscovi E, Solomon F, Groisman EA. Molecular basis of the magnesium deprivation response in Salmonella typhimurium: identification of PhoP-regulated genes. J Bacteriol 1996; 178(17): 5092-9.
[http://dx.doi.org/10.1128/jb.178.17.5092-5099.1996] [PMID: 8752324]

[34] Prasad T, Chandra A, Mukhopadhyay CK, Prasad R. Unexpected link between iron and drug resistance of Candida spp.: iron depletion enhances membrane fluidity and drug diffusion, leading to drug-susceptible cells. Antimicrob Agents Chemother 2006; 50(11): 3597-606.
[http://dx.doi.org/10.1128/AAC.00653-06] [PMID: 16954314]

[35] Hameed S, Dhamgaye S, Singh A, Goswami SK, Prasad R. Calcineurin signaling and membrane lipid homeostasis regulates iron mediated multidrug resistance mechanisms in Candida albicans. PLoS One 2011; 6(4): e18684.
[http://dx.doi.org/10.1371/journal.pone.0018684] [PMID: 21533276]

[36] Pal R, Hameed S, Sharma S, Fatima Z. Influence of iron deprivation on virulence traits of mycobacteria. Braz J Infect Dis 2016; 20(6): 585-91.
[http://dx.doi.org/10.1016/j.bjid.2016.08.010] [PMID: 27755980]

[37] Pal R, Hameed S, Fatima Z. Altered drug efflux under iron deprivation unveils abrogated MmpL3 driven mycolic acid transport and fluidity in mycobacteria. Biometals 2018.
[http://dx.doi.org/10.1007/s10534-018-0157-8] [PMID: 30430296]

SUBJECT INDEX

A

Atta-ur-Rahman (Ed.)

B

C

D

E

M

N

R

S

T

U

V

W

www.ingramcontent.com/pod-product-compliance
Lightning Source LLC
Chambersburg PA
CBHW051144220526
45473CB00003B/652
* 9 7 8 1 6 8 1 0 8 6 3 8 5 *